D0381896

About Island Press

Island Press is the only nonprofit organization in the United States whose principal purpose is the publication of books on environmental issues and natural resource management. We provide solutions-oriented information to professionals, public officials, business and community leaders, and concerned citizens who are shaping responses to environmental problems.

In 1994, Island Press celebrated its tenth anniversary as the leading provider of timely and practical books that take a multidisciplinary approach to critical environmental concerns. Our growing list of titles reflects our commitment to bringing the best of an expanding body of literature to the environmental community throughout North America and the world.

Support for Island Press is provided by Apple Computer, Inc., The Bullitt Foundation, The Geraldine R. Dodge Foundation, The Energy Foundation, The Ford Foundation, The W. Alton Jones Foundation, The Lyndhurst Foundation, The John D. and Catherine T. MacArthur Foundation, The Andrew W. Mellon Foundation, The Joyce Mertz-Gilmore Foundation, The National Fish and Wildlife Foundation, The Pew Charitable Trusts, The Pew Global Stewardship Initiative, The Rockefeller Philanthropic Collaborative, Inc., and individual donors.

About World Resources Institute

The World Resources Institute (WRI) is an independent center for policy research and technical assistance on global environmental and development issues. WRI's mission is to move human society to live in ways that protect Earth's environment and its capacity to provide for the needs and aspirations of current and future generations.

Because people are inspired by ideas, empowered by knowledge, and moved to change by greater understanding, the Institute provides—and helps other institutions provide—objective information and practical proposals for policy and institutional change that will foster environmentally sound, socially equitable development. WRI's particular concerns are with globally significant environmental problems and their interaction with economic development and social equity at all levels.

The Institute's current areas of work include economics, forests, biodiversity, climate change, energy, sustainable agriculture, resource and environmental information, trade, technology, national strategies for environmental and resource management, and human health.

In all of its policy research and work with institutions, WRI tries to build bridges between ideas and action, meshing the insights of scientific research, economic and institutional analyses, and practical experience with the need for open and participatory decision-making.

Frontiers of Sustainability

Frontiers of Sustainability

Environmentally Sound Agriculture, Forestry, Transportation, and Power Production

Roger Dower
Daryl Ditz
Paul Faeth
Nels Johnson
Keith Kozloff
James J. MacKenzie

Introduction by Walter V. Reid and Roger Dower

Foreword by Jonathan Lash

World Resources Institute

ISLAND PRESS

Washington, D.C. ◆ Covelo, California

Copyright © 1997 by Island Press

All rights reserved under International and Pan-American Copyright Conventions. No part of this book may be reproduced in any form or by any means without permission in writing from the publisher: Island Press, 1718 Connecticut Avenue, N.W., Suite 300, Washington, DC 20009.

ISLAND PRESS is a trademark of The Center for Resource Economics.

Frontiers of Sustainability: environmentally sound agriculture, forestry, transportation, and power production/Roger Dower [et al.]; introduction by Walter V. Reid and Roger Dower; foreword by Jonathan Lash.
p. cm.
Includes bibliographical references and index.
ISBN 1-55963-546-0 (alk. paper)
1. Sustainable development—United States.   2. Natural resources-United States—Management.
3. Environmental protection—United States.   4. Economic Development—Environmental aspects—United States.   I. Dower, Roger C.

HC110.E5F77 1997

333.7' 0973—dc21   97-5206
                                                                                                                  CIP

Printed on recycled, acid-free paper ♲
Manufactured in the United States of America

10 9 8 7 6 5 4 3 2 1

CONTENTS

FOREWORD

Sustainable development—the integration of economic, environmental, and social goals—is a goal on which reasonable people may agree in principle, but passionately disagree in practice. Yet, sustainable development is fundamentally a matter of practice, of minimizing the conflicts among these three goals and making choices when conflicts are unavoidable.

In the United States, few question the progress made in fostering both economic and social goals over the past several decades. By most measures, America is much wealthier: Gross Domestic Product increased by 95 percent from 1970 to 1994, while per capita income grew from $12,000 to $18,000. Our wealth was not shared equally, however: about 12 percent of all American families live below the poverty line, an increase since 1976 in both absolute and relative terms.

As for environmental goals, there is much to be proud of, but much more to do. Since 1970, most conventional air pollutants have declined and national water quality has improved significantly. Between 1982 and 1992, the estimated rate of soil erosion on U.S. cropland dropped by about 30 percent.

But growth is partly offsetting progress. In the case of transportation, the U.S. motor vehicle fleet nearly doubled between 1970 and 1995. Emissions standards for new vehicles have grown progressively tougher, yet between 1970 and 1994 nitrogen oxide emissions increased and carbon monoxide emissions dropped only 31 percent. Partly as a result, some 60 million Americans still live in counties that did not meet at least one air quality standard in 1993.

Furthermore, about 40 percent of the nation's surveyed rivers, lakes, and estuaries are too polluted to swim or fish. Supporting the average American takes some 80 tons of material annually, much of which winds up on the trash heap.

Finally, two massive global threats lie ahead: climate change and the loss of biodiversity. Climate change could conceivably flood coastal areas, multiply the stresses on biological systems, and create a series of threats to human well-being. Biodiversity loss diminishes the resilience of the biological systems that support life and the wealth of nature on which agricultural and forest productivity ultimately depend.

Thus, the unfinished agenda is daunting.

For all that, there are some grounds for optimism about the future. The President's Council on Sustainable Development, which I was honored to co-chair, found strong public interest and thriving community innovation in the pursuit of fair and environmentally sound prosperity.

Community residents disturbed by such issues as congestion, pollution, and inner-city job losses are an important resource. Citizens and neighbors can bring diverse interests together, identify and agree on positive goals, and organize for responsive action. Already in many communities across the nation, businesses, households, and government are working together to use their land, energy, and resources more efficiently. They are achieving a higher quality of life with minimal waste and environmental damage.

Some issues can be solved only if nations agree on common goals and responsibilities. The United States was a leader in developing a global agreement to phase out chlorofluorocarbons, the human-made gases that destroy the ozone layer. Inspired U.S. leadership is just as vital in tackling climate change, biodiversity, and other global environmental issues.

The Council's 1996 report marked the start of a national journey toward sustainability. The contributors agreed, for example, that some things must grow—jobs, productivity, wages, capital and savings, profits, information, knowledge, and education—while others—pollution, waste, and poverty—must not. We also agreed that market incentives and consumer power can boost environmental performance at less cost than traditional regulatory approaches.

From the Council's well-received report we can get a general sense of where we should be heading, but not a detailed roadmap. *Frontiers of Sustainability* begins to fill in some of those blanks. It looks at agriculture, electricity generation, transportation, and forestry and describes the kind of actions needed to put each on the path to efficiency, environmental integrity, and growth.

In the power sector, which is now being restructured, policy-makers can reform price signals, ensure that all technologies compete on an equal footing, support private-sector efforts to commercialize emerging technologies, and encourage voluntary corporate leadership.

In forestry, we can learn to plan with the future in mind, start a national network of sustainable forest demonstration sites, reform federal estate tax law to help preserve private forest lands, create new financial incentives to encourage restoration of degraded private timberlands, and use targeted incentives and other measures to protect critically endangered forest ecosystems.

In agriculture, we can make income-support programs more efficiently promote both agricultural and environmental goals, invest in agricultural research that could yield both economic and environmental gains, and encourage private-sector stewardship that will bring sustainability closer to home.

In transportation, electric-drive vehicles, congestion pricing, parking reform, and increased fuel taxes can provide a bridge to futuristic public transportation that takes advantage of startling technological progress to provide clean, fast, convenient service.

Shifting the United States toward sustainability won't happen through a single policy or regulatory fix. Many new policies and incentives are needed. But adhering to a few basic rules of thumb—putting our incentives where our objectives are, inclusion, fairness, and intergenerational equity—will make the challenging transition before us easier and more successful.

For their support of this project we thank the Charles Stewart Mott Foundation, the Nathan Cummings Foundation, the Joyce Foundation, the Winslow Foundation, the Surdna Foundation, Inc., and the W. Alton Jones Foundation.

Jonathan Lash
President
World Resources Institute

ACKNOWLEDGMENTS

The authors would like to thank the many colleagues and friends who contributed to this project. In particular, we would like to thank the members of the Project Advisory Committee (listed on p. 369), who provided ongoing peer review during the project, meeting at several critical stages to offer advice and ideas and to review findings and proposed research. We are grateful to our colleagues at WRI, including Donna Wise, Jonathan Lash, and Allen Hammond for both time and ideas and to Gus Speth, Jessica T. Mathews, Thomas Stoel, Jr., Theresa Bradley, Alan Brewster, Bruce Cabarle, Christy Dobbels, Janet Ranganathan, Bob Repetto and Ann Thrupp for comments on early drafts.

Special thanks are due to reviewers of the individual chapters: Margot Anderson, Martin Bender, Pierre Crosson, Dave Ervin, Elissa Graffy, Jim Maetzold, Kathleen Merrigan, Susan Offutt, Rob Paarlberg (*Agriculture Sector*); Leon Lowery, Larry Owens, Roger Sant, Mary Tucker (*Electricity Sector*); Richard Denison, Alan Durning, Robert Ewing, William Frohnmayer, Carrie Meyer, Dave Panco, Hal Salwasser, Joseph Strickland (*Forestry Sector*); David Burwell, Don Chen, Hank Dittmar, Charles Lave (*Transportation Sector*); Deborah Jensen, Kai N. Lee, Kathleen Merrigan, Susan Offutt, David Richards, Edward L. Strohbehn Jr. (*Overview*).

Our focus group interviews were ably conducted by Tom Cosgrove of MacWilliams, Cosgrove, Snider, Smith, and Robinson. We also thank the participants in our forest sector sustainability roundtable discussion held in Portland, including Robert Ewing, William Frohnmayer, Catherine Mater, Larry Potts, Joseph Strickland, and Henry Whittemore. Special thanks are also due to Michael Tennis of the Union of Concerned Scientists, who did much of the analysis for the electric power sector case study, to John Dunlop, who organized a workshop in Minnesota on that sectoral study, and to the workshop participants: Al Bartsch, Joseph Bizzano Jr., William Blaser, Margaret Donohoe, Betsy Engelking, Bill Grant, Mark Haller, Mary Hirschbeck, Anne Hunt, Marion Kloster, Drew Larson, Mark Laub, Bob Lee, Bill Leeper, Alfred Marcus, Carl Michaud, Jim Nicols, Michael Noble, William Poppert, Paula Prahl, Lola Schoenrick, Linda Sohutz, Michael Tennis, and Audrey Zibelman.

We thank the members of our resource panel, who provided guidance on how to make the findings of this project most useful to various audiences. The members included Matthew Arnold, Tom Bonnet, David Gatin, Ann James,

Bonnie Kranzer, Gary Lawrence, Cathy Lerza, Tracy Mehan, Margaret O'Dell, Julia Parzen, and Ross Stevens.

Finally, we are very grateful to Patricia McGinnis, Alexandra Sevilla, and Karin Wiener, who provided research assistance. Kathleen Courrier deserves thanks for her skillful editing and oversight of the preparation of the book, Maggie Powell and Hyacinth Billings for publication production, and Christy Dobbels, Sridevi Nanjundaram, Eva Vasiliades, and Serene Jweied for their invaluable help with manuscript preparation, administration, coordination, and research assistance.

1.
INTRODUCTION

Walter V. Reid and Roger Dower

As a nation, we have improved environmental quality significantly over the last 25 years. The number of U.S. cities in violation of air quality standards has fallen dramatically. Rivers, streams, and waterways once written off as polluted are now sources of civic pride and renewed recreational and development opportunities. Toxic pollution from industrial facilities has been cut by more than 40 percent in less than a decade. And a half dozen plant and animal species once on extinction's brink have been taken off the endangered species list (U.S. FWS, 1997). Although the gains came at a price—more than $127 billion a year spent on environmental protection measures (U.S. EPA, 1990)—they also return uncounted economic benefits through improved public health and worker productivity, enhanced recreational opportunities, and increased productivity of our renewable resources.

This record of environmental progress stands in sharp contrast to other seemingly intractable national problems, such as poverty, drug abuse, and crime. But the nation's environmental problems are far from solved. Indeed, many of the greatest challenges lie ahead, and many are less visible than air pollution and less controllable than the discharge of toxic wastes from a factory. Indeed, their impact on our health or economy may not be felt for decades. These challenges include the risk of global climate change, the loss of biological diversity, continued decline in resource productivity, and health risks associated with the vast quantity of toxic chemicals we still release into the environment. They include the steps that remain in making our air healthy to breathe and our water fit to drink.

The first phase of our nation's environmental history was a story of resource exploitation and unchecked pollution. The second, just now ending, has centered on pollution abatement—cleaning up the mess left by our blinkered pursuit of economic growth. In the third, more systemic changes will be needed. It's no longer enough to fix economic development's environmental side effects. U.S. policies must tackle the underlying causes of environmental degradation, not just the symptoms. This means preventing, rather than treating, industrial pollution. It means designing whole new systems for moving people, goods, and ideas. It means living within our means—making do with our natural allowance of biological production and water recharge, rooting out waste and pollution. It means linking environmental progress to economic and social progress and ensuring quicker feedback when our actions trigger unanticipated environmental, social, or economic costs. Now, the challenge is to make environmental quality a product of our development, not a casualty; we must find new paths to a future in which we efficiently meet demands for jobs, education, goods and services, and healthy communities without overtaxing Earth's life-support systems. In short, we need to stimulate a transition in technologies, policies, and lifestyles that will place the nation on a sustainable development path. *(See Box 1-1.)*

The obstacles in the path to these new technologies, policies, and lifestyles are immense—indeed, so large that some analysts and much of the public deny its possibility. In focus group discussions undertaken for this study, Americans commented that the goal of sustainable development was "hopeful" or seemed to be "nirvana." One man in Atlanta said:

> *"I think it's impossible. It's conflicting ... You need people to fill the jobs and people need places to live and they need food and durable goods that have to be produced, which takes away from the environment."*

On the other hand, the United States has solved some sustainability problems once considered intractable. A decade ago, scientists discovered that the release of large amounts of ozone-depleting gases including chlorofluorocarbons (CFCs) could imperil Earth's stratospheric ozone layer, and hence all life. As a nation, we decided to safeguard the stratospheric ozone layer by halting such releases even though the use of CFCs was entrenched in the U.S. economy—more than 100 billion dollar's worth of equipment relied on CFCs. The United States signed an international agreement to phase out ozone-depleting substances. On January 1, 1996, chemical manufacturers stopped producing CFCs for U.S. consumption except for a few essential uses. The technological breakthroughs that sprang from concern for sustainability proved far less disruptive and costly than anyone had predicted (Cook, 1996).

Frontiers of Sustainability concludes that, despite progress, the United States is not yet on a sustainable development path. Our current policies, technologies,

BOX 1-1. What Is Sustainable Development?

In 1987, the Brundtland Commission Report defined sustainable development as development that meets the needs of the present without compromising the ability of future generations to meet their needs (World Commission on Environment and Development, 1987). It means improving environmental quality and economic living standards without triggering catastrophic events—in short, pursuing progress toward economic, social, and environmental goals all at once. The broad scope of the sustainable development challenge is well documented in the 1992 Earth Summit's *Agenda 21* and the 1996 President's Council on Sustainable Development's *Sustainable America: A New Consensus for Prosperity, Opportunity, and a Healthy Environment for the Future.*

Only history can judge which development paths are sustainable. Centrally planned economies were not. Fisheries management that paid little heed to scientific information about stock size proved unsustainable for coastal communities in California in the 1940s and, more recently, for communities in Maryland, Massachusetts, Rhode Island, and Maine. The expansion of dryland agriculture into the Plains states during the unusually wet years at the turn of the century proved unsustainable when weather patterns returned to normal, precipitating the Dust Bowl of the 1930s.

But even if time is the ultimate judge of sustainability, it's possible we can figure out which development paths or activities are *not* sustainable and then chart new paths. Our petroleum-based ground transportation system, for example, is threatened by the long-term depletion of global oil reserves. Similarly, electric power produced from fossil fuels isn't sustainable because it releases greenhouse gases into the atmosphere. Policies, technologies, or behaviors introduced to put such basic human enterprises on a better footing are all steps toward sustainability.

corporate practices, and public attitudes and behaviors will not meet the public's environmental goals or ensure continued economic and social well-being. And we have not yet begun to tackle climate change and the loss of biodiversity. Recognizing that it has the wherewithal but needs the will, the United States must make a difficult but historic transition starting in the next few years.

Along with the environmental concerns addressed in this book, economic, social, environmental, and demographic concerns all figure into sustainability, as do social fairness, justice, and intergenerational equity. Real progress toward any of these goals requires progress toward all.

U.S. progress toward sustainability also depends on the progress of other nations. Global environmental, social, and economic challenges all crop up in the context of a global economy and interlinked ecosystems. For example, the volume of greenhouse gases emitted by the United States and the size of overseas private investments made by U.S. companies both influence other nations' pursuit of sustainability. In turn, the United States can neither defuse the domestic threats posed by climate change nor end its natural population growth nor conserve its fish stocks without other nations' cooperation.

But where to begin? Because the challenge can seem overwhelming, we decided to narrow our scope to four important sectors of the U.S. economy. We also limited our research to sustainability's environmental dimensions. We focused on agriculture, electricity generation, transportation, and forestry partly because WRI has built considerable expertise in these fields through past research. *(See Figure 1-1 and Table 1-1.)* Also, though these four sectors represent only about 9 percent of the U.S. Gross Domestic Product (GDP), they account for the great bulk of Americans' demands on biological resources and for the lion's share of U.S. energy consumption. (Cropland, rangeland, and forests make up 77 percent of the area of the contiguous states (Daugherty, 1995).) Two sectors rely heavily on fossil fuel combustion—risky in view of global climate change—and a significant change for this fuel source will be slow and difficult. Forest and agriculture are linked to at least a dozen sustainability challenges—from the productivity of our biological resources, to water pollution and toxic discharges, to species loss.

Our research team zeroed in on sustainability challenges within our borders. But solutions to many—say, climate change, biodiversity loss, and population growth—will require international action. In turn, domestic moves can threaten other countries' sustainability. Too often, sustainable development has been treated by rich countries as if it applied only to the poorer ones. But environmental problems born of affluence are just as difficult to solve as those born of poverty, and if the United States gets its own house in order it can better lead and aid international efforts. For these reasons, the research team looked for "win/win" actions that the United States can take toward sustainable development.

We used a variety of methods in the sectoral studies, but applied the same basic questions in each. How will we tackle the environmental and resource problems that threaten sustainability? What changes in policies, institutions, and attitudes will catalyze economic "re-engineering" so that affordable environmental progress continues? And, perhaps most important, how do we solve these problems at their roots and just what might a sustainable United States be like?

FIGURE **1-1.** **Trends in the Four Sectors**

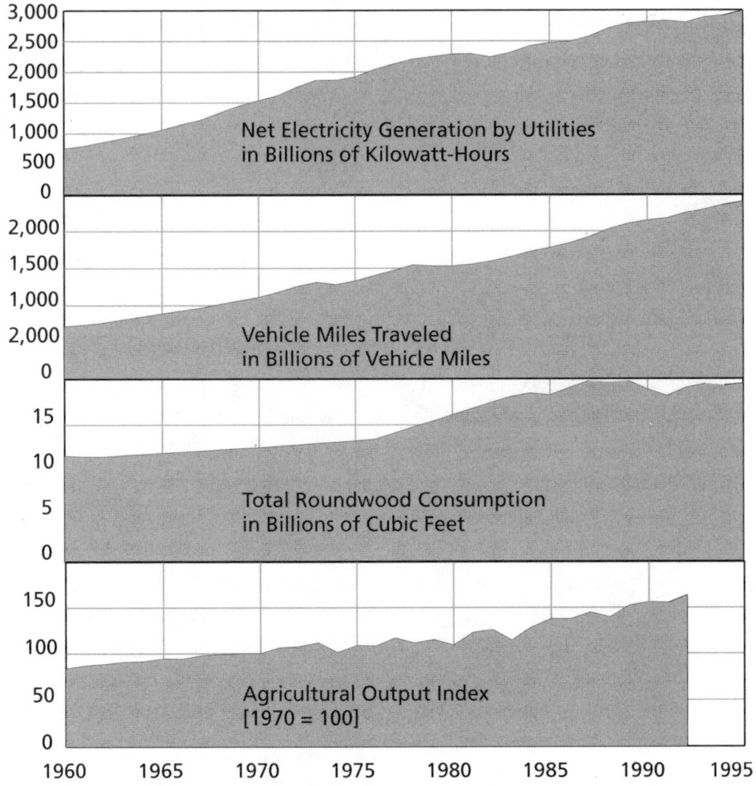

Sources: USDOE, 1996; AAMA, 1996; Haynes et al., 1995; USDA, 1994.

From the four sectoral studies, the environmental dimensions of sustainable development emerge, along with a surprisingly consistent picture of the types of changes needed to make our society and economy sustainable.

WHERE ARE WE NOW AND WHERE ARE WE HEADED?

For the past 25 years, U.S. environmental policy has emphasized pollution control and underscored palpable threats to natural resources. A careful reading of the economic, demographic, and environmental trends shows that while our nation has brought many types of pollution under control while maintaining strong

TABLE 1-1. Sector Profiles

Sector	Economic Output	GDP	Employment	Sector Description
Electric Power	$107 billion contribution to 1992 GDP	1.7%	955,000	Largest single consumer of primary fuels (36%, up from 15% in 1950); 87 percent of total U.S. coal use in 1995. Sources of power: coal (55%), nuclear (22%), natural gas and oil (13%), hydropower (10%).
Ground Transportation	$277 billion contribution to 1992 GDP (all expenditures for moving goods and passengers in excess of $900 billion in 1994)	4.4%	5.5 million	200 million cars, buses, and trucks (1995); 30% of world's motor vehicles (only 4.5% of world population); 4 million miles of roads and highways; 3.7 trillion passenger miles of travel each year.
Forestry and Forest Products	$114 billion contribution to 1992 GDP	1.8%	1.8 million	World's largest producer of wood fiber and forest products; 9th largest industrial sector. Forest land ownership: small private (58%), private firms (14.5%), state and federal government (26.5%).
Agriculture	$72 billion to 1992 GDP	1.1%	3.5 million directly (18 million in related industries)	World's largest exporter (10% of U.S. merchandise exports accounting for 23% of world agricultural exports in 1995). Output more than doubled from 1948 to 1993 while acreage shrank.

economic growth, the job is far from over and certain long-term environmental threats—particularly climate change and biodiversity loss—still pose significant economic as well as environmental challenges.

Economic Outlook

The United States is the world's greatest economic powerhouse. Gross domestic product (GDP), the most widely used (though incomplete) measure of economic performance, increased by 95 percent between 1970 and 1994. We rank first in the world in total economic output and among the top nations in output per person, which has risen by 53 percent over the last 25 years. Personal income has also grown, from around $12,000 per capita in 1970 to $18,000 in 1994. The United States currently enjoys one of the lowest unemployment rates of all developed countries, even while absorbing more than 48 million new workers during the last quarter century. In 1995, some 146,000 jobs per month were added to the workforce, many more than any of our European competitors created. Worker productivity here has grown, on average, by around 1 percent per year from 1973 through 1995. And even though it grew faster during, say, the 1960s, it still outpaces that of most of our global competitors. While Americans literally make a federal issue out of the national budget deficit, which *does* influence the well-being of future generations, we still have the lowest annual deficit as a percentage of GDP among the Group of Seven economic powers—Canada, France, Germany, Italy, Japan, United Kingdom, and the United States. And while ours is falling, others' seem to be on the rise. As a whole, the United States is prosperous—one part of sustainable development.

Can we keep up the good work? While economic forecasts are chronically unreliable, no compelling evidence suggests that the United States economy will stop growing anytime soon. The presence of the fundamental determinants of economic growth and prosperity—an educated, productive labor force; an open, competitive market; and our considerable assets—all augur well for the continuing growth of national wealth. Ups and downs are inevitable and business cycles will always heave us where they will, but it is hard to imagine a plausible economic threat at home large enough to reverse the trend of the last quarter century within the next.

Less rosy than the aggregate economic picture are trends related to how our national wealth is distributed. Our overall economic growth masks a more subtle trend toward relatively greater growth for wealthy families compared to that for middle-income or poorer ones. From 1979 to 1993, per capita incomes increased but the growth occurred in the top 40 percent of families, while the rest experienced a painful decline in income (Council of Economic Advisors, 1996). For the very poor, the situation is worse. The number of families living below the poverty line rose in absolute and relative terms between 1976 and 1994 for all racial groups and currently stands at around 8.1 million—or around 12 percent of all families. This gap could widen without dampening overall U.S. economic growth, but it is neither acceptable nor sustainable because social tensions will probably swell with growing inequality.

Demographic Outlook

The pressures of population growth contribute to virtually all environmental problems related to sustainable development. In 1970, the U.S. population stood at around 205 million people. At the beginning of 1997, we numbered around 266 million. By 2020, the population is expected to reach 326 million, having grown by 0.9 percent per year. *(See Figure 1-2.)* This population growth will occur even though the average family size is already below the "replacement" level of 2.1 children per woman (it was 2.06 in 1992) because of demographic "momentum" (the built-in population growth stemming from the relatively large number of young people now approaching reproductive age) and immigration, which now accounts for slightly more than one-third of U.S. population growth. In the coming decades, the relative importance of population growth from immigration will increase.

Population growth can have some economic benefits, but it increases consumption of materials—both natural and human-made—and emissions of wastes. In particular, U.S. population growth has helped drive up fuel consump-

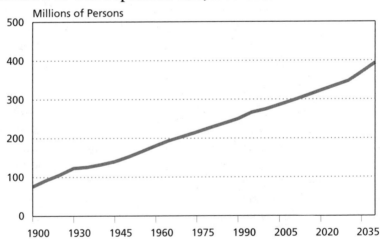

FIGURE **1-2. U.S. Population Size, 1900–2040**

Sources: Council of Economic Advisers, *Economic Report to the President 1996,* Washington, D.C., and the U.S. Bureau of the Census, *Statistical Abstract of the United States, 1996* (116th edition), Washington, D.C., 1996.

tion and CO_2 emissions. With slower population growth, the nation would have used less fuel, had less air pollution, relied less on imported oil, enjoyed greater mobility, and had more time to develop alternatives to oil-powered vehicles. Over the near term, the trick is to minimize the environmental impacts of the growing population by using resources more efficiently, thereby reducing both demand for resources and the production of pollutants and wastes. And over the long term, a stable U.S. and global population is absolutely essential to sustainability.

Stabilizing the U.S. population is no simple matter. Education, new economic opportunities, better pre- and postnatal care, and more family planning services are just part of what is required. The President's Council on Sustainable Development (1996) has recommended that all citizens be assured access to basic reproductive health services and that the public and private sector work as partners to enhance opportunities for women, paying special attention to socioeconomic factors that encourage unintended and teen pregnancies among the disadvantaged. Population equilibrium will also be difficult to achieve while immigration remains high (unless natural population growth falls below the replacement level). For a country comprised largely of descendants of immigrants, ending immigration at the border is almost unthinkable. In fact, pressure for immigration to the United States will end only when global population is stabilized and economic disparities among nations is reduced.

Environmental Outlook: The Good News

Many indicators of U.S. environmental quality have improved steadily for decades. Most of the pollutants targeted in our efforts to control air pollution—carbon monoxide and particulate matter, among others—have declined fairly dramatically in absolute terms, even as economic activity has grown. *(See Figure 1-3.)* Emissions of volatile organic compounds—the catalysts and ingredients of smog—have dropped by 24 percent, from 31 million tons in 1970 to 23 million in 1994. Sulfur dioxide emissions decreased by 18 percent between 1986 and 1995 (U.S. EPA, 1996a and 1996b). Even in the electric power sector, emissions of sulfur dioxide began to decline after 1980 despite rapidly rising power generation from coal. *(See Figure 1-4.)*

National water-quality trends, though harder to assess, have also improved significantly. Violations of Maximum Contaminant Levels in drinking water had declined by 47 percent between 1986 and 1994. The percentage of the U.S. population served by wastewater treatment plants is among the world's highest, exceeded only by that for Germany and the United Kingdom. The amount of toxic

FIGURE 1-3. **Trends in U.S. Particulate (PM-10) Emissions (Excluding Dust, Wildfires, Wind Erosion)**

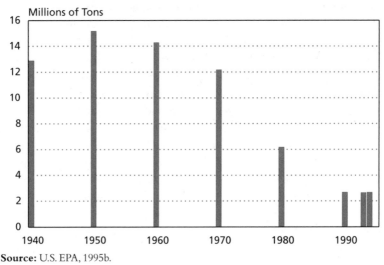

Source: U.S. EPA, 1995b.

chemicals jettisoned into our environment—at least of the 300 or so whose release must, by law, be reported—has also fallen. *(See Figure 1-5.)*

Some of the nation's most striking environmental gains were made in cleaning up ground transportation. Thanks to federal regulation, total lead emissions to the atmosphere dropped by 98 percent between 1970 and 1994. *(See Figure 1-6.)* High levels of lead in the blood have been conclusively linked to neurological damage in young children and to impaired motor-physical development. The average fuel efficiency or miles per gallon (mpg) of our passenger car fleet increased by 60 percent between 1970 and 1993 (U.S. DOE, 1995). And, despite growth in the nation's fleet, carbon monoxide emissions from vehicles decreased by about 30 percent between 1970 and 1994 while emissions of volatile organic compounds declined by over 50 percent (U.S. EPA, 1995b).

While cleaning up many of the visible and smelly forms of pollution, we have also slowed the loss of some of the natural resources underpinning our agriculture and forest sectors. Between 1982 and 1992, estimated soil erosion on U.S. cropland dropped by about 30 percent. The annual rate of wetland conversion to agriculture slowed from roughly 600,000 acres per year between 1954 and

FIGURE **1-4. U.S. Electric Utility SO₂ Emissions and Power Generation from Coal**

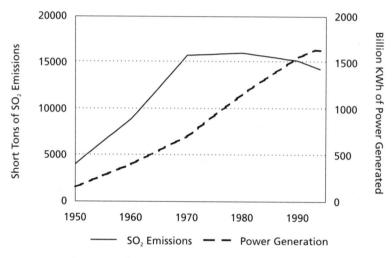

Sources: U.S. Department of Energy, 1994; U.S. EPA, 1995a.

1974 to about 29,000 acres per year in 1987-1991. Forest area (including both natural forest and plantations) has remained stable since 1920, even though the country's population has grown by 150 million. The release of water pollutants by pulp and paper mills declined from 210 pounds per ton of production in 1945 to less than 10 pounds per ton today. The use of recycled fiber has grown from 8 percent in 1962 to nearly 35 percent today. And the area of legally protected forest has grown from less than 3 percent of total forest area in 1960 to about 6 percent in 1992.

Environmental Outlook: The Bad News

On the other side of the environmental ledger, some problems have persisted or grown, and longer term threats could pose far greater economic challenges than those conquered so far. The nation's overall level of materials consumption by weight has remained fairly constant since 1975 (Adriaanse et al., in press). While it's good news that economic growth is outpacing our "material flow," the scale of that flow is not. To meet the needs of the average U.S. citizen for a single year, more than 80 tons of material is processed or moved through our economy.

FIGURE 1-5. **Trends in Toxic Chemicals from Industry**

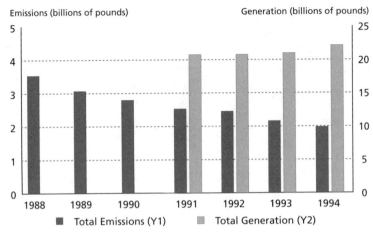

Emissions (billions of pounds)   Generation (billions of pounds)

■ Total Emissions (Y1)   ■ Total Generation (Y2)

Source: U.S. EPA, 1996b.

The United States still falls short of meeting its air and water quality goals. Emission levels of NO_x have barely budged since 1980. High levels of nitrogen oxides contribute to smog and ozone formation and to the acidification of rainfall, rivers, lakes, and soils. Even more telling, though air quality has improved overall, 60 million Americans live in counties that did not meet at least one air quality standard in 1993 (U.S. EPA, 1993). Meanwhile, we have not yet succeeded in controlling water pollution from sources besides large factories or other easy targets—such as farms or run-off from rain on city streets. Over the past 30 years, the use of both chemical fertilizer and pesticides has doubled (by weight of active ingredients). Today, water quality problems keep Americans from using 44 percent of the nation's river miles, 57 percent of lake acres, and 44 percent of estuary waters.

Environmental degradation of the sort we tend to worry about most—toxic chemicals in the environment—continues to pose health risks. Although the sectors examined here did not include many of the largest sources of toxic wastes (such as the chemical and primary metals industries), the research team did find health risks associated with the growing use of pesticides in agriculture and the continuing release of pollutants from the pulp and paper industry (which ranks third in total quantity of toxic chemicals released).

FIGURE **1-6. Trends in U.S. Lead Emissions**

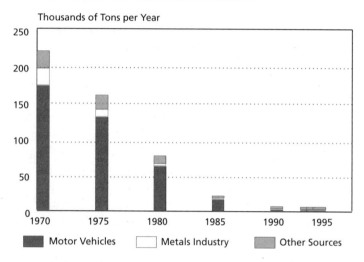

Source: U.S. EPA, 1995a.

One of the most striking conclusions of our sectoral studies was that many of the most serious environmental problems—those whose solutions will require dramatic changes in what we produce, how we produce it, and what we consume, and those that threaten the nation's economic well-being—receive far less attention than more traditional environmental concerns. *(See Box 1-2.)* Of these, two stand out as potential time bombs: climate change and biodiversity loss.

The threat posed by climate change to the U.S. future is essentially irreversible in any timeframe of interest. The Intergovernmental Panel on Climate Change (IPCC) projects that by the year 2100 the Earth's global mean temperature could rise 1.6 to 6.3 degrees Fahrenheit and sea level could rise 13 to 94 cm if nothing is done to limit greenhouse gas emissions (IPCC, 1995). Two of the sectors examined in *Frontiers of Sustainability*—electric power and transportation—are the largest contributors (35 and 30 percent, respectively) to carbon emissions in the United States (U.S. DOE, 1996). While U.S. CO_2 emissions are leveling off, the country remains the leading source of emissions, exceeding the second highest country—China—by more than 80 percent *(See Figure 1-7.)* And two of the sectors—agriculture and forestry—number among those that stand to lose the most if climate changes rapidly.

Agricultural productivity is threatened by changes in precipitation and in the frequency of extreme weather events that human-induced global climate change could trigger. Because scientists cannot yet accurately predict the regional impacts of climate change, the net impact on U.S. agriculture remains unknown. Under some scenarios, increased carbon-dioxide concentrations could aid plant growth and breeders could have ample time to cope with the need for crop varieties adapted to new climates. But the risk of disasters on a par with the Dustbowl of the 1930s are just as real.

Forest productivity faces similar challenges. But, forests aren't the quick-change artists that modern agriculture is. The rotation cycle for trees can be decades (or even centuries) instead of seasons. Rapid climate change could profoundly alter forest productivity, both through direct impacts on tree growth and through changes in disease or pest populations.

To reduce the risk of climate change to forest and field, we must cut carbon emissions from the electric power and transportation sectors to below current

---

BOX 1-2. Principal Sustainability Issues in Four Key Sectors

Electric Power Generation: Electric power generation is the single largest U.S. contributor to the risk of global warming, accounting for 35 percent of U.S. carbon emissions. Global warming, in turn, threatens both agriculture and forests. The energy sector is also hard pressed to reduce SO_x and NO_x emissions.

Ground Transportation: Ground transportation is a major contributor to the risk of global warming, accounting for 30 percent of U.S. carbon emissions. Any set-back thus puts other sectors at risk. In its own right, today's transportation system is unsustainable because congestion is increasing and because new fuels will have to be introduced over the next half century as global oil production peaks and begins to decline. Until that shift occurs, reducing NO_x emissions will be difficult.

Forestry: Both timber production and the forest products industry face sustainability challenges. Declining productivity growth and the potential impacts of climate change threaten timber production, and current timber practices contribute to the loss of biological diversity. The forest products industry still releases more toxic chemicals into the environment than ecosystems can absorb.

Agriculture. The two primary threats to the sustainability of the U.S. agricultural system are the loss of germplasm and climate change. In addition, current agricultural practices and excessive pesticide use lead to soil erosion and water quality degradation.

FIGURE 1-7. U.S. CO$_2$ Emissions, 1950–92

Source: Carbon Dioxide Information Analysis Center, 1995.

levels. To hold global concentrations of greenhouse gases to no more than a doubling of preindustrial levels—a benchmark but not a risk-free target—would require reducing global carbon emissions over time to about 25 percent of current levels (IPCC, 1995). Particularly since emissions from developing countries are expected to grow, all nations will ultimately have to make drastic reductions. Achieving that goal in the United States will not be easy. The Department of Energy projects that CO$_2$ emissions from power generation alone will increase by one-third from 496 million metric tons in 1995 to 656 million metric tons in 2015.

Biodiversity loss could also undermine sustainability. The nation's agricultural productivity could plummet if the genetic diversity—or germplasm—needed to maintain and increase productivity growth shrinks. This threat can't be cut down to size unilaterally since crops grown in the United States depend on infusions of germplasm from around the world, and efforts to conserve this raw material for plant breeders are falling well short of the mark. One international group of experts estimates that current expenditures on conserving the world collection of 4 million unique samples of germplasm amount to only one-third of the $200 million per year needed (Keystone, 1991).

Other components of biodiversity—our species and ecosystems—are also under threat. Old-growth forests outside of Alaska have been reduced to less than 5 percent of their original extent, and many forest ecosystem types are threatened or endangered. More than 1,000 species are on the U.S. "threatened and endangered" list. Of these, some 69 percent are declining in population size or, at best, holding their own, while only 10 percent of populations are increasing (USDI, 1990). *(See Figure 1-8.)* In the early 1990s, more than half of these listed species lived in or depended on forest ecosystems, and four of the ten U.S. regions with the highest levels of species endangerment include extensive forest areas.

The erosion of the nation's biodiversity has direct economic implications. For example, the extinction of certain salmon varieties in the Pacific Northwest would destroy some commercial and recreational fishing, and the loss of freshwater wetlands with their built-in capacity to absorb floodwaters has magnified flood damage from storms. Such direct economic costs aside, our pattern of economic growth fails to deliver what the American public strongly desires: a way to maintain the nation's biodiversity for future generations. Unlike

FIGURE 1-8. **U.S. Endangered and Threatened Species Listings**

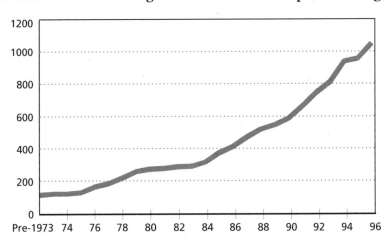

Source: Division of Endangered Species, U.S. Fish and Wildlife Service, 1997.

many forms of pollution which can eventually be corrected, species are lost forever.

In our transport system, unsustainability stems from our nearly total dependence on petroleum. Oil production continues to decline in the United States, and global production could start falling early in the next century as the world's rapidly growing demand overwhelms the finite global supply.

But doesn't history show that substitutes are developed as nonrenewable resources become scarcer and more expensive? Certainly, U.S. petroleum resources came on line *before* U.S. supplies of whale oil had been exhausted; and attractive alternatives such as photovoltaic cells already exist for fossil fuels, so the finiteness of oil supplies places no absolute limit on transportation or heating fuels. Still, because our transportation systems are the arteries of our economy and oil is the lifeblood of that system, and because introducing new power sources is tortuously difficult, heavy reliance on oil puts the sustainability of transportation at risk.

Finally, though permeated with uncertainty, the declining rate of growth in production of the nation's forest lands may signal a serious long-term threat to sustainability. Unlike agriculture, which is increasingly productive, U.S. forest productivity—as measured by the total net annual growth of wood on the relatively constant U.S. forested land area—appears to have plateaued. Productivity increased by 2.7 percent per year between 1952 and 1991, while the highest projections of forest productivity gains over the next four decades are less than one percent per year; and some studies indicate that productivity will remain flat. Stunted growth in forest productivity stems partly from declining forest health—rates of tree mortality rose by 24 percent between 1986 and 1991. Because the causes of this are poorly understood—though air pollution and suppression of natural fire regimes seem involved—and because continued decline in forest health could set back production of timber and many other forest services and products, forest health must be considered a significant long-term sustainability concern.

Striking threads run through these long-term threats to sustainability. All are virtually irreversible once they are allowed to get out of hand. Our climate system requires hundreds of years to "self-correct." Once lost, the wild genetic plant material needed to build new ultra-productive crop species (or to breed varieties that can withstand climate change) is essentially gone forever. Declines in forest productivity take decades or centuries to reverse. Even more to the point, it is hard to imagine substitutes or replacements for our climate system, missing species, or complex ecosystems.

Our greatest environmental achievements in recent decades have been in response to immediate threats to human health, such as lead pollution in the air, or threats that could be readily observed, such as smog or water pollution. In contrast, many new sustainability challenges will influence people's lives a little today and far more in future generations. Yet, the United States is not prepared to meet long-term sustainability challenges.

As individuals, many Americans plan for and invest in the future. But getting a social consensus on the need for anticipatory action is far more difficult than, say, deciding to establish a trust fund for one's grandchildren—witness the current stalemate on how to solve the Social Security crisis expected a decade from now. Impacts in the future are easy to brush aside, especially if uncertain. In fact, a core message from our sectoral studies is that a wide range of decisions on what to consume and when, and what to produce and how much, and what to invest in, are made in the here and now only—as if tomorrow did not matter. In our focus group discussions, one man from Minneapolis commented:

> *"I think this is a society that doesn't really think that much about the future. We don't care. There aren't enough people that really care about the environment, the national debt. A good proportion of Americans think about themselves, and what can I get now."*

VISIONS OF THE FUTURE

The future implied by existing trends is not encouraging, but it is by no means a given. Quite the contrary, any of several sustainable futures seems possible. For each sector we picked one that reverses the unwanted trends we see unfolding now, and—however raggedly—each intersects with the other three visions. None is cast so far into the future that we can't set it in motion today. All are technologically possible or feasible, not dependent on some new process or technology we can't imagine or lay hands on. All help us weather surprise or uncertainty, enhance useful feedback to decision-makers, and reduce vulnerability to poor decisions. All build on notions of flexibility and substitutability to sketch a world with more options and choices. Finally, all spread responsibility for environmental protection among all agents of change—consumers, producers, and governments alike. If we follow these principles on a two- or three-decade tour of agriculture, forestry, transportation, and electricity production, quality of life for ourselves and our children will continue to improve. And the country's definition of "quality" will move beyond bigger houses, better education, and better jobs to include a healthy social and physical environment.

A Vision of an Environmentally Sustainable United States

A sustainable American economy will generate enough jobs to employ the workforce. The proportion of jobs for highly paid skilled labor will also grow, allowing Americans to enjoy growing incomes and to reinvest in the economy, to educate their children better and become lifelong learners, and to save for their own retirement.

As the economy continues to develop, it will also lighten its environmental "footprint." The workforce will find ways to "recycle, reduce, and reuse." The economy as a whole will produce more with less, using new technologies, advanced information systems, and better design procedures to become more productive. Greater productivity will reduce industry's tab for environmental compliance and cleanup.

Wastes generated in the production of manufactured goods will diminish as efficiency improves. For this reason, total natural resource consumption by the average American will go down even as the amount of goods produced grows modestly. Industry's move to recycle and reuse manufactured parts more routinely will further propel this trend.

The U.S. economy will switch from fossil fuels to renewable energy sources, and this domestic energy transition will help ignite a global one. Resource shortages will drive change, but so will the eventual need to address the risk, and perhaps the emergent costs, of global climate change.

In many ways, services—particularly information services—will substitute for goods. Although it's already become a cliché, the "information superhighway" could allow many people to attend meetings, shop, and stay in contact with family and friends "virtually." Using the Internet could help us eliminate excessive travel and meetings. Information, communications, and sectors of the economy that use little or no materials will grow as a proportion of total economic activity.

As we learn to use material more efficiently, fewer pollutants will find their way into the environment. Greater knowledge of chemistry and improved process design will increasingly allow industrial toxic chemicals to be reused or captured instead of released to the environment. Fewer highly toxic chemicals will be released as less harmful substitutes are found.

pushed by the American public. Environmental awareness grows with incomes, as does the desire for healthy communities and better recreational opportunities. A highly productive workforce will spend more time at leisure and more leisure time will be spent out of doors. Revenues from tourism and affiliated industries will rise. As more industries and jobs become free to locate far from today's industrial or financial centers and as the ranks of retirees grow, more and more Americans will choose their hometowns with environmental and other non-work-related issues in mind.

As more and more Americans seek to live and work in areas near such natural amenities as water, mountains, or forests, communities will realize the increasing value of protecting and restoring natural assets to attract economic investment. For this reason, many resource jobs in forestry, mining, and other fields will be supplanted by high-tech industries and service sectors, such as health care.

Changes will transform the four sectors covered in this report. In agriculture, for instance, soil erosion will be contained at levels that don't sap productivity, partly because farmers will be able to carefully monitor soil and crop health. Pesticide use will drop dramatically, as U.S. agriculture comes to rely principally on biological, agronomic, and other mechanisms to control intruding organisms. Nutrients will be applied in ways that minimize losses and in amounts keyed precisely to crop requirements. Economic signals will be used to support conservation objectives. For example, it will become common for "point source" polluters to pay such "non-point sources" as agriculture to reduce "chemical loadings" and to claim credit for these reductions. Agriculture will become a significant supplier of energy, paper, and industrial feedstocks. Perennial crops grown on tens of millions of acres will be harvested to make electricity.

At the core of a sustainable ground transportation system will be emissionless vehicles powered increasingly by renewable energy sources (such as electricity or hydrogen derived from photovoltaic cells, wind turbines, biomass, wastes, or other renewable technologies). Concentrations of urban air pollutants, largely the result of vehicle emissions, will fall as gas-powered cars, trucks, and buses are retired. Greenhouse gas emissions will follow the same downward curve.

A technological transformation will also sweep the electric utility sector as renewables capture a rapidly growing share of energy supply—the sanest approach to coming to grips with climate change. Petroleum, coal, and natural gas won't necessarily disappear, but fossil fuels will reflect the real risks of using them. Electricity grids will still exist, but large central generating stations will provide a smaller percentage of the power our economy needs. Many households and businesses will meet their needs for space heating and hot water with small turbines or hydrogen fuel cells. The fuels for these electric generators will be clean gases, including natural gas (for some time to come), gasified biomass, and hydrogen generated by electrolysis of water using renewable sources of electricity. In this far more sustainable future, photovoltaic cells will be incorporated into roofing shingles and other building materials and accompanied by on-site electricity or hydrogen storage.

Americans will continue to look to the forest sector to provide products used to communicate information, package and store materials, frame buildings, and furnish the home and workplace. But they will increasingly value (and pay for) the basic environmental services forests provide—clean water supplies, flood protection, and habitat for fish and wildlife. Technological changes will substitute for some paper products, eventually depressing overall demand for pulpwood. While the acreage of protected forest will remain relatively stable, strategic additions and land swaps among private land owners and state and federal agencies will allow the United States to maintain a more representative national network of protected forest ecosystems. Air and water emissions for forest product manufacturing will continue their steady decline, and "closed loop" manufacturing processes that keep waste and pollution to a minimum will become the industry norm in the next generation of capital investments. The forest sector will provide fewer logging and sawmill jobs, though some communities will still depend heavily on traditional forest sector economies. However, emerging markets—for specialty wood products, recreation, nontimber forest products, and even such environmental services as carbon sequestration—will afford many people and communities with historical ties to forests new opportunities to make a decent living without uprooting.

Governments will provide incentives for changes such as these, but increasingly will help determine the factors that industry and consumers bring into their decisions, not the decisions themselves. Government will watch performance, not methods, giving businesses the leeway to decide how to control pollution and waste. A gradual move away from taxation of such economic "goods" as income and investment will be supplanted, at least partially, by a move toward taxing "bads," such as pollution.

Americans will call for stewardship by government and corporations and will practice it themselves at home and on the job. Citizens will realize that all environmental problems can't be solved by government, or by industry, and that such quotidian choices as what foods we eat, what cars we drive, and how we drive them all count. Children will be a force for change as environmentally aware offspring assess their parents' environmental performance.

America will come to define security in broader terms than in the past. As the overwhelming threat of nuclear annihilation recedes, less cataclysmic but equally important issues—environmental degradation, migration, and economic disparities—will move to center stage. With the world's population pushing the planet's ecological boundaries, environmental problems and social instability will erupt in more regions and will capture world attention. The United States will find that it is in its own best interest to take a strong leadership role in combating global climate change, resource shortages, and poverty since the only alternative is supplying bread and soldiers once these pressures push populous regions over the brink. Immigration pressures will mount in both rich and poor nations. Political conflict between the "haves" and "have-nots" will overshadow those of political or economic ideologies. The necessity of cooperation and the folly of confrontation will be clearer than ever and will usher a new era of collaborative problem-solving in world politics.

OVERCOMING BARRIERS TO CHANGE

The future described here and in the four sectoral studies is one of many sustainable itineraries, but it is a far cry from the trajectory the United States is on. Change is indicated—change in decision-making, in the make-up of the capital stock of our economy, and in consumption patterns and personal behavior. These changes will not occur without new incentives to innovate; the removal of the hurdles to a smooth economic, technological, and institutional transition; and attitude changes in tens of thousands of communities.

The roots of these barriers run deep. Many U.S. policies—say, subsidy payments to ranchers or the protection we give to the domestic sugar industry from international competition—create powerful vested interests that can effectively block change. In other instances, change may be delayed because we lack the scientific knowledge needed to pin down the best course of action even though we know the current course goes nowhere. Or partisan politics may polarize an issue and block resolution. All of these factors block progress, but our sectoral studies revealed two even more fundamental barriers: (i) technological "lock-in" and capital investments in outmoded technology, and (ii) public attitudes. If new ideas can trump political inertia in these two areas, sustainability will have the foothold it needs.

Technological Lock-in

Resistance to change is literally built into the technology that we use. What makes adopting new technologies so difficult, even when they seem clearly superior to those in use, is the large investment of capital and know-how in the present technology (Arthur, 1990). In all four sectors, the shift to a more sustainable future is slowed by this "lock-in."

As a very simple example of technological lock-in, consider the keyboard on your computer. The sequence of the keys is a throwback to the design of the original manual typewriter, which was intentionally made to be inefficient, with few of the most commonly used letters near the center of the keyboard, to keep typing speeds low enough that the keys wouldn't jam. Now this slowhand "QWERTY" typewriter (named for the first letters in the upper left of the keyboard) has been "locked in" as the technology of choice even though the reason for its ascendancy makes no sense at all in the computer era. Left to market forces alone, we will use the QWERTY typewriter as long as people type, even though much more efficient alternatives exist. In the case of typewriters, this is a small price to pay. In the case of gasoline engines, pesticides, or coal-burning power plants, the price of staying with what you know is the difference between growing environmental threats and sustainable development.

One example of the positive feedback loop associated with technological lock-ins can be seen in the growing quantities of pesticides used in U.S. agriculture. Initially, pesticides reduce pest damage, but eventually pests develop resistance to the pesticides, making heavier applications necessary. Indeed, while U.S. pesticide use has soared over the past decades, losses to pests have not fallen substantially because pests evolve rapidly and the strains resistant to pesticides thrive. U.S. crop losses to insects, diseases, and weeds from 1942 to 1950 totaled about 32 percent; in 1984, about 37 percent. Nevertheless, were we to end pesticide use precipitously, crop losses might temporarily skyrocket since the crops have been bred to depend on them.

Ground transportation faces a similar hurdle. Our transportation system might have locked-in on electric vehicles in 1900, when nearly 40 percent of new vehicles ran on batteries, 22 percent on gasoline, and 40 percent on steam (MacKenzie, 1994). But as gasoline became the fuel of choice, a vast infrastructure of service stations, road networks (rather than electric railways), and even city layouts came about to accommodate gas-powered vehicles, so now the switch back to electric vehicles or to other transportation fuels is going to be expensive and difficult. In a classic Catch-22, people are wary of purchasing less polluting vehicles powered by natural gas because few fuel stations exist to serve them, and these stations are few and far between because Americans aren't purchasing the vehicles.

Resistance to change is also built into investment. Consumers who want to buy a more fuel-efficient automobile have to take into account how much has already been invested in their current cars. If a manufacturer of paper products wants to install a new super-clean production process, some financial advisor is sure to point out the value of the capital investment sunk in its current plant. Replacing one asset for another prematurely—when the cost of replacement exceeds the benefits of new technology—can be expensive. How many of us put off buying a new energy-efficient refrigerator until the old one actually quits? The same concern over recouping the original investment applies in business, at home, and for governments. The value of the fixed capital stock of the U.S. electricity industry (excluding government utilities), for example, was $810 billion in 1992—about 5 percent of the nation's total fixed private capital. *(See Chapter 5.)* That massive investment implies substantial inertia to change.

Public Attitudes

Many of the proposed changes in technologies, policies, and behaviors identified in WRI's sectoral studies have been well-known for years, but the U.S. track record is disappointing. For example, in 1993 a proposal to introduce a gas tax that would have helped to reduce CO_2 emissions became controversial and eventually was reduced to less than 5 cents per gallon—about $25 per car per year. Similarly, how many of us have passed by compact fluorescent lights for the energy-intensive but more convenient alternatives? Since consumers and tax-payers already feel strapped—never mind that today's cost of driving is lower in the United States than any country in Europe and local communities may see greater economic opportunity in developing land rather than protecting it—the notion that sustainability is worth paying for must be driven home before the policies can be introduced to tackle these problems.

Reducing the threat of global warming, for example, will require a change in our economic system as deep as those wrought in our social systems by the civil rights movement in the 1950s and 1960s. Righting the wrongs of discrimination could not begin until the public began to accept that discrimination was both wrong and rampant. Only with that change of heart could new laws and policies be put in place to require long-overdue changes in institutions and behaviors to address the problem, which even to this day is far from solved. The point is simply that we will not tackle the threat of climate change if only scientists and policy-makers know and care about the risks.

Confirmation of the key role of public attitudes was an important conclusion of WRI's sectoral studies, but our research wasn't designed to yield detailed recommendations on how to change them. Greater openness and increased information clearly can be influential, but in an era of information overload and at a time when environmental disputes are already too often oversimplified, information alone isn't enough. Still, reason for optimism that social change will occur lies in the rapid pace of technological and institutional change that Americans now take for granted. It no longer requires great imagination to envision all Americans driving electric vehicles or powering their houses from rooftop solar cells or other renewables. Indeed, we have come to expect that we will be doing things quite differently ten years from now simply because we were doing things quite differently ten years ago. Once the solution is within sight, public acceptance becomes easier to engineer.

Change need not force us to abandon basic American values. Indeed, the visions outlined in *Frontiers of Sustainability* suggest a world of fairly simple virtues—efficiency in converting materials into products and services; ingenuity in designing policies and fashioning technologies to lessen our environmental footprint; stewardship of our wealth of natural resources and capital; cooperation among individuals and institutions to find and act on mutual interests; and choices that enable us to intelligently balance our needs and desires with those of others, now and in the future.

Public attitudes and policies influence each other. To make sure that influence is positive in the transition to sustainability, Americans and their leaders must keep fairness, justice, and common sense in mind, for when logical and necessary changes run aground politically, it is often because the proposals fail one of three basic tests of public support.

The Common Sense Test

As rancorous as Congressional debate on the role of environmental protection in the United States is, the American public is *not* deeply divided on the issue. In fact, again and again national polls confirm that Americans consider environmental protection important and they don't want to give up hard-earned benefits. The evidence is that most Americans take deep pride in our natural heritage and past environmental accomplishments and have at least a glimmering of the challenges to come. Public support for the environment, however, does not translate into unquestioning acceptance of every program, policy, and priority. Overtaxed and feeling at odds with Washington, Americans want environmental initiatives to pass a common sense test. Are the changes or sacrifices required reasonable and understandable? Do they match the gravity of the challenge? Will they work?

A few examples illustrate this "show me" skepticism. The American public fairly solidly supports endangered species protection, but in communities where jobs or opportunities may be put at risk by steps to protect a species, people often want proof that the value of protecting *this* species or habitat is really worth any required economic sacrifices—witness the political storm in the Pacific Northwest over saving the spotted owl and the old-growth trees it calls home. Similarly, most Americans also seem to want automobile pollution reduced, but centralized emissions-testing programs—annual tailpipe tests that usually mean missing work and waiting in line—are unpopular. And corporations constantly complain that environmental regulations themselves are less troublesome than the delays in investment and production decisions that sometimes go with them. In case after case, Americans agree there is a problem, but take exception to the proposed solutions.

The Fairness Test

Nobody likes to make sacrifices or changes that others aren't asked to make, especially if the problems stem from everyone's actions. Firms will fight pollution-control requirements that appear to single them out and ignore other pollution sources, especially if the others are competitors. Strategies to protect and enhance American environmental assets will work best if they are viewed as making everyone who is part of the problem accountable for it. In our focus groups, one woman from Atlanta noted that effective policies must *"work towards universal acceptance, not just a few who must sacrifice. Having lots of money should not mean you don't have to comply."*

This test is consistent with the steps recommended here for the transition to sustainability. Making better decisions about the future requires tailoring solutions precisely, perhaps assigning different players different responsibilities but leaving no one out who has a stake in the outcome. Decentralizing more decisions, giving them to regions, states, or localities, is one way. Another is to chip away at the American public's outdated belief that large corporations cause most of our environmental problems—an excuse for abdicating all personal responsibility. For example, sustaining the positive environmental trends of the past two decades will almost certainly require policies that will make driving cost more. Detroit will have to change, but so will all drivers.

But one dimension of the fairness test works against the acceptance of taxes or additional charges to internalize environmental costs. Fairness also means rewarding positive behavior and penalizing bad. Almost a truism, this guideline hits a political nerve in the United States. The public has never warmed to pollution charges or taxes, for example, mainly because no one likes the notion of new taxes, but also because even efficient or effective taxes are viewed as penalties to be paid by the good and the bad alike. Conversely, much of the public does support subsidies—payments aimed at getting individuals or firms to "do the right thing." Many people consider a per-bag fee for waste disposal unfair because they feel that they have no choice but to dispose of wastes. A more effective approach may thus be to promote a revenue-neutral disposal fee—someone putting out one can would receive a discount while the neighbor who puts out three would pay a tax.

The Justice Test

The justice test asks whether a policy is the "right" thing to do. No U.S. environmental law has been predicated on the notion that wealthier individuals should be able to buy higher environmental quality. Instead, the underlying principle is that all Americans have a right to live in a sound and healthy environment. Yet, for numerous reasons, poor people and minorities today do bear a disproportionate burden of the health risks of environmental degradation. For example, African Americans and Hispanic Americans are more likely than Anglo-Americans to live in areas with poor air quality. *(See Figure 1-9.)* And, a recent study by the Florida Environmental Equity and Justice Commission found

FIGURE 1-9. **Percentage of Population Exposed to Poor Air Quality by Race**

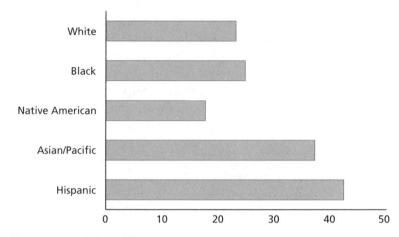

Source: Seitz and Plepys, 1995.

more than two thirds of some 3,000 hazardous waste sites in Florida are located in pockets of poverty where minorities live (Florida Environmental Equity and Justice Commission, 1996).

A strong case has been made by the President's Council on Sustainable Development that economic growth, environmental protection, and social equity are linked (President's Council on Sustainable Development, 1996). Our study supports this conclusion. The nation's first priority for improving environmental quality must be to ensure that all Americans live and work in a healthy environment, and turning this principle into reality means according higher priority to controlling pollution affecting farmers, urban dwellers, and minorities. Nor should any policies for seeding the transition to sustainable development even temporarily add to the burdens on beleaguered groups.

Many seemingly logical policies have been stalled because they seemed to hurt some communities more than others. Bottle bills, designed to encourage recycling by requiring deposits on bottles that would be refunded when the bottle was returned to the store, have sometimes been opposed because the urban poor, with less space to store bottles and no vehicles to ferry them to stores, would pay a disproportionate price. The gas tax introduced in 1993 was initially opposed by advocates for lower income groups on the grounds that the poor must spend a greater share of their income on transportation. The transition to sustainability *will* require individuals who choose products that are more environmentally destructive or who engage in environmentally destructive activities to pay a higher price. But sustainability isn't served if they make the poor worse off.

HOW TO BE MORE SUSTAINABLE

How can the United States get on a more sustainable development path? Step one is to design our vision of the future collectively through dialogue and consensus-building in processes such as those that are already completed globally (*Agenda 21* of the Earth Summit) and nationally (President's Council for Sustainable Development) and those that are now taking place in many U.S. regions and cities. Step two is to form goals based on this vision, primarily increased prosperity, increased fairness, and improved environmental quality. Exactly how depends on the sector, but five basic actions summarize the core lessons from WRI's four studies. This "greenprint" for a more sustainable United States can be applied in other economic sectors, in states, and in communities.

1.   Build the Future into Today's Technology Choices

The four visions of sustainability mapped out in this book involve technologies different from those we commonly use today. Electric vehicles, energy supplies from biomass or photovoltaic cells, and paper products made from grasses represent new ways of meeting demands for economic services while diminishing the environmental footprint. New technologies can expand our choices, create less damaging substitutes for valued products, and help us put our environmental assets to new uses or even make up for lost ecological wealth. Still, technological change is no cure-all. Technology will never deliver good substitutes for certain natural resources, and a wide range of choices for products and services means little without the knowledge needed to choose and use them wisely. But since the dynamic U.S. technological base will change no matter what we do, the challenge is to make it work for sustainability instead of against it.

The stock of technology that we bequeath to future generations is not determined in a vacuum. Although serendipitous discoveries in a laboratory or at a workbench do happen and can profoundly change our lives, complex institutional, legal, economic, and regulatory factors also govern the development of the tools, machines, and production processes that define our economic system.

Tax laws influence the rate and pace of investments in new machines by specifying how long it is profitable to keep old ones and by influencing how much new ones will cost. Patent laws can create or destroy incentives for innovation by defining the terms of ownership. Environmental regulations that tell companies exactly how to comply can freeze technology development and actually work against our environmental goals and objectives. The level of technical training and education of our workforce can mean the difference between creating new opportunities and choices and making do with what we have on hand. Corporate cultures and traditions make one company embrace technological change as an opportunity and another see it as a threat or ignore it altogether. Government largess can, sometimes unintentionally, dictate which technologies companies and individuals try.

Many of the policy recommendations that stem from WRI's case studies rely on the market to stimulate more efficient and locally appropriate solutions to environmental challenges. But market-based approaches can't always overcome technological "lock-in." Typically, the short-term cost of switching to what would in the long run be a more economically desirable system exceeds the short-term benefits of clinging to the status quo. Our sectoral studies identified several ways to spark technological innovation and to break technological lock-in. Three in particular stand out: the use of performance-based standards, government R&D, and public environmental reporting (discussed under Action #3).

Use Performance-based Standards to Stimulate Innovation

No one likes to be told what to do and when to do it, and most Americans think they can find the best way to do something if left to their own devices. Yet, under the traditional approach to environmental protection, government has both set the goals for environmental quality and specified the technologies or processes required to achieve them—wastewater treatment plants, smokestack scrubbers, and so forth. This system often stifles innovation. Judging compliance with environmental standards strictly in terms of the use of a specific technology may kill interest in another much better one. Far more effective is judging environmental *performance* and allowing firms wide technological leeway in achieving it.

Greater reliance on performance and less on technology prescriptions gets the desired environmental outcomes without straining innovation and productivity. The impressive achievement made in phasing out the production of CFCs for use in the United States was a response to a performance goal—no

CFCs—that allowed manufacturers wide liberty in compliance. Similarly, in response to the 1996 California Air Resources Board (CARB) regulations requiring that by the year 2003 one-tenth of all cars offered for sale in California must be exhaust free, automakers the world over are developing battery- and hydrogen-powered vehicles and hybrid cars. CARB's performance goal—zero emissions—has opened the door to the development of many alternatives to the oil-powered car and gives Detroit (and its competitors) the flexibility needed to try to beat the competition. Whereas rigid regulation can create adversaries, flexibility can create partnerships in environmental protection, as well as cost savings and more immediate environmental benefits.

One reason technology standards have long been preferred to environmental performance goals is that measuring performance is difficult. For U.S. EPA to measure air or water emissions from tens of thousands of factories would have been impractical, and the cost for firms themselves to measure some types of emissions would have been far too steep. But technological innovation is dissolving this barrier since new monitoring and information equipment based on biosensors, microelectronics, and remote sensing (among others) now allow technicians to track emissions at lower cost and on a continuous basis. Unfortunately, their potential remains largely untapped so far.

Increase Government Support for R&D Geared to Sustainability Challenges

The development of new technologies rests squarely on the shoulders of the nation's research and development capabilities. U.S. laboratories, workplaces, homes, and even garages (where Apple Computer's Steven Jobs and many other technology entrepreneurs got their start) are the birthplaces of new technologies and new ways of meeting demands for economic services. Research and development (R&D) comes only at a price, however. Funds for greater research on new crop species, on "smokeless" energy-supply technologies, or new batteries for electric vehicles have to come from somewhere, and private investment in R&D has to compete with other investments, most of which pay off faster. Then too, the risks are high: R&D is not guaranteed to produce anything of economic value, especially for the original investor who took the risk.

In a competitive economy, firms are wary of making R&D expenditures if they think their competitors may cash in on the results. These concerns are amplified in the case of "green" technologies since R&D spending on them often faces direct competition from such spending on conventional energy sources or on traditional transportation options that have powerful constituencies. In this instance, government needs to play an aggressive role—often in partnership with business—to fund and stimulate the R&D needed to make the shift to sustainability. Moreover, governments should flex their purchasing power more to create demand for environmentally sound technologies.

These public investments will be easier to make if Americans understand that national security is at stake. For example, the U.S. government has invested heavily in agricultural research and extension since 1864. One result has been the continuous gains in yield that have been a hallmark of U.S. agriculture. Today, both financial and political support for public agricultural research is shrinking. Our research suggests that what support remains should be used to address broad public concerns related to agriculture, including adaptation to climate change, protection of germplasm, and pollution reduction. To be sure, the technologies to cope with these problems will not come quickly from private R&D alone.

We identify a somewhat different approach for public support of new technologies for supplying electric power. One quick way to accelerate development of renewable energy technologies is to create more demand for their use. A good option for creating this demand is to establish a "Sustainable Energy Trust Fund" (SETF), comparable to the Highway Trust Fund that pays for highway maintenance through a gasoline tax. The SETF—which would be run by electric utilities—would obtain revenue through a small fee on electricity sales and use these funds to cover the difference between market energy prices and the price of renewable power.

As the quintessential innovative thinker Albert Einstein put it, "We cannot solve the problems that we have created with the same thinking that created them." With his words in mind, we need to redirect and boost R&D spending by looking for better ways of providing services. Government R&D spending should not prop up an inefficient or outmoded industry; it should catalyze technology turnover. Encouraging greater private sector buy-ins to public-private research partnerships and making a long-term commitment to government R&D will help bring the business world into the sustainability camp.

2. Bring Environmental Accountability "Closer to Home"

Many of the policy recommendations presented create incentives and decision structures that draw the future into our current economic choices. Pricing transportation services to reflect the true costs of driving, making agriculture financially responsible for the pollution it causes, and increasing the price of fossil fuels used in the electric power industry all help to ensure that the decisions made—whether by corporate executives or commuters—give greater weight to the future environmental costs or benefits. The key is widening the range of informed choices at the exact point where economic decisions are made and making it as easy as possible for Americans to take advantage of the options.

One of the biggest opportunities for shifting our development path closer to sustainability lies in "getting the prices right"—that is, internalizing current and future environmental costs and risks in the prices that businesses and consumers pay for products and services. At their core, such policies make those who make decisions about what to buy or what to throw away fully accountable for their choices. In short, accountability shifts from regulators to those making economic decisions.

Policies that bring decisions about our environmental future closer to home depart sharply from the tradition of U.S. environmental policy-making. In early efforts to reduce pollution's toll on the U.S. environment, the strong presumption was that the federal government needed to step in, provide national mandates, and enforce them with brass knuckles. State and local governments and communities retained certain responsibilities and choices, but they mostly had to meet national goals using standardized "one-size-fits-all" approaches. National mandates and standards setting baselines for environmental quality still make good sense since many pollutants cross state lines, industries could otherwise re-locate to states with lax standards, and some broad national objectives—such as protecting biodiversity or priceless natural landscapes—may conflict with narrower local views. Moreover, such a centralized "command and control" approach is largely responsible for the achievements made so far in U.S. environmental quality and remains the best bet when only a few large polluting factories or utilities are involved. But centralizing standard setting and accountability for decisions is not the way to manage countless small and non-point pollution sources or to meet the need to reconcile conservation and development objectives at a local level.

Simply "decentralizing" environmental decision-making is as likely to degrade as to restore the environment. For example, the Air Quality Act of 1967, which required states to adopt their own air quality standards, set off a stampede among states to see who could set the weakest standards so as to attract industry. The better way is to:

1. Set a baseline of state, regional, or national standards (with provision for more stringent standards if local conditions warrant) through social consensus by asking if significant consequences would be felt by others

not involved in setting that standard. (If so, the standard should be set at a higher and more inclusive level.)

2. Decentralize accountability for decisions to the lowest point at which the full consequences of those decisions can be felt (often, by using pricing mechanisms to internalize the full costs of those decisions or through participatory decision-making involving local stakeholders).

3. Establish effective monitoring and enforcement programs to ensure that standards are achieved.

For example, the consequences of a farmer's decision to overuse fertilizer may bypass the farmer and hit individuals using the polluted waters downstream. As a result, many counties and states now regulate non-point source pollution to decrease nutrient run-off. But a far better approach would be to tax farmers' fertilizer use in excess of need (judged by the difference between the amount of nutrients applied and the amount harvested in the crops). The total amount of fertilizer used by farmers would fall to the level that best meets social objectives, while any individual farmer would still be free to choose how much fertilizer to use and where.

Moving environmental accountability closer to home opens the door to surpassing compliance with regulated standards. Why should a factory manager reduce emissions beyond a regulated standard if his firm gets nothing for the extra effort? Only if the company is fully accountable for the costs and benefits of its choices about how much to emit—and can thereby realize greater economic gains by further cutting emissions—will managers have an incentive to go beyond the standards. At an even more localized level, individuals or businesses facing fees for disposing of paper, bottles, and other trash are much more likely to find effective ways to reduce the volume of waste they generate than any local government could. This change in the incentive structure is at the core of sustainable development since it builds progress toward improved environmental quality into sound decision-making and does not force actions that don't make sense.

The need for effective price signals is particularly clear in transportation. Drivers and their passengers currently do not pay many of the costs that they impose on society, such as the costs of combating global warming, the medical costs of people sickened by air pollution, and the military costs of defending overseas suppliers of crude oil. The failure to pay the full costs of driving encourages the overuse of imported oil, excessive greenhouse gas emissions, high emissions of air pollutants, and the underuse of alternatives to driving oil-powered cars, including alternative fuels and alternatives to driving itself. Currently, U.S. drivers enjoy the lowest gasoline prices and lowest gas taxes of any citizens in developed nations. *(See Figure 1-10.)*

Shifting accountability closer to home involves more than getting the price right. In the forest sector, one key is developing regional and state forest-sector management plans with the participation of all who manage, use, or benefit from the region's forests. Traditionally, the question of who wins and who loses in such plans, and whether the environment is improved or degraded, was answered at the state or federal level. Small wonder local opposition to decisions reached paralyzing proportions. By establishing bottom-line state or federal level objectives but setting local stakeholders free to meet them any way they can, government increases the chances that broadly acceptable or win/win solutions will be found.

A more sweeping approach is to re-think our nation's entire system of taxation. Currently, local, state, and national governments raise revenue by taxing the very things that we want to encourage in society—labor, investment, and savings. From an economic standpoint, society would gain wealth if we shifted some of the tax burden from intrinsic goods to air and water pollution, greenhouse gas emissions, wetlands loss, forest degradation, soil erosion, and the like (Repetto et al., 1992). This way, individual decision-makers would be more accountable to society for their environmental decisions and society as a whole gets a "double dividend"—reductions in both pollution and taxes.

3. Inform Our Choices

We cannot make wise decisions about the future without good information on our range of choices and their consequences. That's why the role of information—helping to create options, motivate innovation, and improve efficiency—figures centrally in each sectoral analysis. The publication of data

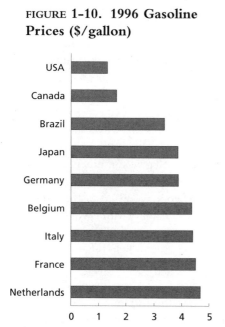

FIGURE **1-10. 1996 Gasoline Prices ($/gallon)**

Source: *Washington Post,* May 11, 1996.

on releases of toxic chemicals to the environment by major U.S. businesses helps communities assess their environmental risks, helps firms measure their environmental performance, and gives government a starting point for reducing unreasonable risks. Information alone can also spark change. Indeed, releases of toxic chemicals reported under the Toxic Release Inventory fell by 44 percent in less than a decade once firms were required to begin reporting their emissions. More generally, our study finds that many of the most important policy changes will not take place until attitudes and behaviors change—and to do that people need ready access to credible and accurate information.

In agriculture, current USDA cosmetic standards for fruits and vegetables encourage excessive pesticide use because grades—and thus prices—are determined largely by appearance, which has nothing to do with nutritional value. Changing this measurement and information system could reduce both environmental and health threats. Information has already helped Americans use energy more efficiently. Many utilities offer consumers free energy audits, and federal law requires labels on major appliances to indicate their relative efficiency ratings. Deregulating electric utilities will redouble the need for consumer information. Because energy users will soon be able to choose one electricity provider over another, they need clear and pointed information on cost, emissions, and the fuels that each electricity provider uses. At the same time, the industry itself needs better information on the likely costs of alternative ways of producing energy so that it can plan "least cost" investments to meet future energy demand.

Information can also help companies shouldering the extra costs of practicing sustainable forest management to recoup their investment in sustainability and stay competitive. Independent (or "third party") certification programs now being set up by the Forest Stewardship Council will give consumers information about which wood products come from sustainably managed forests and which don't. Consumers willing to pay extra for the "good wood" make the sustainable timber practices a competitive opportunity, not a liability. More generally, sharing information on demonstration projects and model forestry planning initiatives like those recommended in the forest sector study can help other communities or regions plan for the future.

Public environmental reporting—like the Toxics Release Inventory—informs public and private choices and speeds the transition to environmentally sound technologies. But in the forest sector it fails to reveal some of the main environmental risks associated with pollution (partly because some chemicals are far more toxic than others, but the TRI measures emissions by weight only), doesn't measure toxics that are disposed of rather than emitted to the water or air *(see Figure 1-5),* and doesn't even touch on other facets of environmental performance. New pulp and paper plants, for example, have modified processes to reduce the release of persistent organic pollutants and lowered the consumption of fossil fuel per unit of production; such changes go unreported, however, so the public can't accurately evaluate progress.

One requirement that would help inform public and private decisions and hasten the transition to sustainability is for factories to report results on "materials balance sheets"—how much resource enters the production process and how much product and waste leaves. This measure of "material flows" would complement the tallies of financial stocks that firms report in their financial balance sheets. These data are already becoming available economy-wide to complete the picture that emerges from national income accounts (Adriaanse et al., in press). They should also be reported at the firm level. Firms regularly report their financial flows—income and expenses—using Security and Exchange Commission criteria to help investors make sound decisions. If they also reported their materials flows, then local communities can do the same. Some companies object that this information could give competitors insights, but financial reporting raises the same threat and key materials information that would reveal trade secrets could be kept confidential.

Looking beyond our four sectoral studies, the indicators now used to gauge economic performance in the United States provide a distorted picture of progress toward sustainable development. Two basic economic measures of national economic status—general economic activity and productivity—keep the worth of our environmental assets and the values Americans place on the environment and our natural resources largely off the books. Using current methods for measuring Gross National Product (GNP), the destruction of valuable natural assets—the loss of a forest or the depletion of a fishery—would be recorded as an economic *gain* (Repetto et al.,1989). And since the GNP is simply the sum total of our nation's balance sheet, the failure to account for the real economic value of our natural resources reverberates throughout our system of economic measures—dangerous because we will underinvest in whatever we can't measure.

Other key economic indicators also yield half-truths. One of the most fundamental measures of the nation's growing economic prosperity is our growing productivity—the rate at which we convert "raw materials" (labor, resources, and capital) into various goods and services. But these much-watched measures include only the "gross" production—they don't subtract the costs of the pollution or environmental degradation stemming from the manufacture or distribution of goods (Repetto et al., 1996). For example, by conventional measures, productivity in the electric power sector declined from 1970 to 1991. But if the economic benefits of substantially reducing emissions over this period are captured in the formula, productivity grew. Clearly, until we begin accounting for environmental costs and benefits in measuring productivity, our decisions won't be well-informed.

Accounting within firms suffers from an analogous problem. Environmental expenditures are seldom traced to their precise cause. Rather, like overheads, environmental costs are spread across all the firm's accounts. Plant managers or chief executive officers reviewing corporate accounts cannot easily tell which lines of their business add most to the environmental tab (Ditz et al., 1995).

If environmental and natural resource values were accounted for more explicitly in measures of economic and financial performance, tracking sustainability would move beyond guesswork. For now, greening our economic and financial accounting systems is itself, as the Nobel Prize-winning economic theorist Robert Solow puts it, practicing sustainability (Solow, 1992).

4. Confront the Biggest Challenges to Sustainability: Climate Change and Biodiversity

Two sustainability challenges stand out both in terms of difficulty and their probable effect on future generations: climate change and the loss of biological diversity. We are currently losing ground on both because they are easy for policy-makers and the public to ignore.

Climate Change

Since energy use is embedded in our economy's warp and weft, climate change affords us no easy outs. Mitigation measures would reduce greenhouse gas emissions or remove some of these gases already lingering in the atmosphere. Adaptation means adjusting technology and management practices to reduce the physical, biological, social, and economic impacts of warming. Either strategy can be adopted before or after the fact of climate change.

Emissions of greenhouse gases can be cut by adjusting energy prices, substituting more efficient technologies for less efficient ones, and changing land-use management, especially in agriculture. In power production, projected CO_2 emissions can be cut by burning fuels more efficiently and by switching from high-carbon fuels (coal) to less carbon-intensive energy sources (natural gas). In electricity production specifically, renewable energy sources (solar, photovoltaic, wind, biomass) can begin to displace fossil fuels. Renewable energy sources would penetrate markets faster if we had a sustainable energy trust fund like the one described here. Many of these are "no regrets" policies: it makes sense to do some fuel switching or conversion to renewable energy, for example, whether or not climate change is in store since all Americans want cleaner air.

Next in line after "no regrets" options come low-cost ones, including major reductions in commercial and residential energy use. Installing new high-efficiency furnaces, air conditioners, lighting, and appliances can cut both energy consumption and CO_2 emissions dramatically. So can putting individual meters in apartments and commercial units. Indeed, together such improvements could halve U.S. energy use. Using the same principles, material and energy inputs in manufacturing can also be cut down to size.

Our transportation sectoral study offers many options for cutting CO_2 emissions. In the near term, carbon emissions can be reduced by improving new-vehicle efficiency (through higher fuel taxes and higher taxes on "gas guzzling" vehicles) and by changing the driving habits of solo commuters and speed demons (through parking reform and congestion pricing). Over the long term, emissions can be cut by reconfiguring cities to reduce the need to drive, by introducing innovative forms of public transport (such as Personal Rapid Transit), and by phasing in electric-drive vehicles powered by batteries, hydrogen fuel cells, or flywheels.

Steps can be taken in agriculture and the forest industry to both mitigate and adapt to climate change. Some 60 million acres of agricultural land are now idled. Millions of these acres are not environmentally sensitive and could be used to grow biomass for electrical generation or liquid fuel production, thereby reducing net emissions of greenhouse gases. Indeed, our forest sector study proposes allowing power plants to offset their carbon emissions by funding forest restoration on degraded land or subsidizing biomass plantations.

As for adaptation to climate change, pace is everything, and changes in global surface temperature could occur 100 times faster than in the past. Room for adaptation must be found throughout the economy, taking natural ecosystems, agriculture, coastal development, energy and water supplies, human health, and more into account. Among the impacts, loss of biodiversity and of the goods and services that ecosystems provide figure prominently, along with the fragmentation of landscapes, which will reduce natural systems' capacity to roll with the punches of climate change.

Three observations about the challenges of adaptation are worth bearing in mind here (IPCC, 1995). First, climate change will add stresses to systems that may already be under assault from other natural or human factors (disease, pollution, etc.) Second, because so many systems are sensitive to climate change, developing adaptation strategies for all will be a major undertaking, and priorities will have to be set. Third, many of the impacts of climate change are still poorly understood and difficult to quantify. That said, opportunities for adaptation abound, and priorities are emerging.

Clearly, one need is to re-think and bolster the nation's research and monitoring programs. Another is to better protect the genetic resources that breeders will need to develop crop and tree varieties suited to new climatic conditions. Also important is fortifying protection of ecologically sensitive areas and reducing habitat fragmentation to maximize species' potential to survive as temperature and precipitation patterns shift.

Biodiversity Loss

A cornerstone of sustainability is protection of the genetic and species diversity that forms the basis of agriculture and natural ecosystems and underwrites the promise of biotechnology. This means avoiding haphazard land-use patterns, overharvesting, and the introduction of exotic species into our ecosystems. Some 1,050 species are on the U.S. list of threatened and endangered species, and many ecosystem types are threatened or already lost. Genetic resources face threats as natural habitats are lost, and the global network of gene banks is too weak and underfunded to ensure their preservation. Like climate change, biodiversity loss is a problem of global scope—some 5 to 10 percent of species will be committed to extinction in tropical forests alone over the next three decades (Reid, 1992). But the U.S. contribution to the problem is significant and the environmental and economic costs to the nation are potentially great, especially as human populations and resource demands grow.

WRI's sectoral studies identified many actions to relieve pressure on biological diversity. Improved forest stewardship and the protection of ecologically sensitive areas would keep remaining U.S. forests healthy. Increased support for the U.S. germplasm conservation network would help us and the rest of the world. Reducing U.S. reliance on fossil fuels for electric power and transportation will reduce the toll of pollution and rapid climate change on biodiversity and ecosystems. Other steps that would help to slow biodiversity loss include investing in far better information on the distribution of species and ecosystems in the United States, expanding protection of ecologically sensitive and biologically diverse regions, increasing financial support to communities that now pay a high price for biodiversity protection, and targeting loans to help build local economies and businesses with a stake in protecting diversity—ecotourism, recreation, fishing, nontimber forest products.

5. Export Sustainability

Our study focussed on the United States as a big producer, a big spender, and a world leader. But facing some environmental challenges requires working with other nations. Moreover, the United States can undermine the sustainability of other nations through its patterns of consumption and investment or, conversely, aid those countries ready to chart their own sustainable development paths.

Energy and Transportation

The sustainability challenges in the energy and transportation sectoral studies tie the United States to other countries in two important ways. First, our oil-powered transportation system depends increasingly on oil supplies from other nations. U.S. crude oil production in the lower 48 states has been declining for more than 25 years, and production in Alaska peaked in 1988 and had fallen 26 percent by the end of 1995. By 2015, imports are projected to account for over 55 percent of our supply, and many of these imports will be from the Middle East, where two-thirds of proven oil resources lie and where political volatility is almost a given.

Second, the greatest sustainability concern related to transportation and electric power production—climate change—is truly global. The United States currently ranks number one in total CO_2 emissions (and in per capita emissions), and so our focus here on policy changes that will reduce emissions at home is well justified. However, the growth in CO_2 emissions in many other countries over the next several decades is projected to outstrip that here. Within OECD countries, emissions are expected to increase some 24 percent above 1990 levels while emissions from the developing world more than double. Since developing countries will account for nearly half of global CO_2 emissions from industrial sources by 2010 (WRI, 1996), even meeting U.S. targets would not save us from serious threats of climate change unless other countries matched our progress.

Agriculture

The two biggest threats to agricultural sustainability cannot be fully addressed within U.S. borders alone. Like mitigating the risk of climate change, conserving genetic diversity for agriculture also requires protection of germplasm and species in many other nations. Global food demand will also determine the sustainability of U.S. agriculture. Worldwide population growth is sure to increase demand for U.S. agricultural products, raise production, and take an attendant environmental toll. The policies recommended here to internalize many of the costs of environmental degradation should largely hold erosion, productivity loss, and pollution to acceptable levels even as production grows. However, recently retired lands—almost one-fifth of our cropland base—by definition had more than their share of environmental troubles, especially soil erosion. (More than 60 percent of the reduction in soil erosion between 1982 and 1992, came about because highly erodible lands were retired under the Conservation Reserve Program.) If these fragile lands are called back into service, environmental costs could rise faster than production increases, reversing progress toward sustainability.

Forestry

Like U.S. agriculture, forestry faces risks from climate change that cannot be addressed simply through U.S. actions at home. But the trade linkages are almost exactly the opposite of those in the agricultural sector. Whereas the United States is a major exporter of agricultural products and exports will grow in coming years, most of the produce of the U.S. forest sector satisfies domestic demand, and U.S. production will not keep pace with growing demand in coming years. Indeed, the actions recommended here will probably slow growth in timber production in the United States. Meanwhile, the combined growth in both U.S. and global demand for wood products is sure to heighten pressure on U.S. forests. As prices rise, many small woodlot owners may be lured into the market, and even if the forest products industry as a whole shifts to sustainable practices, the forest-management practices of these smaller, less well-informed and less well-capitalized players may undermine progress toward sustainability. Nor can the United States "export unsustainability"—and in some cases jobs as well—to other nations by buying wood from other countries using antediluvian harvest practices. The solution is to help other nations learn how to practice sustainable forest management.

Although most actions needed to place the United States on a sustainable path can be taken nationally or subnationally, global responses are needed to deal with such pressing challenges as climate change and biodiversity loss. The United States has much to learn from other nations as well. Fortunately, the worldwide similarities in nations' sustainability problems have a bright side. Technologies developed in the United States to make resource use more efficient and to reduce pollution—renewable energy sources, emissionless vehicles, ultra-efficient industrial processes—will find markets in other countries. And the use of those technologies will, in turn, reduce the risk that our own steps toward sustainability are undermining progress elsewhere.

CONCLUSION

We present in these studies and this overview the broad guidelines and specific policy steps needed to secure a U.S. future in which generations to come have a stock of environmental and natural resource assets at least as good as our own. By focusing on what is really important to Americans, by grounding work in virtues that Americans already profess, by pushing flexible, fair, forward-looking policy and institutional change, and by accenting the how as much as the why, our decision-making framework gives both the present and the future their due. Our sector by sector visions show what the nation stands to gain just as WRI's analysis of where current trends would lead the United States reveals a darker future.

Our findings contradict recent pronouncements that our environmental problems are largely solved or that more environmental protection costs too much. Those who interpret U.S. environmental progress as proof that environmental degradation is no longer an issue fail to grasp the big picture, which shows progress on some fronts, backsliding on others, and the emergence of new threats that will require our collective imagination and will to tame. Yet, the formidable technological, policy, and attitude hurdles that stand between our current development path and a much better one will yield to ingenuity, discipline, and commitment. *That* is the promise of sustainability.

REFERENCES

Adriaanse, A., S. Bringezu, A. Hammond, Y. Moriguchi, E. Rodenburg, D. Rogich, and H. Schütz. In press. *Resource Flows: The Material Basis of Industrial Economies.* Washington, D.C.: World Resources Institute; Wuppertal, Germany: Wuppertal Institute; The Hague, The Netherlands: The Netherlands Ministry of Housing, Spatial Planning, and Environment; Tsukuba, Japan: National Institute of Environmental Studies.

American Automobile Manufacturers Association, *Motor Vehicle Facts and Figures 1996,* Washington, D.C.: American Automobile Manufacturers Association, 1996.

Arthur, W.B. "Positive feedbacks in the economy," *Scientific American,* February 1990, pp. 92-99.

Carbon Dioxide Information Analysis Center (CDIAC), Environmental Sciences Division, Oak Ridge National Laboratory, "1992 Estimates of CO_2 Emissions from Fossil Fuel Burning and Cement Manufacturing Based on the United Nations Energy Statistics and the U.S. Bureau of Mines Cement Manufacturing Data," ORNL/CDIAC-25, NDP-030 (an accessible numerical database) Oak Ridge, Tennessee, September 1995.

Cook, Elizabeth (ed.). *Ozone Protection in the United States: Elements of Success*.Washington, D.C.: World Resources Institute, 1996.

Council of Economic Advisors. *Economic Report to the President 1996*. Washington, D.C.: Government Printing Office, February 1996.

Daugherty, A.B. *Major Uses of Land in the United States.* Washington, D.C.: U.S. Department of Agriculture, Economic Research Service, 1995.

Ditz, D., J. Ranganathan, and R.D. Banks (eds.). *Green Ledgers: Case Studies in Corporate Environmental Accounting.* Washington, D.C.: World Resources Institute, 1995.

Florida Environmental Equity and Justice Commission. *Final Report 1996*. Tallahassee, FL: Florida A&M University, 1996.

Haynes, R.W., D.M. Adams, and J.R. Mills. *The 1993 RPA Timber Assessment Update*, USDA Forest Service General Technical Report RM-259, Fort Collins, CO: USDA, March 1995.

Intergovernmental Panel on Climate Change (IPCC). *Climate Change, 1995, The Science of Climate Change.* Cambridge, U.K.: Cambridge University Press, 1996, p. 25.

—. *Climate Change 1995: Summary for Policy Makers*, Cambridge, U.K.: Cambridge University Press, 1995a.

—. *Climate Change 1995, Impacts, Adaptations and Mitigation of Climate Change: Scientific-Technical Analyses*. Cambridge, U.K.: Cambridge University Press, 1995b.

Keystone Center. *Final Consensus Report: Global Initiative for the Security and Sustainable Use of Plant Genetic Resources.* Keystone, CO: Keystone Center, 1991.

MacKenzie, J.J. *The Keys to the Car.* Washington, D.C.: World Resources Institute, 1994.

President's Council on Sustainable Development (PCSD), *Sustainable America, a New Consensus for Prosperity, Opportunity, and a Healthy Environment for the Future*, Washington, D.C.: U.S. Government Printing Office, February 1996.

Reid, W.V. "How many species will there be?" in T.C. Whitmore and J.A. Sayer (eds.) *Tropical Forests and Species Extinction.* London, U.K.: Chapman and Hall, 1992, pp. 55-73.

Repetto, R., D. Rothman, P. Faeth, and D. Austin. *Has Environmental Protection Really Reduced Productivity Growth?* Washington, D.C.: World Resources Institute, 1996.

Repetto, R., R.C. Dower, R. Jenkins, and J. Geoghegan. *Green Fees: How a Tax Shift Can Work for the Environment and the Economy.* Washington, D.C.: World Resources Institute, 1992.

Repetto, R., W. Magrath, M. Wells, C. Beer, and F. Rossini. *Wasting Assets: Natural Resources in the National Income Accounts.* Washington, D.C.: World Resources Institute, 1989.

Seitz, Fred and Christine Plepys, "Monitoring Air Quality in Healthy People 2000," *Healthy People 2000 Statistical Notes,* Washington, D.C.: HHS/CDC/NCEH, No. 9, September 1995.

Shelby, M. et al. *The Climate Change Implications of Eliminating U.S. Energy (and Related) Subsidies.* Washington, D.C.: U.S. Environmental Protection Agency, 1995.

Solow, R. *An Almost Practical Step Toward Sustainability.* Washington, D.C.: Resources for the Future, 1992.

U.S. Bureau of the Census. *Statistical Abstract of the United States, 1995.* Washington, D.C.: U.S. Department of Commerce, September 1995.

U.S. Department of Agriculture, *Agricultural Resources and Environmental Indicators.* Washington, D.C.: USDA Economic Research Service, Natural Resources and Environment Division, Agricultural Handbook No. 705, December 1994.

U.S. Department of Energy, Energy Information Administration. *Annual Energy Outlook, 1996.* Washington, D.C.: U.S. Government Printing Office, January 1996.

—. *Annual Energy Review 1994,* Washington, D.C.: U.S. Government Printing Office, 1995.

U.S. Department of the Interior. *Endangered and Threatened Species Recovery Program.* Washington, D.C.: USDI Fish and Wildlife Service, 1990.

U.S. Environmental Protection Agency. *Brochure on National Air Quality: Status and Trends.* Washington, D.C.: U.S. Government Printing Office, 1996a.

—. *1994 Toxics Release Inventory.* Washington, D.C.: U.S. Government Printing Office, June 1996b.

—. *National Air Pollutant Emission Trends, 1900–1994.* Washington, D.C.: U.S. Government Printing Office, 1995a.

—. *National Air Quality and Emissions Trends Report, 1995.* Washington, D.C.: U.S. Government Printing Office, 1995b.

—. *National Air Quality and Emissions Trends Report, 1994.* Washington, D.C.: U.S. Government Printing Office, 1994.

—. *National Air Quality and Emissions Trends Report, 1993.* Washington, D.C: U.S. Government Printing Office, 1993.

—. *Environmental Investments: The Cost of a Clean Environment.* Washington, D.C.: U.S. Government Printing Office, November 1990.

U.S. Fish and Wildlife Service, Division of Endangered Species, unpublished data, January, 1997.

World Commission on Environment and Development. *Our Common Future.* Oxford, U.K.: Oxford University Press, 1987.

World Resources Institute. *World Resources 1996–1997.* Washington, D.C.: World Resources Institute, 1996.

2.
SUSTAINABILITY AND U.S. AGRICULTURE
Problems, Progress, and Prospects

Paul Faeth

OVERVIEW

Agricultural sustainability is not a new concern, as the age-old adage "don't eat your seed corn" demonstrates clearly. Still, history records numerous examples of communities, cultures, and even civilizations that have died out because they failed to recognize or to heed sustainability threats.

In the past, most sustainability concerns were in the here and now because the scale of human activity was small. Local actions could address problems, or the deleterious effects could be absorbed by the environment. When land resources were plentiful, many cultures simply abandoned depleted areas and moved on, letting the degraded areas recover naturally over time. But since population densities have increased and land resources have become constraining, fallowing is now used in few places around the world, while agricultural land use has intensified dramatically.

Sustainability concerns evolve along with production patterns and practices, economic development, social changes, and knowledge. What is considered sustainable 50 years from now may include concerns not yet imagined, just as today's issues differ from the crises of the Dust Bowl and the Great Depression. As our influence on the physical environment increases, the issues we must face grow broader and more complex, and the cost of failure looms larger.

The nation's current concerns about agricultural sustainability depend on where we think we are now and where we want to go. For this reason, this chapter has three parts: a vision of a sustainable future; an understanding of our current position in relation to that future; and a path of action to make that future a reality.

A Vision of an Environmentally Sustainable United States

A Vision of a Sustainable Agricultural Future

A sustainable agricultural system will produce food and fiber to balance demand at prices society considers acceptable. Everybody, including the poor, will have access to enough food to meet all nutritional needs.

Americans will respond to dietary cautions and eat healthier, consuming fewer fats and less sugar and meat. Average caloric consumption will decline, as will the average American waistline. A new focus on grains and vegetables, instead of meat and dairy products, will make the use of grain production more efficient. As a result, relatively less land will be needed for agriculture, and environmental impacts from our patterns of domestic consumption will be less harmful. Reduced per capita consumption of agricultural commodities (volume, not value) in the United States for health reasons will be balanced to a certain extent by increased demand abroad, but crop prices will continue their long-term decline, allowing the poor to spend a smaller cut of their income on food. At the same time, the domestic market for organically grown products will become mainstream, as consumers' awareness and incomes increase, and farmers, realizing the cost savings and the benefits to their own health, expand into this field. Price differentials will decline but will still allow producers to benefit. Biological pest and soil-management techniques developed for organic production will transfer to conventional production.

Agriculture will become a significant supplier of energy, paper, and industrial feedstocks. Perennial crops grown on tens of millions of acres will be harvested for electricity. Renewable energy from agriculture will help to mitigate global climate change. New crops will be grown for their fiber as a source of pulp for paper. Chemical industries will find many new uses for conventional agricultural commodities to replace fossil fuels.

Agricultural practices will be highly productive, but will not deplete the natural resource base, nor overwhelm ecosystems' ability to assimilate pollutants and to recover from past damages. Soil erosion will be contained, so productivity won't fall off. Pests will be managed principally through biological, agronomic, and cultural control mechanisms. Pesticides will be used as a last resort in strategically small amounts that will break down quickly into harmless components. Essential plant nutrients will be applied in ways that minimize losses while meeting crop requirements precisely. As farmers and researchers learn to squeeze more production out of fewer inputs, agricultural productivity will continue to climb.

Rural communities will depend less on agriculture, and rural economies will become more diverse and thus more resilient. Educational and infrastructural investments will attract various industries, along with higher paying jobs, so the tax base will support continued economic development. Advancements in computing and communication will allow people to live in rural areas but to work for information industries in distant locations.

Agriculture will become independent of government subsidies. Yet, agricultural income will grow. Farmers will become much more sophisticated at business operations, and new financial tools and institutions will support them. As government steps aside, new businesses will emerge to help them insure against losses, sell their crops, manage pests, develop long-range business plans, and conserve their natural resources. Many of these services will be available on the computer consoles in their tractors.

Economic signals will be used to support conservation. Conservation objectives will be achieved through contracting arrangements that government regulation encourages but not at the government's expense. It will become common for dischargers to reduce emissions beyond requirements and to trade emissions reductions with each other, paying least-cost sources like agriculture to reduce loadings and achieving significant water quality gains relatively inexpensively. Taxes on excess nutrient applications using simple mass-balance calculations and pesticide taxes based on potential environmental harm will encourage efficient farming and pay for the development of technologies needed.

Knowledge of how ecosystems operate will improve substantially as technologies, from computer modeling to satellite monitoring, develop and improve. Farmers will have an intimate understanding of how their decisions affect their own operations, the watershed they live in, and the larger environment. New technologies will provide farmers with the information needed to optimally manage their resources. Simple tests to monitor management indicators—such as soil and crop health, nutrient-use efficiency, and pollutant runoff—will improve efficiency and profits. As these changes take hold, more and more of the nation's lakes, rivers, and estuaries will begin to recover from the pressures of agricultural and other sources of pollution, and more people will enjoy the economic and recreational benefits of clean water.

Greater understanding of molecular biology and genetics will give researchers the tools to develop plants that are more highly customized for climate zones, soil types and pest threats, especially as the stock of genetic resources recovers. Researchers will have cataloged and described the genetic codes of many plants and will have easy access to the required genetic resources. The time required to develop new varieties will shrink significantly.

Precipitation will become significantly more variable as climate changes. Yet, American farmers will cope. Weather forecasting will become more precise and accurate over longer periods of time, allowing farmers an extended planning horizon. Healthier soils will hold more moisture, crop varieties will resist drought better, cropping patterns will change, and irrigation will become more common and more efficient. Prices for consumers will remain low and stable.

Where Do We Want to Go?

The vision put forth here of a sustainable agricultural system includes basic goals and objectives that don't change with time. It also includes subjective and ever-debatable preferences for what we would like to have happen and what we would like to avoid. Some of the technological aspects of this admittedly optimistic vision are probably the closest to realization. (For example, farmers already have access to Global Positioning Systems that use satellites to tell their tractors how much fertilizer to apply to a field.) The social aspects may be most difficult to realize, since almost by definition established institutions resist change.

Where Are We?

How do we chart progress toward the vision stated here (or some alternative vision)? The simple answer is that we need to define and track indicators of sustainability. In practice, this isn't so easy. The choice of indicators follows from the definition of sustainability used, and in agriculture such definitions abound. The extensive literature on agricultural sustainability includes a variety of books and journals and numerous definitions of the term have been put forward by various authors and organizations.[*] There is even a legal definition included in the 1990 U.S. farm bill:

---

[*] See Allen and Sachs (1993) and Hoag and Skold (1996) for summaries of definitional discussions.

> *Sustainable agriculture is an integrated system of plant and animal production practices having site-specific application that will, over the long-term: satisfy human food and fiber needs; enhance environmental quality and the natural resource base upon which the agriculture economy depends; make the most of nonrenewable resources and on-farm resources, and integrate, where appropriate, natural biological cycles and controls; sustain the economic viability of farm operations; and enhance the quality of life for farmers, ranchers, and society as a whole (U.S. House of Representatives, FACTA, 1990).*

Broadly defined, sustainability means that economic activity should meet current needs without foreclosing future options (World Commission on Environment and Development, 1987). Textbook definitions of income take this notion of sustainability into account: income is defined as the maximum amount that can be consumed in the current year without reducing potential consumption in future years (Edwards and Bell, 1961; Hicks, 1946). Basically, this means that we should live off the interest generated by natural and human–made assets, not the capital.

Economist Herman Daly (1991) has defined three necessary conditions for sustaining the physical resource base:

◆ The rate at which renewable resources are used should not exceed their rate of regeneration.

◆ The rate at which nonrenewable resources are used should not exceed the rate at which sustainable renewable substitutes are developed.

◆ The rate of pollution emission should not exhaust the environment's assimilative capacity.

Box 2-1 features a selected set of indicators for the agricultural sector that follow from these various definitions of agricultural sustainability. The indicators emphasize certain *necessary conditions* to provide humans with food and fiber without impinging unnecessarily on other sectors and resources. Those have been separated here from other *desirable conditions* that are often considered part of sustainability, but in fact are social objectives.

Indicators are useless without interpretation, of course. Accordingly the following criteria help reveal whether a given concern is a sustainability threat and, if so, how serious it is.

Criterion 1. Direction of Trend: Is the Situation Getting Better or Worse? An assessment of long-term quantitative changes in a natural resource, environ-

BOX 2-1. Agricultural Sustainability at a Glance

How would we know if agriculture were on a sustainable path? Among other things:

- soil would not erode at a rate that harmed soil productivity;
- nutrient applications would roughly balance plant uptake;
- rivers, lakes, and estuaries could meet their designated uses;
- concentrations of pesticides and nutrients in ground and surface waters and the atmosphere would be below damage-causing levels;
- pesticides would be targeted more carefully to specific pests;
- the stock of agricultural genetic resources would be stable or increasing;
- the diversity of crop rotations and crop varieties would be increasing;
- yields and production would increase at least as fast as population growth;
- agricultural output growth would increase faster than input growth;
- climatic variability would be stable or declining;
- agriculture would be more financially sustainable, with income growth but declining dependence on government support; and
- commodity prices would be stable and consumers, especially the poor, would spend a smaller share of their income on food.

mental condition, or other factors related to the productivity of natural resources and ecosystems is essential to problem definition. Sustained improvement in a trend lessens concern about sustainability while deterioration hastens it.

Criterion 2. Irreversibility: Are the Consequences Permanent? Irrevocable losses of a desired natural resource, impairment of an ecological process, or degradation of an environmental condition warrant special attention. Few processes except species extinction are irreversible in an absolute sense. But some, such as climate change, may be irreversible in all but geologic timeframes. Some problems may be reversed if practical constraints of money, time, and social or political support can be overcome.

Criterion 3. Substitutability: Are There Viable Alternatives? Some processes can be replaced through technology, management, or changes in demand, which can be fostered by changes in policy. For others, substitutes may be impractical, too costly, or simply unavailable. When practical responses exist, sustainability concerns will be less pressing than in situations where they don't.

Criterion 4. Risk: What's at Stake if We Make the Wrong Decision? Our knowledge of the consequences of change is far from perfect. Some concerns, such as those about soil productivity, have been addressed by voluminous research, while for others—such as pesticides and health or the nature and pace of climate change—information gaps remain enormous, so it is harder to understand the consequences and size up the risks. In any case, significant potential impact and substantial uncertainty elevate sustainability concerns.

Table 2-1 provides a crude summary of the sustainability assessment spelled out later in this report. The intent is to gauge and prioritize, in a practical and reasonably objective way, the various threats to agricultural sustainability in the United States and to size up the net impact on production related issues.

Lowrance and colleagues (1986) looked at sustainability from the viewpoint of hierarchies or scales—an approach we adapt here to the physical scope of impact. The scale of effects vary from relatively small (fields and watersheds) to very large (bioregions, climate zones, global). The greater the impact on net agricultural production and the broader the scale of impact, the likelier it is that food supply will be disrupted and higher production costs will be imposed on farmers and higher food costs on consumers. Similarly, the broader the environmental impact, the greater the costs imposed on other economic sectors for clean-up or mitigation.

The direction of the current trend indicates whether the problem is getting better and roughly how long it will take to correct it. For economic, social, or physical reasons, some problems can never be corrected. Long timeframes for correction (or irreversibility) matter less if there are technical and policy options available to mitigate the problem's effects.

Of course, the less we know about what causes a problem, what its effects are, and how effective potential responses might be, the more difficult it will be to deal with it. Thus, the bottom line is the last column—the "sustainability threat"—which integrates the information in the other columns.

In this analysis, *soil erosion* is deemed a small threat to sustainability because it can be readily corrected with well-understood techniques at acceptable costs. Sedimentation's off-farm effects are significant and the problem is extensive, but it is also local—derived from each field and addressed there—and major institutions and policies have been created to combat erosion.

Nutrient runoff and *pesticide pollution* have no effect on agricultural production, but can have major impacts on the environment. Nutrient runoff can upset the ecological balance of rivers, lakes, and estuaries, and it can increase the cost of treating water for urban and industrial use. Pesticides can harm the health of applicators and other farmworkers. Significant attention and research have probably reduced these problems, and numerous technical and policy responses

TABLE 2-1. Summary Assessment of Sustainability Threats to U.S. Agriculture. (Some of the more conventional concerns will likely be less of a problem than the emerging issues of climate change and agricultural germplasm loss.)

Sustainability Concern	Nature of the Problem	Potential Effect on *Net* Agricultural Production	Scope of Environmental Concern
Soil Erosion	soil productivity; surface water pollution; sedimentation	small	field, watershed
Nutrient Runoff	surface and ground water pollution; eutrophication	none	watershed, aquifer
Pesticide Pollution	ecosystem and human health	none	watershed, aquifer
Wetland Losses	flooding, habitat loss	small	watershed, bioregion
Water Supply for Agriculture	loss of irrigation capability	small	field, aquifer
Farmland Loss	urban runoff; landscape values; specialty crops	very small	field, watershed
Declining Farm Numbers	rural character and viability	probably none	probably none
Agricultural Germplasm Loss	crop and livestock development; disease	large	bioregion, international
Global Climate Change	changes in temperature, precipitation and variability	unknown	climate zones, globe

are now in place. Recent studies have raised new concerns about pesticides' effects on animal and human hormones and immune systems, though these effects are so far poorly understood. Because of their fairly broad impacts on ecosystems, these problems are considered "medium threats."

Wetlands losses through conversion to agriculture has been a significant problem in the past, but the rate of conversion to agriculture has declined dramatically over the last few decades. Conversion to urban uses is now more significant. Still, the fact that much of our stock of wetlands has been converted remains a problem for flood control, fisheries, and wildlife management. Besides stopping the conversion of wetlands, active restoration is also necessary. Except for occasional flooding, wetlands loss seldom affects agricultural production and

		TABLE 2-1 (cont.).		

Direction of Trend	Timeframe for Remediation	Certainty of Fundamental Understanding	Availability of Responses	Sustainability Threat
getting much better	years to decades	high	many	small
probably getting better	decades	medium	many	medium
probably getting better	decades	low	many	medium
getting slightly worse	decades	high	some	small
getting worse	decades to centuries	high	many	small
getting worse	practically irreversible	high	many	probably not a threat
getting worse	likely irreversible	high	few	probably not a threat
getting worse	irreversible	medium	some	large
probably getting worse	centuries	very low	some	large

any such effects are fairly localized. Further, responses are being developed to address this problem, and some new funds are being devoted to wetlands restoration, so this is a small sustainability threat.

Declining farm numbers and farmland loss probably pose no significant threat to sustainability, even though both problems are most likely irreversible and, by most people's lights, getting worse. However divisive the issues, declining numbers of farms probably do not hurt either agricultural production or the environment. For some time, the United States has overproduced agricultural commodities, so sustainability is a concern only vis-à-vis the production of some specialty crops. However, many urbanizing communities consider the issue very important since conversion of farmland to urban and suburban uses can increase water pollution. While trying to keep people on the farm is probably impossible since longstanding social and economic pressures both push and pull them off, there are many ways to address farmland loss at the local, state, and federal levels.

Since most of the nation's agricultural land is rainfed, the issue of *water supplies* is relevant in only a few areas. Within these, however, the economic value of production is enormous. Agriculture in California would all but cease without irrigation. Elsewhere, irrigation has allowed the production of surplus crops better suited to other areas of the country—whether cotton in Arizona or corn in the Mountain states. The net impact on production is small because a large volume of water is used so inefficiently on low-value crops. Adjustments could be made fairly quickly if policies were corrected to encourage conservation, so this problem can only be considered a relatively small threat to sustainability.

Agricultural germplasm loss is a major threat to the sustainability of U.S. agriculture. Historic episodes of major crop losses due to crop uniformity—in the United States and elsewhere—have demonstrated that agriculture is vulnerable to genetic losses, though this aspect is the smaller of two related concerns. The larger is that the extinction of genetic stocks irretrievably closes off future opportunities for crop and livestock improvement and the development of genetic responses to emerging pest and climate threats. Genetic extinction is forever and it represents an unknowable but undeniably large opportunity cost. Conversely, genetic improvements have resulted in huge production gains over extensive areas.

The net effect of *global climate change* on agricultural production is hard to gauge because physical phenomena could be countervailing and because scientists cannot predict the nature, extent, or rate of climate change. But we know with certainty that our climate is already warmer than it otherwise would be because of heat-trapping "greenhouse" gases (such as carbon dioxide) in the atmosphere and that the release of greenhouse gases from fossil fuel combustion and other sources is accelerating as the world economy grows. Recent evidence suggests that the most deleterious effect may be the increased variability of precipitation, not changes in average temperatures or precipitation levels. While climate change could be reversible, the world economy and the physical environment might take centuries to adjust. Also, once released into the atmosphere, some greenhouse gases last only a few decades while others can linger for as long as 50,000 years. Climate will not simply move to some new steady state. It would change continually, making human adjustments difficult. If climate change is slow and steady, farmers and researchers can probably adapt, at least in wealthy countries like the United States. But, if the changes are sudden and dramatic—a distinct possibility—then major disruptions could follow. No other issue could affect us at so large a scale. For these reasons, global climate change merits a ranking as a large threat to sustainability.

How Do We Get There?

On any trip, the traveler must know the destination, current position, and route. When traveling to a place where one has never been before, some sort of map or directions are also essential to avoid getting lost or encountering major problems. It's possible, though much more difficult, to get somewhere by trial and error, setting off with nothing more than a general sense of direction. The same is true with sustainable development. It *might* be possible to achieve it without a shared vision, assessments of progress, or a plan, but the journey could be bumpy and costly.

This chapter articulates a destination. It also outlines our current compass reading. First, long-recognized issues are discussed, then emerging threats. Next, the supply side is evaluated and the agricultural factors that could contribute to sustainability are assessed. Finally, a road map for sustainability is drawn. Its milestones are actions that will make the greatest difference and get to the heart of the problem fastest.

CONVENTIONAL THREATS TO PHYSICAL SUSTAINABILITY

Some threats to agriculture's basic objectives have long been identified and have been addressed through various policies and programs. These "conventional" threats include soil erosion, water quality, wetlands loss, water supply, and toxics in the environment.

The Natural Resources Conservation Service recently surveyed its clients, mostly farmers, to find out their top four natural resource concerns. The results *(Figure 2-1)* showed that most consider water quality first, with soil erosion, sustainability, land conversion and pesticide management rounding out the top five (USDA/NRCS, 1995b). (Respondents were not given the option of choosing global climate change and the erosion of agricultural genetic stocks.)

Soil Erosion

A large segment of the U.S. landscape is in agriculture. The total area of the lower 48 states is almost 1.9 billion acres, more than half of which is used for cropland, rangeland, and pasture (Daugherty, 1995). Most of the remainder is used for

FIGURE 2-1. Most people consider water quality to be the top natural resource issue related to agriculture. Soil erosion, sustainability, land conversion, and pesticide management round out the top five.

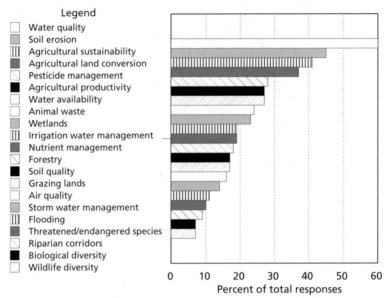

forestry, but much of this acreage is grazed. Soil erosion is of significant concern, then, simply because agriculture covers so much of the land base.

Changes in policy and production practices over the last two decades have reduced soil erosion and attendant losses of productivity. The threat that became apparent during the Dust Bowl period is subsiding as highly erodible land has either been retired or managed with conservation in mind. Erosion is dangerous mainly at the levels of the field and the watershed because it can lower crop yields and because sediment, nutrients, and pesticides from eroding soils are major causes of water quality problems.

There appears to be very little, if any, trade-off between current and future generations regarding soil erosion's effect on productivity. But inter-sectoral equity concerns are significant because the rest of the economy pays to correct both soil erosion and nutrient and pesticide pollution from agriculture.

More than 60 years ago, the federal government created the Soil Conservation Service to give farmers the technical support needed to adopt soil conservation

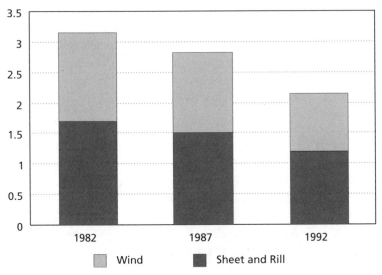

FIGURE 2-2. **Annual Soil Erosion from Cropland**

measures. The government also established cost-share and land retirement programs and regulations to address these problems. The results can be said to be partially successful and improving steadily as total gross erosion declines, though many parts of the country are still experiencing erosion above rates thought to harm soil productivity.

Between 1982 and 1992, estimated soil erosion on U.S. land dropped by about 30 percent. *(See Figure 2-2.)* Most of this gain (700 million tons per year) came from retiring highly erodible lands under the Conservation Reserve Program. Also at play were the adoption of conservation practices put forward through technical assistance programs (300 million tons per year) and the mandated adoption of conservation-tillage practices through federal commodity programs (100 million tons per year).

Conservation tillage practices, which can reduce soil erosion by up to 90 percent, have been shown to increase productivity and profits (Faeth, 1995) and farmers have begun to adopt these practices widely without government subsidies. Nationally, conservation tillage is now used on more than one-third of all cropland. In some Indiana counties, up to two-thirds of all land is planted without any tillage simply because it saves farmers money. The prospects for further adoption of conservation tillage and other measures to improve soil productivity appear bright.

Even so, much more remains to be done. One commonly used measure of the sustainability of soil resource use is the "T-value," which indicates the maximum rate of erosion that can be tolerated to indefinitely maintain soil productivity.[*] Figure 2-3 shows that in 1992, some 35 percent of all cropland was eroding at rates higher than the T-value (Alt, pers. comm., 1995). While this is a concern, it is also a mark of improvement—in 1982, the comparable figure was 44 percent (Alt et al., 1989).

Water Quality

Although soil erosion influences soil productivity, erosion's off-farm effects on local and regional water quality are far more important. By one recent estimate, the value of soil productivity losses due to reduced yields total about $200 million per year, while the off-farm economic costs of soil erosion from cropland alone run about $2 billion per year, lowering the productivity of other water-dependent sectors (Faeth, 1995). Run-off from soil tillage, fertilizer and pesticide use, and dairy and other concentrated feeding operations, and other agricultural

FIGURE 2-3. **Cropland Erosion Compared to Tolerances**

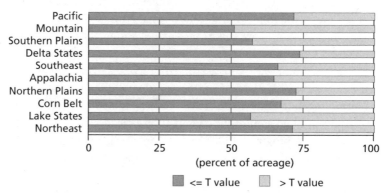

---

[*]   T-values are different for each soil and are a function of soil depth and formation processes. T-values are not known with any precision and for evaluative purposes a value of five tons per acre per year is most often used. While T-values are basically a rough guess based on the judgment of scientists, no better indicator exists.

BOX 2-2. The Clean Water Act

Federal water pollution control law attempts to protect water quality by controlling the discharge of pollutants. Congress passed the Clean Water Act in 1972 with the basic objective "to restore and maintain the chemical, physical and biological integrity of the Nation's waters." The Act sought to have "fishable and swimmable waters" by 1983, "zero discharge" of pollutants by 1985, and to eliminate the release of "toxics in toxic amounts." The Act made the EPA responsible for setting national standards for the discharge of effluents on an industry-by-industry basis, considering both the capabilities of pollution control technologies and the costs of application. This legislation makes it illegal to discharge pollutants to surface waters without a permit. The legislation specifically focused on point sources. The Act was extensively modified in 1977 and 1987 to expand EPA's powers and to address non-point pollution through voluntary programs.

Adapted from: Adler et al., 1993, and Arbuckle et al., 1993.

---

activities pollutes the nation's lakes, streams, rivers, and estuaries. Siltation from soil erosion is a principal cause of water pollution, but nutrients, organic matter, and pesticides also contribute. This pollution costs municipalities, industries, and government agencies sizable sums by increasing water-purification costs, reducing water-storage capacity, inhibiting water flow, damaging fresh and marine fisheries, and limiting recreational opportunities (Ribaudo, 1989).

Significant reductions in the control of "point source" emissions—those coming from industrial and municipal pipes—have been made over the last 20 years through federal and state regulation and cost-sharing programs. *(See Box 2-2.)* Major federal, state, and local investments in primary and secondary water treatment have been made to improve water quality too. Yet, today 44 percent of the nation's river miles, 57 percent of lake acres, and 44 percent of estuary waters aren't clean enough to support their designated uses—a legally defined indicator of the sustainability of our freshwater resources. Recent surveys undertaken by the states show the quality of lakes and rivers to be declining.[*] EPA goals call for

---

[*]   Water quality should be improving because of past investments, yet surveys show the opposite. Newer surveys may not be strictly comparable with older surveys because more waters are being surveyed, standards may be getting stricter, or other compensating factors such as elimination of wetlands or stocks of pollutants remain even though pollution has since stopped (Adler et al., 1993).

TABLE 2-2. Agriculture and Water Quality Impairment in the United States			
Source of impairment	**Percentage Impaired**		
	Rivers	**Lakes**	**Estuaries**
Agriculture	72	56	43
Hydro/habitat modification	7	23	10
Storm runoff/sewers	11	24	43
Land disposal	not available	16	not available
Municipal/industrial	22	21	76
Cause of impairment			
Siltation	45	22	12
Nutrients	37	40	55
Pathogens	27	8	42
Organic enrichment	24	24	34
Pesticides	26	9	7
Suspended solids	13	6	11
Salinity	12	<1	7
Metals	6	41	4

Source: USDA/ERS, 1994c, pp. 60 and 61.

a further 14 percent reduction in the discharge of pollutants from sewage systems, water-treatment plants, and industrial sources at a cost of some $15.5 billion (U.S. EPA, 1993).

As other sectors have made environmental gains, agriculture has become the main source of water quality impairment. While so-called point sources, where pollution is emitted from a single point such as a pipe, have come under increasingly tight regulations, non-point agricultural sources remain largely unregulated except for the biggest operations.

Siltation and nutrient run-off are the two most important single causes of impairment of rivers and lakes. Between 1963 and 1978, agricultural use of inorganic nutrients doubled and then stabilized at about 20 million tons per year. *(See Figure 2-4.)* Although increases in nutrient application rates have helped fuel

FIGURE 2-4. **Inorganic Nutrient Use**

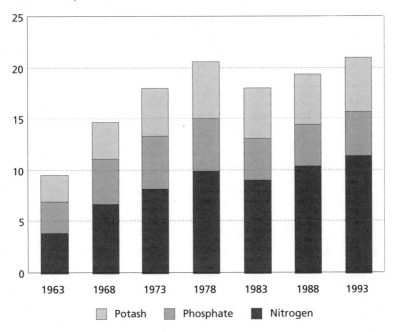

Million Tons per Year

Potash   Phosphate   Nitrogen

production increases, many farmers apply more nutrients than plants can use or do not account for nutrients that are applied as manure, fixed by plants, or derived from precipitation. Nutrient surpluses can pollute ground and surface waters. The National Research Council's Committee on Long-Range Soil and Water Conservation (1993b, p. 445) has estimated national residuals[*] for nitrogen and phosphorous. These residuals—ranging from 33 to 40 percent of all inputs (6.7 to 9.1 million metric tons) for nitrogen and 63 percent for phosphorous (2.9

---

[*] Calculation of residuals is based on a mass-balance approach. The residual is the difference between the inputs and the outputs. Inputs are derived from commercial fertilizer, manure, fixation by legumes, and the content of crop residues. Outputs include harvested crops and crop residues (National Research Council, 1993b).

TABLE 2-3. Of all the nitrogen and phosphorous applied in the United States, one-third of the nitrogen and almost two-thirds of the phosphorous is not used by crops and can become the source of local and regional water quality problems.

Inputs and Outputs	Nitrogen (medium scenario)		Phosphorous	
Inputs	million metric tons	percent of total inputs	million metric tons	percent of total inputs
Fertilizer	9.4	45	3.6	79
Manure	1.7	8	0.7	15
Legumes	6.9	33	na	na
Crop residues	2.0	14	0.3	6
Total	20.9	100	4.5	100
Outputs				
Harvested crops	10.9	51	1.3	29
Crop residues	2.9	14	.3	6
Total	13.5	64	1.6	36
Net Balance (Residual)	**7.4**	**36**	**2.9**	**63**

Source: National Research Council, 1993b, p. 445.

million metric tons)—gauge potential losses to the environment and opportunities for efficiency gains. *(See Table 2-3.)* Fortunately, many currently available technologies and management strategies can make nutrient use more efficient. But with fertilizer so cheap, costs and uncertainty make farmers wary: new technologies entail adoption costs; heavy rains can wash nutrients away, so farmers tend to overapply; animal manures are considered a waste-disposal problem instead of a resource, and their nutrient content is often uncertain.

Government Approaches to Agricultural Water Quality and Conservation Issues

Since agricultural sources of pollution remain largely unregulated, progress so far in reducing polluted run-off from agriculture is owed to federal subsidies. Until this year, conservation-compliance provisions tied to commodity price subsidies required farmers on highly erodible land to adopt conservation practices, though program participation itself is voluntary. Past agricultural commod-

BOX 2-3. Commodity Programs and the Environment

Until 1996, farmers received price supports for seven commodities: corn, wheat, sorghum, barley, oats, rice, and cotton. Under USDA rules, participating farmers had to establish *acreage bases* for each crop; price supports were paid only on acreage in the program. The rules forbade planting one crop on another's base, so the commodity program historically locked farmers into a certain cropping pattern, rules rewarded farmers for growing program crops, and punished them for growing non-program crops. Prior farm legislation discouraged the use of crop rotations that included non-program crops that can help farmers to break pest cycles, reduce soil erosion, and manage soil fertility.

A study by the World Resources Institute (Faeth, 1995) looked at the economic and environmental benefits of increasing farmers' flexibility in planting. The study showed that if farmers were allowed to plant whatever they wanted on half their base acreage but received no price supports, while program rules were maintained on the other half and price supports were given, then farmers would use more varied crop rotations. Fiscal costs of the farm programs would drop by $5 billion per year and soil erosion damages would decline by 14 percent. Crop prices would increase as some unsubsidized land would go into other uses, and farm income would decline by just 1 percent.

ity programs did offer some incentives for conservation too, but the negative effects of controlling crop-planting decisions far outweighed the benefits. Most such provisions were eliminated with the new legislation—the Federal Agricultural Improvement and Reform Act (FAIR) passed in 1996. *(See Box 2-3.)*

The new legislation gives farmers much more planting flexibility. They are free to plant any crop except fruits and vegetables on contract acres, and the acreage-idling programs have been scrapped. Payments are based on "production flexibility contracts" that require farmers to comply with established conservation plans and wetland-planting restrictions and to keep their land in agricultural uses. Farmers receive payments based on their previous participation in the commodity programs and share allocations of funds provided by Congress by crop.

To mitigate agricultural production's negative environmental impacts, Congress approved five conservation provisions or programs:

1. the swampbuster provision and the Wetlands Reserve Program;
2. the conservation compliance provision;

3. the Conservation Reserve Program;

4. the Environmental Quality Incentives Program, and

5. the Wildlife Habitat Incentives Program.

Swampbuster and the Wetlands Reserve Program

The swampbuster provision sets rules and penalties to discourage inadvertent or purposeful drainings of wetlands. Some land can be converted if another drained wetland is restored. Certain violations trigger the loss of all federal benefits.

Under WRP, government shares with farmers the costs of returning farmland to wetlands and provides funds to lease or purchase that land. Wetlands farmed or converted before 1985 and current wetlands or lands adjacent to wetlands are eligible for enrollment. (USDA/NASS 1992a). The WRP was maintained in the FAIR Act and capped at 975,000 acres.

Conservation Compliance

This provision discourages continued production on highly erodible cropland without an approved conservation plan. Producers may become ineligible for federal benefits if they have no locally approved plan developed and implemented. Land is considered "highly erodible" if its erosion potential is greater than eight times the rate of soil formation. Farmers can self-certify their plans and get out of compliance if they can prove economic hardship will result (Osborn, 1996).

Conservation Reserve Program

The CRP is by far the largest and most important U.S. conservation program, covering 36.4 million acres at average yearly rental rates of about $50 an acre (USDA/NASS 1993b). Participating producers get paid to remove from production for 10 to 15 years highly erodible land or land that significantly lowers water quality. Farmers must establish a protective cover of grass or trees.

Updated in 1996, the CRP is as much a supply-control program as it is a conservation program. Enrollment has concentrated in major crop-producing areas, not where soil erosion has high off-site costs. A major benefit of the CRP appears to be its contribution to avian habitat (Allen and Sachs 1993; Johnson and Ekstrand 1994), and beginning with the tenth contracting period, the environmental performance of this program has improved with targeting (Osborn 1991; Cook 1994). The CRP was capped at 36.4 million acres under the 1996 legislation. The law also allows farmers to get out of CRP contracts after

TABLE 2-4. Benefits of USDA Conservation Programs on Erosion 1992. The Conservation Reserve Program is responsible for the largest share of erosion reductions.

Program	Benefit
Total Erosion Reduction (million tons per year)	1,139 (total)
Conservation Reserve Program	672
Conservation Compliance	103
Agricultural Conservation Program	30
Conservation Technical Assistance	298
Annual Acreage Reduction Program	36

Source: USDA/NASS 1993b, p. 36.

five years if the land covered is environmentally sensitive. New enrollments will have to meet more stringent criteria (Osborn, 1996).

Environmental Quality Incentives Projects

This program combines several programs created by prior farm bills—the Agricultural Conservation Program, the Great Plains Conservation Program, the Water Quality Incentives Program, and the Colorado River Basin Salinity Control Program. Through education, technical assistance, and financial assistance, EQIP encourages farmers to adopt environmentally beneficial practices. Funding levels are set at $200 million per year, with half the amount allocated to livestock activities for the first time. The new legislation explicitly calls on the United States Department of Agriculture to develop this program so as to maximize environmental benefits (Osborn, 1996).

Table 2-4 shows some of the benefits of the conservation programs outlined above or of their predecessors. Look particularly at CRP. As points of comparison, 1992 erosion from all cropland was 2.13 *billion* tons (USDA/NASS 1994b).

Wetlands

Converting wetlands to agricultural production has contributed to water-quality problems. Wetlands filter silt and nutrients from water flow, thus protecting lakes, rivers, and estuaries from pollutant's harmful effects. Wetlands also help protect lowlands from flooding, among other factors. *(See Table 2-5.)* More than half of the 220 million acres of wetlands found 200 years ago in the lower 48 states have been converted to other uses. In recent years, the annual rate of conversion of wetlands to agriculture has slowed dramatically, from roughly 600,000 acres per

TABLE 2-5. Functions, Related Effects, and Corresponding Social Values of Wetlands		
Hydrologic		
Short-term surface water storage	Reduced downstream flood peaks	Reduced damage from floodwaters
Long-term surface water storage	Maintenance of base flows, seasonal flow distribution	Maintenance of fish habitat during dry periods
Maintenance of high water table	Maintenance of hydophytic community	Maintenance of biodiversity
Biogeochemical		
Transformation, cycling of elements	Maintenance of nutrient stocks within wetland	Wood production
Retention, removal of dissolved substances	Reduced transport of nutrients downstream	Maintenance of water quality
Accumulation of peat	Retention of nutrients, metals, other substances	Maintenance of water quality
Accumulation of inorganic sediments	Retention of sediments, some nutrients	Maintenance of water quality
Habitat and food web support		
Maintenance of characteristic plant communities	Food, nesting, cover for animals	Support for furbearers, waterfowl
Maintenance of characteristic energy flow	Support for populations of vertebrates	Maintenance of biodiversity

Source: National Research Council, 1995, p. 35.

year during 1954-1974 to about 29,000 acres per year during 1987-1991. *(See Figure 2-5.)* Although agriculture's share of wetland conversion is now dominated by losses to urban uses, more than 100,000 acres per year of this valuable resource are still being lost (USDA/ERS, 1994c, p. 12). As an indicator of the sustainability of wildlife habitat and water quality past losses of wetlands are worrisome. However, the dramatic slowdown is a positive sign. Many communities now recognize the value of their wetlands, and states have instituted regulations and protection mechanisms. But it is not enough to simply stop wetlands losses to preserve the benefits that wetlands produce—restoration is also essential.

Water Supply

Water use in agriculture has been recognized as a sustainability issue. While the United States has enough water overall, regional disparities exist. In the East, consumptive water use is less than 1 percent of supply, while in the Colorado

FIGURE 2-5. **Wetland Conversion by Cause, 1954–91**

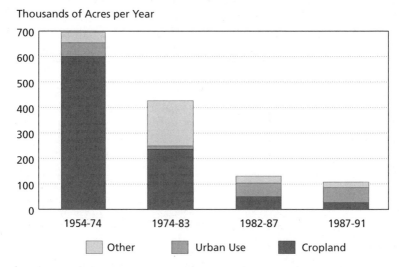

Thousands of Acres per Year

Other   Urban Use   Cropland

River, Rio Grande, and Great Basins, consumptive use is 94, 64, and 49 percent of supply, respectively (USDA/ERS, 1994c, p. 45).

Increased demand for water from municipal consumptive uses and from such nonconsumptive uses as recreation and wildlife habitat maintenance can be expected to continue to increase. In the past, increased demand has been met through dam construction, the transfer of water from other areas, and groundwater. But many of the best sites for water impoundment have been used or are the subject of intense public concern for environmental reasons, and groundwater resource development is approaching its natural limits. Since agriculture is by far the major consumer of fresh water, accounting for 42 percent of all uses (Solley et al., 1993), demands from other sectors will inevitably reduce agriculture's supply. *(See Table 2-6.)* Much of agriculture's water use is for relatively low-value commodities that can—and should—be grown in other areas without irrigation, so that water for more valuable uses would be available.

Currently, about 52 million acres of cropland are irrigated, up from 38 million acres in 1972. Two-thirds of all irrigated land is used to grow six crops: corn for grain, hay, cotton, wheat, rice, and soybeans—with corn and hay accounting for 38 percent of irrigated acres. The average subsidy rate for federally supplied water

TABLE 2-6. Agriculture is by far the largest user of fresh water in the United States, accounting for 42 percent of all uses nationally in 1990. About three-fourths of agricultural water use is lost, primarily through evaporation, transpiration, and conveyance.

| Top Five States | Total Use (fresh) | Agricultural Use (Irrigation and Livestock) | | | Irrigated Acres (thousands) |
		Total Ag Use (M gal/day)	Ag Use from Groundwater (M gal/day)	Total Ag Use as Percent of Total Use	
California	35,100	28,311	10,905	81	9,480
Texas	20,100	8,718	5,683	43	6,220
Idaho	19,700	19,260	7,180	98	3,410
Illinois	18,000	141	136	<1	287
Colorado	12,700	11,762	2,582	93	3,557
Total U.S.	339,000	141,500	53,690	42	57,200

Source: Solley et al., 1993.

TABLE 2-7. In 1986, much of California's irrigation water was used for relatively low-value agricultural commodities.

Crop	Water Use (million acre-feet/year)	Crop Value ($ million)	Ratio of Crop Value to Water Use ($/ac-ft)
Pasture	4.19	93	0.22
Alfalfa Hay	4.09	570	1.39
Cotton	3.42	843	2.46
Rice	2.62	204	0.78
Grapes	1.59	1,412	8.89
Almonds & Pistachios	1.08	633	5.82
Tomatoes	0.65	485	7.45
Oranges & Lemons	0.55	848	15.31
Total/Average	18.19	5,088	2.80

Source: Reisner and Bates, 1990, p. 31. Ratios are derived.

BOX 2-4. Recovery of the Ogallala Aquifer in Texas

The Texas High Plains, one of the country's richest agricultural regions since the early 1950s, depends almost entirely on the drastically over-drafted Ogallala Aquifer for its irrigation supply. Precipitation is only about 12 to 16 inches per year. In places, the water table was dropping as much as five feet per year, making it certain that sooner or later the Ogallala would be too depleted to pump economically.

When energy prices doubled in the late 1970s and predictions suggested that the Ogallala had only 20 or 30 years more life, the region's farmers came to their own rescue. After ten years of diligent, self-financed conservation efforts, water use was cut back by 25 to 40 percent across the region. Farmers replaced unlined ditches with pipes, made water delivery more precise, recovered tailwater for reuse, and monitored the soil to determine moisture needs. Thanks to these and other improvements, the water table in the High Plains began to stabilize. In 1986, the water table actually increased by six inches—the first rise since 1941.

Source: Excerpted from Reisner and Bates, 1990.

through the Bureau of Reclamation is $54 per acre served, with some payments reaching $150 per acre served. (USDA/ERS, 1994c).

California has more irrigated acres than any other state. As Table 2-7 shows, 60 percent of the water used is for relatively low-value commodities, largely because the cost of water is heavily subsidized by government agencies. The amount of water applied to pasture, alfalfa, rice, and cotton in California is enough to supply 70 million people. These crops contribute 15 percent to California's agricultural income and just one-quarter of 1 percent to the state's total income (Reisner and Bates, 1990).

Irrigation often relies on essentially nonrenewable assets. While irrigation from surface water relies on precipitation—an obviously renewable resource—it is also made possible by human-made reservoirs that eventually fill with sediment and lose storage capacity. Replacement reservoirs must be located on less acceptable sites and thus cost more to build.

Groundwater, which is recharged from the surface or from lateral movement, is being used at rates beyond recharge in twelve states, including all the major irrigation states. Some areas of Idaho, Texas, and Kansas have reported water table declines of more than four feet per year. Texas, Kansas, California, and Nebraska each have more than 2 million acres of irrigated land subject to at least six inches of decline per year (USDA/ERS, 1994c). Where groundwater is not recharged, or is recharged only very slowly, resource depletion is inevitable and can merely be prolonged by conservation. Where recharge rates are higher, conservation can preserve the resource's economic value by avoiding depletion. *(See Box 2-4.)* As the water table declines, the cost of lifting the water increases rapidly until the resource eventually becomes too expensive to use. Incremental costs may be small—one estimate puts the average at $2 per acre or less—but these costs are often cumulative and permanent (USDA, 1994c).

Toxics in the Environment

Once released into the environment, pesticides can find their way into watersheds, aquifers, and the air—where they injure human and ecosystem health. Pesticides are widely used because they make agriculture more stable and productive by mitigating losses to insects, weeds, fungi, rodents, and other pests. Albeit imperfectly, U.S. policy has addressed the use of toxics in agricultural production and their role in human cancers. A new fear is that these chemical compounds may cause hormonal problems and suppress the immune system in humans and animals. Human health concerns in the United States are greatest

FIGURE **2-6. Pesticide Use on Crops**

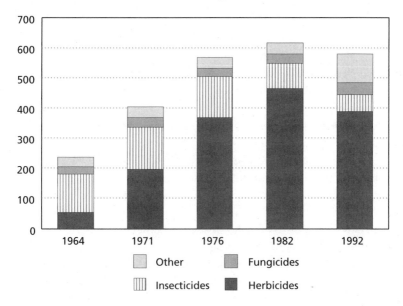

for farmers, farmworkers, homeowners, and those who work in pesticide factories.

Although blemished in many respects, the U.S. regulatory system developed over the last three decades effectively controls the release of the "biocides." Most newer pesticides achieve results with much smaller amounts of active ingredients that are less persistent in the environment, though some may also be more immediately toxic.

Pesticides have been an important source of productivity gains since World War II. Farmers now depend heavily on pesticides because they can reduce the risks associated with pest outbreaks and they cost less than other production inputs (Conway and Pretty, 1991). Pesticides allow the continuous production of high-value crops that otherwise would have to be grown in rotation with crops of lesser value to break pest cycles. Over the last three decades, total pesticide use has doubled, and herbicide use has increased sevenfold. *(See Figure 2-6.)*

More pesticides are used on corn than on any other commodity: 58 percent of the herbicides, 36 percent of the insecticides, and 42 percent of the total. Cotton is the second heaviest user of pesticides, accounting for 10 percent of all applications (USDA/ERS, 1994c, p. 90). Fruits and vegetables have the highest rates of active ingredient application per acre, at 49 and 21 pounds per acre, respectively, compared to 3 pounds per acre for corn (USDA/NASS, Agricultural Chemical Usage, 1991-94).

Unfortunately, the use of pesticides has not eliminated or even reduced pest losses, which have stayed constant as a proportion of yields. The average crop loss to insects, diseases, and weeds in the United States during 1942-1950 was about 32 percent; in 1984, losses to the same pests totaled about 37 percent (Conway and Pretty, 1991, p. 23).

One reason for the constant rate of crop loss is that pests adapt and grow resistant to pesticides. Fifty years ago, only 44 insect species were known to be resistant to insecticides. In 1986, the National Research Council placed this number at 447, three-quarters of them agricultural pests. Weeds can also possess or develop resistance. Resistant weeds may flourish when herbicides are applied, creating a secondary problem. Fungal resistance to fungicides was not a problem until the mid-1970s, but by the late 1980s, resistance had been observed in all major fungicide classifications (Ware, 1989).

The use of pesticides in agricultural production remains a serious environmental problem because of the sheer volume of chemicals used and their broad spectrum of effects. Consider atrazine, currently the most commonly used herbicide. Some 72 million pounds of active ingredient are applied to more than 63 million acres of cropland (Gianessi and Anderson, 1995). Atrazine is highly toxic to fish, crustaceans, bees, molluscs, aquatic insects, amphibians, and soil invertebrates, and somewhat to moderately toxic to mammals, displaying long-term effects as a carcinogen, mutagen, and immunotoxin. Atrazine is moderately persistent in the environment and is soluble in water (Briggs et al., 1992).

Chloropyrifos is the most commonly used insecticide. Annually almost 15 million pounds of active ingredient are applied to more than 1.5 million acres (Gianessi and Anderson, 1995). It has medium to high immediate toxicity to mammals and displays long-term fetotoxic and neurotoxic cumulative effects. Chloropyrifos has been known to cause sterility and impotence in bulls, and it has very high immediate toxicity to birds, molluscs, fish, amphibians, crustaceans, and aquatic insects (Briggs et al., 1992).

Pesticides have become common in our environment. A 1990 survey by EPA estimated that about 10 percent of community water systems and 8 percent of rural domestic wells in the country contain one or more pesticides, though in each case slightly fewer than 1 percent of these cases exceeded the federal guidelines for lifetime exposure. A similar study of rural wells in Iowa found that 13 percent of the wells contained one or more pesticides, most commonly atrazine. In northwestern Iowa, the incidence was greater than 23 percent. The study team estimated that 5,400 Iowans were consuming water with pesticide concentrations above the federal guidelines (Kross et al., 1990). Some pesticides are even found in rainwater now (Richards et al., 1987).

Fortunately, researchers have developed and some farmers have implemented a control strategy—Integrated Pest Management (IPM)—that can reduce pesticide use. Ecologically and economically based, IPM includes a variety of options, such as crop rotation, tillage, water management, biological control, and pesticide use, to manage pests below a level where they cause financial losses. Results are impressive. For example, IPM has been shown to reduce pesticide use by 75 percent in cotton, 80 percent in peanuts, and 69 percent in almonds while maintaining adequate protection against pests (National Research Council, 1993b).

Households may be inadvertently exposed to pesticides from water and foods, but they also subject themselves and the environment to significant doses. The EPA (1994) estimates that 69 million U.S. households use pesticides, typically at higher rates than farmers do (Pimentel et al., 1991). On lawns, homeowners apply 3.2-9.8 pounds of pesticide per treated acre per year on average—up to

2.6 times greater than agricultural use rates. And while 1.3 million farmers and pesticide applicators have taken classes to learn how to use, store, and dispose of pesticides, consumers typically buy and use them without any training whatsoever (U.S. EPA, 1994). These users may be subjecting themselves to significantly higher cancer rates for the sake of perfect lawns; in households that use home and garden pesticides, children face a risk of leukemia that reportedly is 6.5 times greater than in other households (Lowengart et al., 1987).

Cancer has been considered pesticides' main threat to human health and regulation appears to be effective against the danger, though opinions on this issue diverge sharply. The National Research Council (1987, p. 65) estimated that pesticides in the food system cause the risk of cancer to increase from about 25 percent in a lifetime, to 25.1 percent for the average person— a minimal increase compared to other risks such as smoking or obesity.

As conventionally measured, pesticides on food may not be a significant health problem for the general population. In 1992, the USDA's Agricultural Marketing Service tested almost 6,000 samples of fresh fruits and vegetables for residues of 14 fungicides, 6 herbicides, and 30 insecticides. Of those, 61 percent contained detectable residues, mostly from post-harvest applications of the 500 possible pesticide/commodity combinations, and 477 showed residues lower than 1 percent of legal tolerances. The largest average residue relative to the tolerance was for benomyl on bananas, where the residues averaged 12 percent of the tolerance (USDA/ERS, 1994c, p. 102, 103).

This risk may be considerably higher for other groups of people, however—especially those who apply and manufacture pesticides. Surveys of mortality patterns show that farmers tend to have a lower incidence of heart disease and all cancers than other occupational groups. Yet, they suffer a higher incidence of such malignancies as non-Hodgkin's lymphoma, leukemia, soft-tissue sarcoma, and brain cancer—all associated with pesticide use (Blair and Zahm, 1991; Hoover and Blair, 1991). Workers who manufacture pesticides also suffer a greater incidence of these diseases (Hoover and Blair, 1991).

Recent studies suggest that conventional measures of pesticides' health impacts may understate health risks. New (though highly contested) data suggests that even in small amounts pesticides may harm immunological and reproductive systems because they can mimic human hormones. For instance, health may be at risk where pesticide levels are below current drinking water standards since low concentrations over very short periods have been found to do chromosomal damage in the ovary cells of hamsters (Biradur and Rayburn, 1995). Various agricultural chemicals may act as so-called endocrine disruptors, disturbing the prenatal and early postnatal development of organs that respond to endocrine signals. The effects include "abnormal thyroid function in birds and fish; decreased fertility in birds, fish, shellfish and turtles; demasculinization and feminization of male fish, birds, and mammals; defeminization and masculinization of female fish, gastropods, and birds; and alteration of immune function in birds and mammals" (Colburn et al., 1993a, p. 378). These problems went unnoticed until the 1950s, but are currently observable in many areas where human-made chemicals have been used (Colburn, et al., 1993a; Conway and Pretty, 1991). At least 8 widely used herbicides, 19 insecticides and 8 fungicides have been reported to have reproductive and endocrine-disrupting effects (Colburn et al., 1993b, p. 379). Pesticides may contribute to unexplained cases of human breast cancer (Davis and Bradlow, 1995) and declining male fertility too (Danish Environmental Protection Agency, 1995; Sharpe and Skakkebaek, 1993), though these linkages are still being debated (Safe, 1995) and numerous scientists question their validity.

Evidence also suggests that pesticides may suppress the immune system and contribute to increased rates of morbidity and mortality. Repetto and Baliga (1996) summarized the available evidence from many "animal studies" that show that pesticides can alter the immune system's structure, disturb immune responses, and lower resistance to infection. Other direct and indirect studies point to similar responses in people. The authors concluded that the preponderance of data point to a significant but unrecognized health threat, particularly for exposed populations in developing countries where most people's health is already stressed by poor living conditions and where many have no access to medical care. In the United States, many pesticides now in use have been shown to affect the immune system, but the net effect on health is unknown. Because the focus has been on cancer for so long, regulatory testing of immunotoxicity is very limited as a requirement for pesticide registering or re-registering.

NEW THREATS TO SUSTAINABILITY

Two emerging threats to agriculture's sustainability are fundamentally different from those of the past—loss of genetic resources and global climate change. The scale of these potential threats is much larger, so broader social consensus and actions are required to address them. Soil erosion can be battled by each individual farmer in the field, where productivity is mainly affected. But to combat the root causes of genetic resource loss and global climate change, farmers can do little on their own except adapt to some extent. These threats hit factors of production for which substitutes are much more difficult to find, and adaptation is expensive.

The impacts of erosion of the genetic stock and global climate change may be irreversible in a human timeframe. While climate change occurs over decades and certain "greenhouse" gases last for many thousands of years, species loss is irretrievable over any human timeframe. Scientific understanding of the consequences of these changes is sketchy at best, but suggests potentially severe outcomes. For this reason, the risk associated with a failure to find out more and to take appropriate action is high.

Erosion of Germplasm

The ability to develop new plants to increase productivity or produce adaptations for changing environmental pressures depends upon whether suitable genetic stock—plants that possess the sought-after trait—is available. Scientists create new varieties by crossing plants that have the desired characteristics. While this can now be done much more quickly with bioengineering methods and traits can be transferred between species, nothing has obviated the need for the basic genetic stock.

Since the 1940s, genetic improvements have accounted for about half of all increases in crop yields (National Research Council, 1993a). Much of this improvement has come about through the crossing of wild plant varieties with varieties already used in crop production. For example, the major breakthroughs in the "green revolution" for both wheat and rice were the crossing of relatively low-yielding dwarf varieties with high-yielding tall varieties. The combination made for a "miracle" high-yielding variety that would not fall over when heavily irrigated and fertilized.

New finds that have proven valuable continue today. A new "super rice" based on a wild relative has fewer tillers, or vegetative stems, so the plant can put more energy into grain production and can be planted more densely. Expectations are that when this new cross is available, yields could increase by 25 percent without any additional use of irrigation water and only minimal additional use of fertilizers. In development at the International Rice Research Institute in the Philippines, this plant could produce an additional 100 million tons of rice per year to feed 450 million people (Consultative Group on International Agricultural Research, 1995). Plants can also be bred to resist pests and other stresses.

Humans have used as many as 5,000 species as food, but only 150 have become commercially important and fewer than 20 provide the bulk of the world's food supply. Three crops—rice, wheat, and maize—account for about 60 percent of the calories and 56 percent of the protein that humans derive from plants (Wilkes, 1983; Frankel and Soulé, 1981).

Except for the sunflower and Jerusalem artichoke, none of the major crops grown in the United States originated here (Reid and Miller, 1989).

One double-edged threat to domesticated varieties is monocultural production and the drive to make varieties more uniform so they can be easily shipped and stored. Fruit and vegetable varieties have been particularly hard hit. As many as 96 percent of the commercial vegetable varieties known to the USDA in 1903 are now extinct. Eighty-six percent of the 7,000 apple varieties used during the nineteenth century have disappeared, along with 88 percent of the 2,683 pear varieties (Fowler and Mooney, 1990). Just three apple varieties now comprise the bulk of U.S. apple production.

A pest or pathogen can cause widespread destruction if it encounters a genetically uniform crop grown over a large area. Yet, nearly all South American coffee trees descend from a single tree that was grown in the Amsterdam botanical garden 200 years ago. Currently, coffee production in South America is threatened by a virulent new disease. Little resistance to this disease has been found, perhaps partly because most of the forests in Ethiopia where the tree originated have disappeared (Fowler and Mooney, 1990).

Similar examples exist in the United States. In the early 1970s, a corn blight epidemic took 15 percent of the entire crop. Hybrids with a certain characteristic were used in more than 70 percent of the corn varieties grown then, making these varieties susceptible to a new race of fungus. Wine production in the Napa and Sonoma areas of California is seriously threatened because 70 percent of the crop is grafted onto a rootstock under attack by a new variant of a graperoot pest. A relative of this pest decimated the European wine industry during the nineteenth century, destroying four million acres of vineyards in France alone (NRC, 1993a).

While genetic uniformity has decreased for wheat and corn since the 1970s and genetic adaptation can be used to overcome certain pests, genetic improvement remains a constant concern requiring years of research and breeding. The International Rice Research Institute released its first high-yielding variety, IR-8, in 1966. Within three years a viral epidemic reduced yields and a new variety, IR-20, was released. This variety was attacked by a different virus and an insect (the brown planthopper) and was replaced in 1973 by IR-26. A new variant of the insect arose to damage IR-26, so a resistant variety, IR-36, was developed

BOX 2-5. The Irish Potato Famine

By the mid-1800s, the potato had become the staple of the poor in Ireland—in 1840, the average adult was eating between nine and fourteen pounds per day. In the summer of 1845, there were numerous reports of an abundant crop in the offing, but when the harvest began in September an awful reality unfolded. A newspaper reported on September 11th of the same year the presence of a "cholera" on all the potatoes harvested: "...the tubers [were] all blasted and unfit for man or beast." The potatoes turned black and rotted, filling the countryside with an awful stench. During the first year, the weather was blamed.

The next year was no better. While three-quarters of the land was devoted to cereals, nearly all of this was exported to England or used to pay rents. Peasants could not afford to keep the cereal they grew and lived on potatoes.

The winter of 1847-48 saw corpses lying in the streets. By the spring of 1849, the toll had become staggering. One county of 5,000 saw more than 700 people die within a two-week period. A total of one to two million people lost their lives and as many migrated to North America.

Weather was not the cause of the failed potato harvest during the famine; it was *Phytophtora infestans,* a potato blight. The genetically limited potatoes grown in Europe could not resist this disease. Also, the social structure of the time was, of course, as much to blame for the famine as the blight. Eighty percent of Ireland's agricultural land, controlled by English landlords, remained in pasture while the peasants starved. The British opposed giving food to the starving lest it encourage the idle poor.

Fortunately, resistance was identified among the potatoes grown in the Andes and Mexico.

Excerpted from Fowler and Mooney, 1990, pp. 43–45.

and released. Yet another biotype of the brown planthopper damaged IR-36, requiring the use of IR-56 in 1980 (National Research Council, 1993a).

Governments have established seedbanks to store and protect many varieties of the world's principal commodities so that they will be on hand when a problem arises. But, the safety of these seedbanks is debatable. Many stored varieties are not cataloged—so breeders can't find them, storage facilities may be inadequate, and the seeds may not be grown out and replaced before they die. The Administrator of the USDA's Agricultural Research Service has estimated that 90 percent of all seed samples collected from outside the country before 1950 have been lost (Raeburn, 1995). Thousands of varieties held in U.S.

germplasm banks—including large collections of soybeans from China, sugar cane from Papua New Guinea, and a unique international collection of maize—have literally been lost or thrown away.

Important genetic collections have deteriorated because they have not been replanted and regenerated. Only one-third of the seed samples collected in the 1940s survive, and just 11 percent of those collected before the 1930s are viable. While these seedbanks have been responsible for billions of dollars worth of crop production, the USDA allocates just $30 million per year to protect them (Raeburn, 1995). In developing countries, often the origins of genetic diversity, neglect is even more pronounced.

The loss of a few thousand varieties may not seem important, but a single plant may hold enormous genetic value. For example, wheat collected in Turkey in 1948 was described as "...a miserable-looking wheat, tall, thin-stemmed, lodges badly, is susceptible to leaf rust, lacks winter hardiness... and has poor baking qualities." Yet, this same specimen was found to have immense value when stripe rust took hold in the northwestern United States because it proved resistant to "four races of stripe rust, 35 races of common bunt, ten races of dwarf bunt" and tolerant of flag smut and snow mold. Similarly, when barley breeders were looking for resistance to barley dwarf yellow virus, only one variety out of 6,500 had the desired trait (Fowler and Mooney, 1990, p. 69).

Crop improvements come not only from seedbanks but also from the wild relatives of domesticated plants. The loss of these potentially invaluable wild relatives is certainly irreversible; their only protection is the habitat in which they live. Unfortunately, for many of the world's most important crops, these habitats are being converted to other uses.

In situ conservation (conservation of species in their natural habitat) is an important element of maintaining genetic stocks because some traits of wild plants may not yet be recognized. This type of conservation also allows nature to develop new genetic traits in plants as they are exposed to selective pressures. Some researchers think that if wild relatives are not protected *in situ* they may succumb, when planted, to diseases that have adapted while the plants were in storage (Raeburn, 1995). *In situ* conservation is particularly essential for trees because they take so long to mature.

Climate Change

More than any other environmental characteristic, climate determines the nature and type of agricultural production. The levels, timing, and variability of temperature and precipitation are first-order constraints, as every farmer knows. In the United States, most agricultural land is rainfed. Much of this land, particularly east of the Mississippi River, is highly productive because rainfall and temperature are very suitable for agriculture.

A central difficulty with climate change is how little we know about it. We do not know how or when climate will change, though most experts now agree that global average temperatures will increase 1.5 to 4.5 degrees Celsius (2.4 to 7.2 degrees Fahrenheit) in the next 50 to 100 years. We do not know the pace or the nature of such changes. Both are critically important. If climate change occurs slowly and is limited to changes in average temperature and precipitation, then adaptation in the agricultural sector may not be a major problem. But, if climate changes in some surprising and dramatic way, disaster could ensue. Further, if the average level *and* variability of precipitation increases, then adaptation will be significantly more taxing. These factors combine to present a potentially large threat to agricultural sustainability.

There are four main time scales over which climate change and variability can be distinguished (Oram, 1989):

1. Short-term natural disasters beyond human control—cyclones, hurricanes, typhoons, and large-scale flooding.

2. Inter-annual and intra-annual variability.

3. Cyclical patterns of climatic anomaly across several years, but within a basically unchanged situation affecting the world as a whole (e.g., the 1940-1970 cooling trend) or geographic regions (e.g., the deteriorating rainfall in the Sahel that began about 1950).

4. Long-term climatic change over several centuries or longer.

The impacts of changes in the *means* of temperature, precipitation, solar radiation, and other climate variables have been assessed in terms of the last two time scales listed above. (*See edited collections and summaries* including Downing, 1996; Houghton et al., 1995; Rosenzweig and Iglesias, 1994; Kaiser and Drennen, 1993; Reilly and Anderson, 1992; Oram, 1989.) But researchers don't agree on the impact of climate change on crop yields. Some results show that warmer temperatures, perhaps coupled with drier growing seasons, would reduce crop yields (Terjung et al., 1984; Warrick, 1984). Other studies suggest that these adverse impacts could be offset by plant fertilization from higher levels of carbon dioxide in the atmosphere (Peart et al., 1989; Robertson et al., 1987). An overview of these assessments done by the Intergovernmental Panel on Climate Change

(Houghton et al., 1995, p. 2) concluded that "[g]lobal agricultural production appears to be sustainable under climate change as expressed by GCMs [global circulation models] under doubled CO_2 equilibrium climate scenarios." The report also notes, however, that "crop yields will vary considerably across regions" and that "the pattern of agricultural production is likely to change in a number of regions."

One assessment (Reilly et al., 1994) expanded the economic impacts of climate change using an economic model of global agricultural trade. The total effects on human welfare range from an increase of $7 billion to a decrease of $38 billion. For the scenario based on the most pessimistic assumptions, the outcome was a decline in total welfare of $250 billion. The countries hardest hit will be the poorer countries with less capital, knowledge, and institutional ability to adapt.

For livestock, the results depend on location—cooler areas fare better while warmer areas fare worse (Baker et al., 1993; Klinedinst et al., 1993). Globally, diseases for both livestock and crops are expected to increase (U.S. EPA, 1989). One study suggests that ticks will become a much greater pest in northern North America, as far north as Nova Scotia, and less troublesome in hot southern zones (Smith and Tirpak, 1988). Other pests from farther south could also move north. Malaria, schistosomiasis, lymphatic filariasis, dengue fever, and other vectors of human disease have been identified as being highly or very likely to spread as climate changes (Weihe and Mertens, 1991).

Most agricultural production analysis of climate change has been based on climatological models that assume a doubling of carbon dioxide. Such models tell us little about daily weather. Impact studies have had to rely on monthly averages, even though in crop production variability is relevant over days, not months (Wilks and Riha, 1996). Further, the atmospheric buildup of greenhouse gases will not stop at a doubling and climate will not automatically achieve some new equilibrium state; indeed, it will continually change, making continual adaptation necessary (John Firor, pers. comm., 1996).

The effect of variability on crop production has been largely ignored in the context of global climate change (Wilks and Riha, 1996). While crops are not highly vulnerable to changes in mean temperatures and precipitation, some crops are very susceptible to changes in variability. For example, when warm days are followed by cold nights, stomates can fail to function. Similar problems can occur when temperatures are unusually high. Variability in temperature can interfere with crop development, flowering, and dormancy. Warm temperatures in late

FIGURE 2-7. **Annual U.S. Climate Extremes Index**

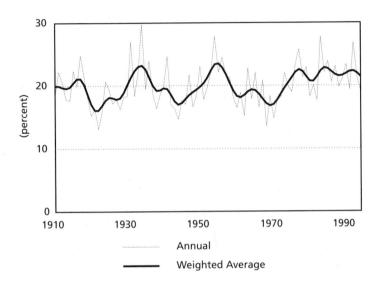

Annual

Weighted Average

winter can break dormancy too soon, and if a warm spell is followed by cold temperatures, new growth can be killed.

Since soils can hold moisture, the effects of variability in precipitation are less extreme, but are not inconsequential. Water shortages can limit photosynthesis and decrease maturation time, lowering yields. Brief water shortages during prime developmental stages can limit the number of seeds that plants produce, severely cutting into yields. Too much water restricts the access of air to roots, constraining growth (Wilks and Riha, 1996). A few recent studies have pointed out the importance of variability in crop production and the need for a much deeper understanding of how climate change could affect variability (Katz, 1996; Mearns et al., 1996; Riha et al., 1996).

Recent evidence shows that climate in the United States has become more extreme in the last 20 years and that this change is most likely related to the greenhouse effect. Thomas Karl and his colleagues at the National Climatic Data Center in North Carolina (1995) developed a Climate Extremes Index using data going back to 1910. This index includes maximum and minimum temperatures, drought and moisture surplus, extreme one-day precipitation, and number of days with precipitation. *(See Figure 2-7.)* While there have been spells of extreme weather in the past, notably in the 1930s and 1950s, the extreme weather of the last two decades has lasted longer. In further analysis, the developers of the index conclude that there is a 90–95 percent probability that these extremes stem from global climate change.

Some observers raise the specter of surprise, defined as an event that "...occurs when perceived reality departs qualitatively from expectations" (Schneider, 1996). One type of surprise could be in the structure, composition, and functioning of ecosystems. Climate change may harm some species and benefit others, giving the winners an advantage and increasing their range and population. In turn, those species higher in the food chain will also be affected.

The way that climate itself changes could also hold surprises. The Gulf Stream carries warm water northward, where it eventually cools and drops some 5,000 meters to the ocean floor, driving ocean currents and weather patterns. Recently, the temperature near Greenland has not been low enough and the water has only sunk about 600 meters, "almost turning off." Major changes to global ocean currents could radically alter climate in unforeseen ways. Certainly such changes have been associated with extreme climate in past ages (Global Environment Change Report, 1996).

Assuming that climate is changing, how big will the impact be on yields, supply, and prices, and how can farmers adapt? One response to changing temperature is to use varieties that have a different growing season; increased temperature means that production could expand northward. Increased temperature also means more heat stress on plants. If precipitation is sufficient, though, a change in varieties can address the problem.

If precipitation becomes more variable, however, adaptation gets harder. Drought at critical periods can devastate yields, and floods are never welcome. Some observers of the 1993 weather pattern—drought in some areas and flooding in others—warn us to expect more of the same if climate changes as expected. Production during the 1993 crop year was down significantly, as was trade and farm income, while prices were up.

Adapting to climate change appears to be within the realm of the possible, but also herculean and expensive. Farmers may have to change cropping patterns (for example, growing more wheat which has smaller water requirements). They may have to use more irrigation and raise animals in climate-controlled environments. Such adaptations could mitigate much of the negative impact (Darwin et al., 1995).

The suitability of many areas for agricultural production is very likely to change temperature and precipitation patterns. Roy Darwin and colleagues of the Economic Research Service (Darwin et al., 1995) coupled global land suitability maps with economic analysis to evaluate changes in land and water resources, concluding that major land use changes would be necessary. For the United States, the area of land suitable for agriculture and forestry may increase, but most of the land opened up will be in Alaska. In the Corn Belt and the Southeast, more land will become less suitable for crop production.

Other disruptive changes can be expected too. As competition for cropland gathers momentum, tropical forests loss may accelerate. Ecosystems threatened by conversion may face additional stresses from temperature and precipitation changes. In some areas, the only viable option will be to shift to lower value production systems that demand less water or to abandon agriculture altogether. Competition for water resources, already contentious in some areas, is likely to increase, with attendant economic and ecologic consequences (Darwin et al., 1995). Farm communities in such areas may be hurt economically.

Fortunately, the United States' extensive agricultural assets and options mean that it can adapt to climate change, if necessary. The Council for Agricultural Science and Technology (1992) studied this issue and identified ten categories of agricultural assets ranging from land and water to human resources and world markets. *(See Table 2-8.)* It also identified key policy responses to increase what it saw as the essential element necessary for adaptation—flexibility.

Oram (1989) of the International Food Policy Research Institute suggests several approaches to dealing with climate change. *(See Box 2-6.)* Most are aimed primarily at developing countries, but there are relevant messages for the United States. In the short term, countries can increase their food security by expanding international and intra-regional trade. In the short to medium term, improved meteorological forecasting could help farmers adjust their production decisions. Over the medium to long term, irrigation investments can guard against climate risk and may encourage diversification and higher yields. Research, particularly on climate change risks in agriculture, has not been well funded, but it could enormously improve understanding and agriculture's ability to adapt.

WHAT ABOUT DEMAND?

Most literature on sustainable agriculture focuses on production and on adaptation to maintain productivity and minimize resource impacts—that is, the supply side. However, the demand side of the issue is at least as important.

	Value for adapting to	
Asset	**climate change**	**Policy steps to increase flexibility**
Land	Extensive cropland across diverse climates provides diversity for adaptation	Reform agricultural policy to encourage flexible land use.
Water	Water, which already limits farming in some areas, is crucial for adaptation if climate becomes more dry.	Change institutions to encourage more prudent use of water. Raise the value of crops produced per volume of water consumed.
Energy	Reliable energy supply essential for many adaptations to new climate.	Improve the efficiency of energy in food production. Explore new biological fuels and ways to stash more carbon in trees and soil.
Physical Infrastructure	Facilitates trade and input flows when market signals change.	Maintain and improve input supply and export delivery infrastructure.
Genetic Diversity	Source of genes to adapt plants and animals to new climates.	Assemble, preserve, and characterize plant and animal genes. Conduct research on alternative crops and animals.
Research Capacity	Source of knowledge and technology for adapting to climate change.	Broaden research to encompass adaptation to climate change. Encourage private research on adaptation. Find farming systems that can be sustained in new climates. Develop alternative food systems.
Information Systems	Provide information needed to track climate change and adapt to it.	Enhance the nations systems that exchange information. Encourage the exchange of agricultural research information.
Human Resources	Provide pool of skills enabling farmers and researchers to adapt to climate change.	Make flexible skills the hallmark of agriculture's human resources. Strengthen rural education systems, particularly continuing education.
Political Institutions	Determine policies and rules that facilitate or hinder adaptation to new climates.	Harmonize agricultural institutions and policies.
World Market	Enables trade to mediate shifts in farm production and sends price signals that eventually adjust production to new climates.	Promote freer trade and avoid protectionism.

TABLE 2-8. Assets, value, and policy steps to address global climate change in the United States as put forward by the Council on Agricultural Science and Technology. Many of the recommendations focus on flexibility of response.

Source: Council for Agricultural Science and Technology, June 1992, p. 7.

BOX 2-6. Adapting to Climate Change

People have used various mechanisms to reduce or spread out climate risks, including:

- *Mobility.* Nomadism, seasonal movement of livestock between uplands and lowlands, shifting cultivation, and migration (as a last resort).
- *Spreading risks.* Large holdings, stratification across ecological zones, fragmentation of holdings, communal land sharing, farming cooperatives, and crop insurance.
- *Manipulation of the environment.* Irrigation, drainage, soil amendment, terracing, contour farming, mulching, and other soil- and moisture-conservation practices, fallowing, fertilizer application, and shelter belts.
- *Diversification of production systems.* Introduction of new crops or cultivars, crop rotation, mixed cropping, multiple cropping, alley cropping, agroforestry, dual-purpose livestock breeds, and mixed livestock.
- *Manipulation of farming systems.* Plant and animal breeding, changing the crop calendar (i.e., time of sowing or harvesting, etc.), adjusting livestock breeding practices (breeding seasons, supplementary feeding, hormonal treatments), mechanizing critical time-dependent operations, and using chemicals for crop protection or as substitutes for machine or human labor (i.e., no tillage).
- *Reduction of post-harvest losses.* Storage, drying, fumigation, processing, and refrigeration.
- *Trade and aid for food security.*
- *Weather wisdom.* Folklore and experience (in traditional societies), meteorological recording, analysis, forecasting, early warning systems, and research on climate and factors that influence the weather.

Source: Adapted from Oram, 1989.

Global Food Demand

The question of balance between food supply and demand is perennial. Until Thomas Malthus, most people looked only at the short term, though stories going back at least as far as the Egyptians tell of attempts to plan for and manage droughts. Malthus, though wrong in many respects, aptly described the real tension between population growth and increased food production. These concerns persist today amid the most rapid population buildup ever.

Over the past 30 years or so, the incidence of famine throughout the world has declined relative to earlier periods. While there are notable exceptions—particularly the death of 30 million people during the Great Leap Forward in China, and the recent famines in Africa—these events owe much more to political turmoil than to the limits of agricultural production.

In India, famine was a fairly regular event before the Green Revolution, occurring almost once a decade. Since the mid-1960s, however, famine has all but vanished from the Indian subcontinent. Other areas of the world have experienced similar benefits from dramatic increases in agricultural production.

In the future, increased agricultural demand will stem not only from absolute increases in population, but also from increased demand per person as incomes rise and demand for more input-intensive commodities, such as meat, increases. In China, India, and other parts of Asia, increases in the demand for meat and wheat are expected to skyrocket as per capita incomes increase and dietary patterns change.

A debate rages, though, over the world's ability to maintain the rate of increase in production needed to keep up with expected demand. In many areas, the Green Revolution technologies are played out, all the suitable agricultural land has already been brought under the plow, and the resources that support agriculture (in particular irrigation works) are being degraded as fast as they are created or renewed. Asia has essentially no arable land left, and increases in production will have to come from increased productivity. In Africa, civil strife, unwise policies, poor infrastructure, ineffective research and extension, and resource degradation all constrain production, resulting in per capita declines in agricultural production over the past decades. Meanwhile, North America, Latin America, and Europe are surplus producers that could produce even more than they do now.

Projections by the International Food Policy Research Institute to the year 2020 show developing countries increasing their imports of wheat by 120 percent, rice by 1,500 percent, and corn by 180 percent. For soybeans, developing countries could shift from a small net export position to a large net importer (Rosegrant et al., 1995, p. 46). For wheat, this increase roughly equals current U.S. production, while for corn the increase is roughly 15 percent of current production. Food demand in developing countries is expected to rise between 1.67 and 3.40 percent per year, with cereals on the lower end and meat products on the upper. In contrast, demand for these products may grow in the developed countries by only 0.37 to 1.10 percent per year (Rosegrant et al., 1995, p. 9).

Despite the swell of international demand for agricultural commodities, prices may continue their long-term decline. As Table 2-9 shows, prices are expected

TABLE 2-9. Although international demand for agricultural commodities is expected to boom, prices may decline significantly.

Commodity	Change in Real Prices 1990–2020 (percent)	Change in World Trade 1990–2020 (percent)
Wheat	−12	61
Rice	−22	139
Maize (Corn)	−23	37
Other coarse grains	−25	62
Soybeans	−11	121
Roots and tubers	−18	29
Beef	−6	187
Pig meat	−10	64
Sheep meat	−4	85
Poultry	−10	227

to drop between 4 and 26 percent, while trade could more than triple for some commodities. For cereals, the net decline in prices is 19 percent; for meats, 9 percent.

With the work done by Lester Brown of the Worldwatch Institute a notable exception, assessments of future world food supply and demand[*] generally show that supply will more or less keep up with demand. Assessments of future food demand uniformly show major increases expected as population grows, but enormous divergence appears in agriculture's projected ability to maintain production increases. Brown, more pessimistic than most about the sustainability of world production,[**] points out various resource problems that could seriously inhibit developing countries' production capabilities. In their assessment (1994), Brown and Kane conclude that having enough food ultimately means stabilizing the world's population.

With an enormous stock of agricultural resources, preeminent research institutions, well-educated farmers, and extensive infrastructure, the United States is in an enviable position. U.S. supply potential greatly exceeds domestic

---

[*]   For a more detailed summary of these issues see *World Resources Institute* (1996).
[**] See Crosson (1995) for an alternative viewpoint.

market demand. Approximately 60 million acres, almost one-fifth of the cropland base, was retired—until recent changes in farm legislation—to help the government balance supply and demand (USDA/ERS, 1994c). These retirement programs have helped reduce soil erosion and polluted run-off from agricultural lands, made the maintenance of fragile lands easier, and provided habitat for wildlife, particularly birds.

For the United States, the major agricultural concerns stemming from worldwide population growth are its impact on export demand, increased domestic production, and its environmental costs. Past environmental gains could be forfeited if retired land comes back into production only to be managed poorly.

Managing higher demand for U.S. agricultural production in an environmentally sensitive way requires increasing both agricultural output and the efficiency of input use. Because the use of some production inputs, especially nutrients, seems inefficient (NRC, 1993b), room for improvement may be large. In addition, the off-farm impacts from surplus land—in general, more subject to soil erosion and so more likely to degrade water quality—need attention.

To meet increased demand, a final necessary element will be to target federal conservation spending to induce the same environmental benefits on less land. One example is enrolling only the most erodible sections of a field or those portions along the banks of streams and rivers. The U.S. General Accounting Office (1995) has estimated that the water-quality benefits of the CRP, currently with an enrollment of 36.4 million acres, could be realized on just 6 million acres.

U.S. Consumption Patterns

The social costs of U.S. consumption patterns are probably much greater than the environmental costs. Heart disease alone kills one-third of all Americans and two-thirds of these cases are associated with poor diet. About 20 percent of deaths from heart disease could be avoided by dietary changes. The direct health care costs and productivity losses due to heart disease have been estimated at $56 billion per year (Frazao, 1995, p. 2).

Cancer takes more than 500,000 lives and costs $104 billion each year in the United States. Diets high in fat and low in fiber are associated with colorectal, breast, and prostate cancer. It is possible that one-third of all cancer cases could be prevented through dietary changes (Frazao, 1995).

The USDA's guidelines for a healthy diet focus on reducing cancer and heart disease. Implicit in its "Food Pyramid" is the need to eat less fat and oil and more fruits and vegetables.

Much of the land used to produce crops in the United States is devoted to the production of meat and dairy products. Corn alone, grown largely as a feedgrain for meat and dairy production, accounts for some 70 million acres. To produce one pound of beef requires seven pounds of feedgrain, but one pound of chicken requires just two pounds of grain. Pork requires about four pounds per pound of meat, while cheese requires three and eggs 2.6 (Brown and Kane, 1994, pp. 66-67). These ratios imply producing a pound of beef also requires more fertilizer and pesticides. Although the USDA recommendations suggest greater meat consumption, if that meat were chicken, as opposed to beef, the environmental impacts need not be worse.

A typical American consumes much more grain than most other people in the world—about 800 kilograms (1,760 pounds), mostly to produce animal products. In comparison, an Italian consumes 400 kilograms (882 pounds), a Chinese 300 kilograms (661 pounds), and an Indian 200 kilograms (441 pounds) (Brown and Kane, 1994).

Sugar is Americans' most overconsumed food. Sugar production is centered in ecologically fragile areas, such as the Florida Everglades and Louisiana. Reducing sugar consumption to recommended levels could prove beneficial, depending on the alternative land uses, though removing sugar-import quotas could have an even greater impact.

Production of fruits and vegetables currently involves intensive use of pesticides and irrigation water. If production systems don't change, environmental harm could increase with production. Demand for organically grown fruits and vegetables has grown rapidly, though as a share of all production this market is tiny (about 3 percent).

What Can We Expect in the Future?

The net environmental impacts from future increases in demand and productivity, coupled with possible changes in consumption patterns and production practices are not at all obvious. In a simple scenario analysis of the possibilities *(see Table 2-10)*, WRI looked at changes in domestic consumption, export demand, changes in consumption patterns in the United States, the adoption of Integrated Pest Management, and more efficient use of fertilizers. For each scenario, demand

TABLE 2-10. Scenarios of population and income, healthier eating, and more efficient use of pesticides and fertilizers. Increases in population and income, and dietary changes recommended by the USDA, lead to more land in production and more input use, but increased adoption of IPM and better fertilizer management can offset these changes.

| | | 2005 | | | |
| | | Current Consumption | | Food Pyramid | |
Crops (million tons)	1995	Current Pesticide and Nutrient Practices	75% IPM with Nutrient Use Efficiency	Current Pesticide and Nutrient Practices	75% IPM with Nutrient Use Efficiency
Feed Grains	732	837 (14)		873 (19)	
Oil Seeds	120	125 (4)		122 (2)	
Wheat	144	158 (10)		150 (4)	
Fruit	63	81 (29)		100 (59)	
Vegetables	65	73 (11)		79 (22)	
Inputs					
Acreage (million acres)	215	224 (4)		223 (4)	
Pesticides (million pounds)	570	595 (4)	473 (-17)	655 (12)	523 (-8)
Nitrogen (million tons)	6.2	7.0 (13)	4.8 (-23)	7.2 (16)	4.9 (-21)
Phosphate (million tons)	2.7	3.0 (11)	2.1 (-23)	3.0 (11)	2.1 (-22)

was estimated and then the impact on crop acreage, pesticide applications, and nutrient use was projected.

The values for 1995 are an average for 1995 and the four preceding years for five major crop categories. Feed grains are by far the largest category, with 732 million tons produced (mostly for animal consumption). Domestic production and production for export are counted. Crops produced outside the United States but consumed within are not included here, nor are the inputs needed for production. Other land uses for agricultural production, such as grazing land and land used to grow crops not considered here, are excluded.

The first scenario for 2005, "Current Consumption," assumes that per capita consumption patterns will resemble today's. Demand is adjusted to reflect increases in domestic population, domestic incomes, and trade. These figures were taken from a USDA analysis of consumption and trade patterns (USDA, 1995a; Blisard and Blaycock, 1993). The greatest increase in demand—29 percent—is for fruits, followed by feed grains—14 percent. Vegetable demand increases by 12 percent, wheat by 10 percent, and oil seeds by 5 percent.

Yield projections reflect the assumption that current yield trends for each of the crop categories will continue. Although growth in demand increases significantly, crop yield increases buffer demand for agricultural land, so crop acreage increases by only about 4 percent.

In the "Current Pesticide and Input Patterns" scenario, current patterns of pesticide and nutrient use are assumed. Pesticide use is considered primarily a function of crop acreage, while nutrient use is tied to production. Pesticide use increases by about 4 percent and nutrient use by 13 percent for nitrogen and 11 percent for phosphorous.

Fortunately, technology is not static in the real world. Pesticide use can be cut significantly if Integrated Pest Management techniques are applied and if nutrient management improves by soil testing and improved application methods such as banding. Estimates from Pimentel (1995) were used to project opportunities for environmental improvement if an aggressive IPM program were adopted. If 75 percent of agricultural production came under Integrated Pest Management—a year 2000 goal of USDA—and all farmers used nutrients more efficiently, major reductions could be achieved. Under "Current Consumption" patterns, pesticide, nitrogen, and phosphate use could decline by 17, 23, and 22 percent, respectively.

As for the implications of eating healthier, would dietary changes recommended by the USDA to promote better health have secondary environmental benefits? Apparently not, at least if there is no shift away from beef eating. The Food Pyramid recommendations, if followed by all Americans, would result in a 56 percent increase in consumption of fruits, a 12 percent increase in that of vegetables, and a 17 percent increase in meat consumption. The intake of sugar and sweeteners would fall by 72 percent and the consumption of fats and oils by 56 percent (Kantor, pers. comm., 1995). Following the Food Pyramid guidelines for the crop categories examined here would lead to significantly increased pesticide use and slightly increased nitrogen use compared to the 2005 estimate under current consumption patterns. Fruits and vegetables are the heaviest users of pesticides per acre, and with increases in demand, pesticide use could follow.

Since inputs are used more efficiently in Integrated Pest Management, any negative impacts from adopting the Food Pyramid recommendations could be mitigated. Pollution would be reduced, for instance, and improved nutrient management could also reduce excess nutrient use.

SOCIAL ASPECTS

Agriculture is not simply a production system; it is also a vital element of our national heritage and social institutions. Rural communities continue to face changes that many people consider undesirable—the concentration of agricultural production, the relative decline in rural populations, and loss of economic vitality. For these and other reasons, definitions of agricultural sustainability have often included various social elements.

Demographics and Economic Development

The most obvious social aspect of agriculture is the dramatic change in demographics since the 1930s. Farm numbers peaked in about 1935 at 6.8 million and have since declined to just over 2 million—holding steady during the 1990s. The most dramatic era of change was the period after World War II up to the mid-1970s, with the rate of decline slowing during the 1980s. The amount of land in farms peaked about 1950 and has declined by about 17 percent since then. Average farm size across the country has tripled since the 1930s.

As farm size has grown, rural population has shrunk. Population losses during the 1980s were most severe in counties that are economically dependent on farming and mining, with 75 and 73 percent, respectively, of these counties losing population. In contrast, of the non-metro areas primarily dependent on industry or a natural amenity (climate, lakes, or other water, varied topography, or low elevation), 43 and 24 percent, respectively, lost people (USDA/ERS, 1994f, p. 154). These shifts have come about as the importance of natural resources has changed from commodities to exploit to magnets for working families, retirees, and vacationers (McGranahan, 1994).

Other rural economic indicators worsened during the 1980s and early 1990s. Rural unemployment has risen; rural per capita income and earnings per job fell farther behind urban areas; and rural poverty rates have increased (Ghelfi, 1994, p. 122). As rural populations and income shrink, the delivery of public services through schools, hospitals and other institutions becomes more expensive (McGranahan, 1994).

These structural changes raise two principal issues related to sustainability. One is the linkage between farm size and environmental stewardship. Many people view the protection of mid-sized family farms as a linchpin of agricultural sustainability. Family-owned and operated farms, they say, are more ecologically friendly than large-scale agriculture. However, a review of studies on farm structure and stewardship done by Atwood and Hallam (1990) suggests that larger farms are at least as conservation-oriented as smaller ones. Higher levels of education and managerial skills are associated with larger and more complex forms of farming, and the two most consistent indicators of the adoption of conservation practices are educational attainment and income. Another study on the adoption of an environmentally friendly post-management technology found that farmers who tried the techniques operated significantly larger farms, and have higher levels of education, greater knowledge of pesticides, and the ability to apply their own pesticides. If anything, the meager research done to date suggests that larger farms may have an environmental advantage, though the issue remains controversial.

A second issue is the role of farming in rural economic development. Many proponents of federal farm-support programs justify these expenses on the grounds that they support farm income and thus economic development. The reality is that agriculture plays an increasingly small part in rural economies. Nationally, agriculture contributes about 15 percent to the incomes of rural economies. The remainder comes from other manufacturing and service industries. While farm jobs have declined by 22 percent between 1975 and 1989, employment in industries related to agriculture has increased by 5 percent and employment in industries weakly related to farming has risen by 60 percent. Non-metro areas gained 11 off-farm jobs for every farm job lost, and just two of these off-farm jobs were farm-related. Since 1975, the number of counties where farming accounts for more than one fourth of employment has dropped from 750 to 335. Most of these counties are concentrated in the Northern Plains and western Corn Belt (Majchrowicz et al., 1994). Investments in education, infrastructure, good social institutions, and adequate financial capital would do much more for rural economies than commodity price supports. Recent agricultural legislation sets a good precedent, allocating $300 million for the support of rural development.

Conversion of Farmland to Other Uses

Many people rate agricultural land conversion as a top natural resource priority. *(See Figure 2-1.)* Without doubt, loss of agricultural land to built-up uses is a quality-of-life issue in many urbanizing areas. But it is less clear that conversion of agricultural land is a priority for sustainability.

Americans place great value on scenic farm vistas. When farmland begins to disappear from an area, the character of that area changes. Development of farmland eliminates open spaces and wildlife habitat in favor of houses, malls, industrial parks—and the traffic congestion that goes along with all (American Farmland Trust, 1993).

Urban sprawl can have economic impacts too. In California, growth and development in California's Central Valley is projected in a study by the American Farmland Trust (1995) to lead to $100 billion worth of cumulative losses in agricultural product sales and employment to the year 2040. The Central Valley has six of the top ten counties in the nation in value of agricultural production. With sales from these counties totaling $13.3 billion per year, this valley accounts for 8 percent of U.S. agricultural sales on just 0.5 percent of the nation's farmland. The expansion of residential and commercial areas is estimated to account for the conversion of 15,000 acres of farmland in the Central Valley each year, and

---

BOX 2-7. Farmland Preservation Categories

Three types of farmlands have been identified as important to preserve:

- *Prime farmland* is best suited for agricultural production. Due to inherent natural characteristics such as level topography, good drainage, adequate moisture supply, favorable soil depth, and favorable soil texture, this land consistently produces the most output with the least input. Some environmental impacts, such as soil erosion, may be lower on prime farmlands.

- *Unique farmland* is not prime, but it has a special combination of soil quality, location, topography, growing season, and moisture supply necessary to produce high yields of specialty crops such as fruits and vegetables. Since these characteristics are geographically fixed, they cannot be reproduced if the land is converted to other uses.

- *Statewide important farmland* is of particular importance in a given state for agricultural production. Generally, these farmlands are nearly prime and may produce high yields.

Source: American Farmland Trust, 1993.

conversion will very likely accelerate since California's population is expected to triple by 2040 (AFT, 1993).

Unique agricultural areas may merit protection *(see Box 2-7)*, but the conversion of farmland is not likely to keep the United States from producing enough food for a very long time, if ever. The United States had approximately 60 million acres of agricultural land idled in 1993, including 36 million in the Conservation Reserve Program and 23 million in annual set-aside programs (USDA/ERS, 1994c, p. 4). For much of the last 60 years, chronic oversupply of agricultural commodities has been the central policy issue addressed by agricultural programs. The government established price guarantees because supply consistently outstripped demand. The government also forced farmers to take land out of production to limit supply and push prices up. Even long-term retirement programs (such as the CRP) were justified as supply-control measures, more than as environmental programs.

Some agricultural areas may be considered unique because considerable investments have been made in infrastructure, particularly irrigation. As one example, the Central Valley was unfit for most agriculture before the government financed large-scale irrigation. But as development proceeds, urban demand for water will lead to water allocation reform and water markets will redirect water to higher value uses. Low-value crop production, such as pasture or alfalfa—major crops in the Central Valley—will face pressure as water prices rise, driving down agricultural income and further accelerating conversion. These lower-value crops can be grown in other parts of the country, so high-value crop production will probably come to dominate agriculture in California.

The central issue of farmland protection is the quality of life in developing areas. Communities have many options for protecting the open spaces and scenic amenities that many people value. Reduced land assessments for agricultural areas, zoning to concentrate development in certain corridors and avoid sprawl, buffer zones between agricultural areas and residential or commercial areas, purchases and retirement of development rights, outright purchases of land, and right-to-farm laws to avoid nuisance complaints against farmers from new developments that are near farms are but a few examples (AFT, 1993). In addition, a recent agricultural law sets aside $35 million in cost-share funding to protect prime farmland.

THE FACTORS OF SUSTAINABILITY

Many facets of the U.S. agricultural sector bear on the choice of policies for promoting sustainability. The few that follow are particularly important.

U.S. Agriculture Has the Potential to Be Flexible

Agriculture is somewhat unique among the economy's production sectors insofar as investment decisions are seldom made once and for all. Many of the investment decisions made by farmers and ranchers are of a short- and medium-term nature, so opportunities for reconsidering investment strategies are frequent. The most basic decision that any farmer makes—what to plant—is made annually for most crops and the farmer has to live with the decision only for the length of the crop's life. Post-harvest, the farmer can reconsider the planting strategy. While some planting decisions may require new equipment, decisions about how much to plant, what varieties to plant, and how much of various inputs to use can be readily adjusted. For livestock, stocking and destocking decisions occur on a somewhat longer timeframe, but herd turnover times are still only a few years. (Orchard production for fruit is a significant exception since investments cover decades.)

Agricultural commodity production does not rely upon large manufacturing facilities, so it is less constrained than other sectors. With the important exception of land, most of the major investments related to agricultural production, such as those for equipment, machinery, and buildings, have shorter lifetimes than in other sectors, where investment decisions may lock a firm into a particular production strategy for decades.

Another salient feature of agriculture is the decentralized production of readily tradable commodities. Regional, national, or international disruptions seldom affect the availability of food for the American consumer, though they may jostle the price.

These advantages of flexibility and adaptability have often been constrained by agricultural policies promulgated by national governments and international treaties. In the United States, the federal government has restricted farmers' crop-production choices through commodity programs intended to boost crop prices and farm incomes. Until the 1996 farm bill ended this system, planting decisions for program participants were determined by production histories and government fiat. Importation of some important commodities, such as dairy products, sugar, peanuts, and tobacco, are still restricted to limit supply and raise prices. Many other countries have also instituted such policies to protect farmers and secure domestic food supplies.

U.S. Agriculture Can Be Highly Receptive to New Technologies

Technological advancements in agriculture have been the source of large and swift productivity gains since World War II. Mechanization, breeding, fertilization, pest control, irrigation, and farm management have all helped enhance production and efficiency. Total output per unit of input has increased by about 116 percent in the last 45 years (Ball, 1995).

Technology has also helped solve environmental problems in agriculture. For example, alternative soil-tillage methods (such as reduced, mulch, and no-till) have helped farmers cut expenses while minimizing soil erosion. Integrated Pest Management techniques that rely on crop rotations, pest scouting, the release of sterile pests, and economic thresholds for spraying have helped farmers reduce pesticide applications, leading to both financial and environmental gains.

When the economic advantages of a given technology are clear, adoption can be swift. The shifts to modern high-yielding varieties, inorganic fertilizers, chemical pest control, and concentrated poultry production all illustrate this point. The concentration of production in the pork industry and the use of bovine somatotropin (a hormone) to increase milk production are current examples.

In agriculture, as in any industry, technology development occurs in response to economic forces as well as scientific advances. Agriculture was mechanized in response to increasing labor costs as the industrial revolution advanced. Reduced tillage technologies were originally developed to keep energy costs down, not to conserve soil. Later, its use on highly erodible land was made a condition of federal commodity subsidies, so adoption took less than five years.

These gains have flowed largely from major investments made in agricultural research and extension. The United States expressed its wish to support agricultural research and extension as far back as 1864, when the Morrill Act established the Land Grant Universities, and in 1890 when this institutional investment was further extended to support extension to minorities. The American agricultural research establishment is truly impressive, with an enormous stock of research infrastructure and human capital. *(See Box 2-8.)*

The history of technology development and adoption suggests that U.S. agriculture is not just highly responsive to technology but is driven by technology. Further, some of the pressing problems of agricultural sustainability lend them-

---

BOX 2-8. Technological Change in Agriculture

Today, we take technological change in agriculture as a given, but it was not always so. From the 1870s to 1910s, the average yields of corn, wheat and cotton did not change. To supply ever increasing demand for food, feed, and fiber, the nation expanded the area in production, nearly tripling cropland acreage from 126 million acres in 1870 to 347 million acres in 1910. The ratio of the productivity of outputs to that of inputs barely changed from 1880 to 1910, from 100 to 102. In contrast, productivity doubled between 1948 and 1992.

In 1800, farmers depended primarily on the ax and fire to clear land; on a primitive plow drawn by animals to till it; on the hoe to control weeds; and on the scythe for harvesting. The work was done by farmers and their families, or by slaves. Because of its vast area, labor was scarce relative to land in the United States. The history of agricultural technology in the 1800s is largely an account of the development of machines and implements to overcome the relative scarcity of labor. The cotton gin developed by Eli Whitney, the cast-iron plow developed by Charles Newhold and commercialized by John Deere, and the harvester developed by McCormick are a few of the best-known examples. The 1880s and 1890s saw the development of the gasoline powered tractor, which was coming into widespread use by 1910. This was a major innovation with vast long-term implications not only for labor productivity but also because it freed tens of millions of acres previously used to provide feed for animals.

While labor productivity increased dramatically during 1910-1930, land productivity did not. Production increased in step with cropland harvested during the period. Yield averages for 1912 and 1930 were the same. The increase in cropland

---

selves to technological and managerial solutions if the incentives are right and economic and environmental policies are better integrated.

Unfortunately, U.S. agricultural history also includes the use of significant subsidies that distort the true economics of resource use, masking information that would promote greater sustainability. Commodity-program restrictions have discouraged farmers from rotating their crops and encouraged them to apply more pesticides. Federally subsidized water projects invite the production of low-value crops in dry regions thirsty for water while below-cost leases of federal rangeland induce ranchers to mismanage grazing lands. Extensive exemption of agriculture from the "polluter-pays principle" is tantamount to a subsidy to pollute the nation's water.

BOX 2-8 cont.

productivity evident by 1940, and continued thereafter, began abruptly in the mid-1930s.

A number of reasons account for this break in productivity, including the rediscovery of Mendel's laws of heredity in 1900, work on mutation and gene theory, and significant investments in research and extension institutions. By the 1920s, research began to pay off in the form of wheat varieties with greater resistance to drought and stem rust. The most dramatic advance came with the development and introduction of hybrid corn, which could double or triple the yields obtained with the traditional varieties then in use. Although these developments took place in the 1920s, they were not widely adopted until the mid-1930s, probably because fertilizer prices were still too high.

Beginning in the 1930s, production patterns shifted from labor-saving and land-using to land- and labor-saving, primarily because land prices began to increase. Farmers invested more in land drainage and irrigation, and also dramatically increased their use of commercial fertilizers as fertilizer-responsive varieties became available and fertilizer prices dropped. The amount of fertilizer applied per acre roughly doubled each decade between 1940 and 1970. From the mid-1950s to 1980, improved crop varieties accounted for slightly less than half of corn-yield increases. The rest trace back to heavier use of fertilizers and pesticides and to better management.

Source: Adapted from Crosson, 1991.

Environmental Problems Are Diffuse and Difficult to Gauge Precisely

Because agricultural production occurs on so much land, the associated environmental problems are diffuse. Except in concentrated livestock operations or processing plants, most agricultural pollution is generally "non-point." Pollution of ground and surface water, for example, results mainly from the accumulation of relatively small amounts of pollutants draining off many fields when it rains or when snow melts. How much any single field contributes is extraordinarily difficult to measure compared to the emissions that are generated from a smokestack or tailpipe, especially when the damage may take years to register.

The impact of land-use changes are also diffuse. The conversion of wetlands and wildlife habitat to agriculture has taken place piecemeal over many decades.

Each conversion has a minor impact, but cumulatively large areas are fragmented and major ecosystems impaired.

When pollution or land conversion problems are regulated, society as a whole benefits, but those who make the changes incur the costs. If the benefits to society exceed the costs of remediation for the farmer, the move is economically justified. But, because the benefits of correcting environmental problems traceable to agriculture have been so difficult to determine, while the costs can be estimated only too precisely, justifying agriculture regulation economically has been hard.

It has also been politically difficult to regulate agriculture. Rural states with small populations but large concentrations of farmers have the same number of Senate votes as every other state. And because the benefits of remediation are diffuse but the costs are concentrated, agricultural producers are generally better organized to represent their interests than are consumers and taxpayers. Environmental gains, such as they are, have come about largely through subsidies to polluters to correct harmful actions, not through regulation. Commodity subsidies have been tied to environmental compliance. Cost-sharing programs have been instituted to share remediation costs.

Sensitive lands have been leased from farmers by the government and taken out of production. This approach—shifting the burden from polluters to taxpayers—has been politically acceptable to most parties because the cost to taxpayers has been relatively small and farmers have enjoyed the income boost or the avoidance of remediation costs. Indirectly, the federal budget also benefited from land-retirement programs: when commodity production was reduced, crop prices rose and the government's price-support expenses fell. Unfortunately, how well this policy approach works depends on how much government funding is available, so making environmental progress has to compete with efforts to reduce fiscal deficits, with competition from other priorities, and with voters' strong desire for tax cuts.

The Biggest Issues Require International Cooperation

The two biggest threats to agricultural sustainability are not strictly matters of agricultural or even domestic policy. Although agricultural producers, industries, and researchers will adapt to global climate change, these adaptations could be costly and divert attention and resources from the resolution of other problems. Avoiding the need for adaptation would make more sense, but since the United States is not the sole contributor of greenhouse gases (though at present it is the largest) it can't solve the problem alone.

Similarly, the problem of genetic resource losses cannot be approached single-handedly by the United States. Most important U.S. commodities originated outside our nation's boundaries. Protecting the ecosystems that produced the ancestors of our most highly valued agricultural plants and animals is outside U.S. jurisdiction. The United States has much to lose if these threats to sustainability are not contained, and our only option lies in cooperating with other nations to mutually address these common concerns.

RECOMMENDATIONS FOR ACTION

The recommendations provided here address the problems examined in this chapter. Aimed at federal, state, and local governments, as well as farmers and agricultural industries, these recommendations, are offered in the hope that they can bring the United States to a future that we want to live in. All fall into four broad categories. The first comprises policies intended to support farm income—a central thrust of most agricultural policy, and the programs that draw the most money. Since most programs have been economically inefficient and environmentally unsound, recent advances in this area are encouraging. But more can be done.

Conservation benefits are public goods, so government has a legitimate role to play in ensuring that broad conservation goals are met. Thus, the second set of recommendations focuses on what government should do more of, less of, or better. Because of the international nature and long-term importance of genetic conservation, national and international governing bodies must play a prominent and strengthened role in conserving this critical resource. Closer to home, conservation subsidies have been key to achieving environmental gains, but we could get much more for the money spent and some subsidies should be scrapped.

The third set of recommendations are aimed at promoting sustainability through research. Agriculture has seen major advances traceable to research investments, and the potential for further gains appears great. Technologies that both increase farmers' profits and make the use of inputs or natural resources more efficient benefit everyone. Also, some climate change appears inevitable, so cost-effective adaptation strategies and ways for agriculture to contribute to its solution will be important. Further knowledge of how pesticides affect human and ecological health is important as we learn how to manage these inputs, improve them, and avoid unintended problems.

The fourth category covers what can be done outside of government. Sustainability must come about through a broad consensus, so industry, trade associations, and consumers all have important roles to play.

Farm Income

1. *The industrialized countries should eliminate barriers that limit trade in agricultural commodities.* Several studies note that greater trading flexibility could buffer climate change's deleterious effects. There are positive economic and environmental benefits as well. Yet, governments have long used trade barriers to increase domestic production and prices for favored commodities. Some countries use them to protect their farmers, thus hoping to increase food security and promote rural development. Such barriers also increase food costs to consumers, limit markets and incomes for developing-country producers, and create economic distortions that damage the environment.

By intent, international barriers protect farmers who are uncompetitive and inefficient. Sugar production in the Everglades is one example. Import quotas on sugar protect a few very wealthy and politically powerful sugar companies but help fuel the widespread degradation of a unique ecosystem. For punishing the Everglades, these producers are paid in excess of a billion dollars a year more for their sugar than American consumers would pay if they bought the sugar on world markets (USDA/ERS, 1994d, p. 370). For ten protected commodities, including barley, beef, butter, cheese, fluid and nonfat dry milk, poultry, rice, and sugar, U.S. consumers pay more than $9 billion per year more in higher food prices (Nelson, pers. comm., May 31, 1994).

While there would be some painful adjustments in the transition, studies repeatedly show that liberalizing agricultural trade would be good for American agriculture overall because of its overall competitive advantage in such major commodities as corn and wheat. More flexible trading regimes would also help to address climate change. If weather variability increases, how well we minimize the economic impacts will depend on access to supplies from regions in other weather belts, and the more that weather variability can be mitigated, the less food prices will fluctuate.

2. *Federal funds previously dedicated to price supports and now spent on direct income payments should be means-tested and used to promote rural development through investments in education, training, and infrastructural development.* In the bargain that made commodity program reform possible, guaranteed prices and uncertain

fiscal exposure for the government were replaced by fixed and declining direct payments to farmers. Federal programs have historically supported just seven agricultural commodities to boost farm income, and the crops that a farmer could plant were limited so the government's fiscal burden wouldn't be too crushing.

In 1996, Congress passed legislation that radically changed farm income-support programs. Instead of basing government payments to farmers on production, payments are based on prior program participation. Program funds are allocated according to what Congress is willing to spend, and these funds are apportioned according to the acreage that farmers have enrolled in the past. Unfortunately, the historical inequity of the programs remains. In recent years, the top 2 percent of all farmers—who have incomes beyond $500,000 per year—received roughly the same number of federal dollars as the bottom 69 percent (U.S. Department of Commerce, 1993, Table 1106). This inequitable subsidy allocation is an artifact of income support based on acres under production—the more farmers produce the more income support they get. If income support were based on need, instead, it would be far more equitable and much cheaper.

The United States has a legitimate interest in promoting rural economic health and well-being. But farm income-support programs are not the right way. Taxpayer dollars would be much better spent, and sustainability better served, if farm income support were paid directly to those who needed the help and the savings put into such job-creating investments as education and rural infrastructure.

The 1996 agricultural legislation allocates $300 million to rural development and $36 billion to direct farm-income support. This new program for rural economic development is a move in the right direction, but the amount is small relative to the problem and the overall spending level. If farm-income support were means-tested, more funds would be available to help the economy as a whole.

Conservation

3. *Congress, the Federal government, and international aid agencies should boost financial and technical support to the domestic and international institutions that protect agriculturally important germplasm.* Seedbank programs at home and abroad are grossly underfunded relative to their value. In recent years, the USDA and many international institutions have faced budget cuts. Some of these international institutions, such as the International Agricultural Research Centers (IARCs),

house seedbanks for major crops and also conduct some of the world's most important breeding programs. Although current food-supply problems result mainly from poverty and political unrest, it would be grossly shortsighted to let this invaluable research and genetic capacity atrophy. Many national centers, also important repositories of germplasm, struggle with technical and financial limitations, barely keeping their programs going. While the United States cannot shoulder the entire burden for maintaining these centers, it should at least increase funding so that seed stocks within its own control can be protected and opportunities for international training and technical advancement provided.

Programs that protect resources *in situ* also deserve much greater support from aid agencies. Highly threatened areas of great biological diversity should be protected from further encroachment and important species should be cataloged.

New partnerships are emerging between developing countries, pharmaceutical companies, and other businesses that use genetic material. In exchange for the right to survey and use genetic material, companies are creating profit-sharing arrangements, training in-country staff, and providing funds for protecting endangered areas. Such arrangements—in everyone's best interest—should be encouraged by governments and shareholders alike.

4. *Congress and the federal government should remain committed to conservation subsidies, but streamline and improve the cost-effectiveness of these programs by consolidating, prioritizing, and retargeting these subsidies.* Conservation programs have been successful but inefficient. Soil-erosion reductions through the Conservation Reserve Program have been greatest in regions where off-farm damages from this form of pollution are the least serious. At the same time, other environmental goals have been underfunded. In recent years, the USDA addressed these concerns head on, but much more remains to be done. Conservation dollars would go much farther if the programs were consolidated, if priorities and criteria were clearer, and if funds were channeled to the highest priority problems and areas. Additionally, when local priorities for public conservation funds are set, decision-making should include a much broader spectrum of the general public—not just farmers and ranchers—so that wildlife, water-quality, and recreational values are taken into consideration.

Eliminating trade barriers (Recommendation 1), particularly sugar quotas, would also allow government conservation funds to go farther. Congress has recently approved funds to purchase and retire land north of the Everglades to protect that unique ecosystem. Land-purchase prices are determined by income generated from production activities—the higher the income, the higher the land value. Because quotas limit sugar imports so as to raise prices and land values, government is forced by its own programs to pay an artificially high price for the land it wishes to retire. Eliminating quotas would allow taxpayers to get environmental benefits at a fair price.

5. *All federal and state agencies should phase out irrigation-water subsidies.* Water subsidies were initially instituted to help settle underpopulated areas in the West. They continue only for political reasons. Subsidies that encourage resource misuse are economically and environmentally costly—anathema to the idea of sustainability.

Many western states are struggling to supply water to rapidly growing urban populations at premium prices while water subsidies to agriculture support the production of low-value crops. Water-supply problems stem largely from gross inefficiencies of use because the resource is undervalued. In addition, water subsidies distort production patterns across the country, providing a windfall to farmers in regions with access to cheap irrigation water and hurting regions without it.

6. *Agricultural production should be subject to minimum environmental standards that are not tied to any subsidies.* Except for the very largest livestock operations, agriculture has remained largely free of the regulation that most other economic sectors have faced over the last 30 years—the Clean Water Act affects only about 5,000-10,000 (less than 1 percent) of 1.1 million farming operations, for instance. Yet, as other sources of pollution have come under regulatory standards, agriculture has emerged as the major polluter of the nation's surface waters. Environmental performance reflects federal subsidies—from performance standards tied to price supports to paid land-retirement programs to cost-sharing for practices or infrastructure that avoids pollution. In most instances, the "polluter-pays" principle has been in force, but in U.S. agriculture, the polluter gets paid; so if federal funds for conservation subsidies dry up, environmental performance in agriculture will be threatened.

In fairness to other industries that invested heavily to improve water quality, agriculture should now be subject to minimum standards to avoid the worst problems. These standards would encourage what many responsible farmers are already doing: soil testing to account for the nutritive value of animal manures and applying fertilizers in amounts that do not exceed plant requirements; managing plant-residue to minimize soil erosion; dealing responsibly with animal wastes; establishing production setbacks near streams and rivers; creating fences to keep animals out of waterways; and using appropriate tillage practices on steep slopes.

7. *States should tax fertilizer and pesticide pollution to encourage efficient use of fertilizers and agro-chemicals.* Fertilizers and pesticides are valuable production inputs, but when they are overused or used improperly they can cause significant environmental damage. In the Netherlands, fertilizer taxes are charged on excess use, based upon a simple "mass balance" approach that an accountant calculated. A similar approach could be applied in the United States. (For each farm, excess nutrients could be calculated as the difference between applications from fertilizers, manure, and leguminous fixation and the actual nutrient levels used by the crops, with some small allowable surplus to account for uncertainty.) Farmers who applied only the necessary nutrients would pay no pollution taxes. Such taxes should also apply to other major users, such as golf courses.

Similarly, pesticide taxes should encourage the use of more benign chemicals and discourage the use of more harmful ones. Pesticides that are more toxic, dispersed more readily into the environment, and more persistent would be taxed the most. The safest would go untaxed. (Such a tax has precedent: the tax that the United States currently imposes on chlorofluorocarbons (CFCs) is based on a given chemical's ability to damage the ozone layer.) Already, several states have implemented small taxes on agricultural inputs to support research. Funds from pollution taxes could be put to a similar use, enabling states to reduce pollution problems and support education and research at land grant universities.

Such taxes won't be popular, but they are justifiable because those who apply fertilizers and chemicals use a public good—the environment—to dispose of their wastes. Taxes are one way to recoup the clean-up costs that these activities impose on everyone else. Pesticide and fertilizer taxes could also be part of a broader tax shift in the U.S. economy, where activities that cause economic harm (i.e., pollution) are taxed instead of activities that result in economic gain, such as investment and employment.

8. *The USDA should reform its grading standards for fruits and vegetables to reflect nutritional quality and value, as opposed to appearance, and educate the public on the health and environmental value of the changes.* Current grading standards for produce

encourage excessive pesticide use by determining grades, and thereby product price, largely on the basis of appearance. These standards should instead reflect nutritional value and quality. Additionally, the USDA should launch a broad educational program with the cooperation of the food industry, and of consumer health and environmental organizations, to educate consumers on the benefits of changing the standards.

9. *States and municipalities should develop, test, and implement tradable permitting schemes for nutrient emissions to improve water quality.* While the Clean Water Act has greatly benefitted water quality, it is approaching its limits as a regulatory solution. A new approach is needed. In many cases, economic and environmental gains could be had by allowing those whose emissions are limited by permit the flexibility needed to make the necessary cuts by purchasing reduction credits from others who can make the cuts more cheaply. Trading of nutrient emission reductions could be allowed between point sources and other point or non-point sources, such as agriculture or urban run-off sources. One township in Connecticut is sharing the costs of a new wastewater-treatment plant in an adjoining town so it can take credit for nutrient removal beyond permit requirements—saving both towns money.

Such schemes have been applied in a handful of places with limited success. The economic and physical potential appears enormous, though so are regulatory and information hurdles. Pollution trading—used effectively to control lead from gasoline refineries and sulfur-dioxide emissions—could be used by states and localities to help improve water quality too. Pilot schemes should be encouraged to test trading approaches to find out what sorts of systems work best.

Research

10. *Publicly funded agricultural research in the United States should be directed more toward resolving natural resource management problems, adapting to sustainability threats, and developing technologies that both reduce production costs and improve environmental performance.* Much, if not most, agricultural research funds are based upon political considerations and historical allocations. Few of those that government manages are allocated competitively with the public interest in mind. As a result, research institutions are somewhat ossified and are not nearly as responsive to current concerns as they should be. Addressing a dwindling client base and problems of the past, these institutions now find their financial and political support declining.

Greater competition, clarity of purpose, and a refocused mission could help them reshape themselves in ways that would benefit the public and earn public support.

While there are many competing demands for funds and never enough to go around, public research dollars should first and foremost be used to address broader public concerns, not narrow private interests. Public research funds should go to work that private investors will ignore, but that nevertheless entail enormous social benefits. Adaptation to climate change, protection of germplasm stocks, and pollution reduction all fall into this category.

Further, research should support technology development that helps farmers simultaneously reduce their production costs and their environmental impacts. Research that achieves both stands a much better chance of being broadly adopted by farmers.

11. *Federal agricultural, environmental, and health agencies should join with pesticide companies to develop a joint research program to deepen understanding of the human and ecological health impacts of pesticides.* Significant gaps remain in scientists' understanding of the hormonal and immunotoxic effects of pesticides. In the United States, very little epidemiological research has been conducted on these effects, even though preliminary evidence indicates that the problems may be significant. Because the issues are poorly understood, they have been highly controversial.

Pesticide companies have a responsibility to ensure that the products they sell do not pose threats to human or ecological health. To their credit, many of the major companies—among them, Ciba-Geigy, DuPont, and Monsanto—have made corporate commitments to product safety and responsible life-cycle product stewardship. In that spirit, agricultural chemical manufacturers should support and join programs to design, carry out, and disclose health research and to design safer crop protection strategies.

12. *Federal, state, and private researchers should together develop economically and environmentally sound ways to use surplus agricultural lands to provide substitutes for fossil fuels.* The agricultural sector is at great risk from climate change, but it also has the potential to respond in a way that can improve farm income while helping to mitigate the threat. Some 60 million acres of agricultural land are now idled for supply or conservation purposes. Much of this land is not environmentally sensitive and could be used to provide biomass for electrical generation or the production of liquid fuels, thus helping to reduce the emission of the gases

believed to cause climate change. Problems remain to be worked out because such fuels are not currently competitive without large and inappropriate government subsidies. But advances in biotechnology may produce more efficient methods for converting feedstocks to fuels, and new crops may prove more suitable for production on easily erodible land.

Private-Sector Actions

13. *Agricultural corporations should find ways to promote sustainability.* U.S. agriculture has great economic importance to the country and a major impact on the nation's environment. Those industries and organizations involved in natural resource management have more opportunities than most—as well as more intimate knowledge and experience—to make the most difference. If the vast resources of the private sector were brought to bear on the questions of sustainability, major progress could be made without government intervention.

In many ways, businesses can benefit from making sustainability part of their corporate objectives. Some leading firms offer cash prizes to employees who find ways to reduce emissions, saving money and avoiding regulatory headaches in the process. The examination of long-term questions can help executives rethink corporate strategies, products, and markets, making the firm both more competitive and more sustainable. Forward-looking firms can set positive examples and help to determine the nature and content of regulatory discussions, make their own intentions seem more credible, and earn the trust of government administrators and the public. Firms can also develop or adopt codes of conduct that specify manufacturing processes, performance standards, and corporate values keyed to high environmental performance.

Industries that purchase agricultural commodities can promote sustainability by setting production standards for the products they buy. Many large food companies—from poultry to pork to popcorn producers—enter into contracts for the production of a variety of commodities. These contracts could specify the use of environmentally friendly production practices, such as Integrated Pest Management and conservation tillage, or the appropriate handling of animal wastes. Some large manufacturing companies, such as Patagonia, even specify organic production of the cotton they purchase. Contracting arrangements like these create both awareness of and markets for environmentally sound practices.

14. *National, state, and local farm and commodity organizations should promote awareness of sustainability among their members.* Some of those organizations have already

taken steps to encourage stewardship—giving awards for highest yields and recognizing farmers who use conservation-tillage methods. Additional awards could recognize organic farmers and farmers with extraordinary environmental performance or innovations. Some farm and commodity organizations also now make information available on technical practices, such as how to prevent pesticide drift, how to reduce soil erosion, and how to minimize water pollution. Some farmers share experiences in adopting "sustainable" farming practices through "T-by 2000" clubs that promote the attainment of soil erosion tolerances by the year 2000 and through other "sustainable ag" groups.

15. *Homeowners, municipalities, golf course owners, and others who manage lawns and turf for scenic and recreational use can help relieve environmental pressures and protect their own health by restricting or eliminating their use of pesticides and fertilizers.* Many homeowners struggle mightily in pursuit of the perfect lawn—often applying relatively large doses of pesticides and fertilizers and threatening not only their own health, but also that of their human and wild neighbors. But nature fights back against imposed uniformity, and once the chemical pressure is removed, diversity reasserts itself.

With some compromises, beautiful lawns can be had with little or no use of pesticides and fertilizers. The first is to tolerate a few weeds and give up the notion of the "weed-free" lawn. Many long-used methods can be used to maintain healthy turf without artificial inputs. Choosing the appropriate grass suited to the climate and soil is a good start. The right grass will grow more vigorously and deter weeds.

Fertilization can be minimized and soil built by letting the clippings fly; this also takes a load off landfills. Interseeding lawns with such nitrogen-fixing plants as short clovers can provide all the nitrogen that turf needs while deepening the color and fighting weeds—a technique long-used on large estates. When fertilization is necessary, slow-release fertilizers or composted sewage sludge keep grass stronger than the fast-release types that readily pollute nearby waterways. Simply raising the height of the mower can help suppress weeds by shading them out. For some grasses, a longer stem also means a deeper root and greater drought tolerance. For many smaller lawns, weed control can be handled the old-fashioned way—by pulling them out. Finally, homeowners can save their backs and beautify their neighborhoods by planting some areas to perennial groundcovers or wildflowers that need little or no maintenance.

CONCLUSION

Far and away, the main emphasis of many who worry about agriculture's sustainability has been on the greening of production systems to make them more environmentally friendly while maintaining profit margins for the farmer. This focus, which has led to significant improvements in agriculture's financial and environmental performance, was appropriate when issues such as soil productivity were an obvious threat.

While this and other past concerns still remain, many are being effectively addressed through research and policy, and many conservation practices are becoming the norm. In both a relative and an absolute sense, traditional concerns such as soil erosion matter much less now than they once did because efforts to respond appear to be reasonably successful and our base of understanding is now large.

As this chapter has shown, current sustainability interests related to agriculture are much broader than soil and water conservation; the most important issues will affect us globally and require international cooperation to resolve. Yet, this fact seems to have escaped not only those who make agricultural sustainability their business, but also the broader agricultural community. As this book goes to press, sustainable agriculture research agendas are being prepared in government agencies while the threats of germplasm loss and climate change are largely ignored. Academic journals are filled with comparisons of productions systems, reinforcing what we already know, but it's harder to find work on the conservation of agricultural biodiversity, unconventional health threats from pesticides, or the effects of increased weather variability. Leading farm and agribusiness groups recently urged the United States to withdraw support for a binding climate convention until "there is a stronger consensus from the scientific community...." The groups threatened to go to Congress if the Administration did not soften language in the climate convention that "blames" agriculture for its share of the problem. These groups did not express any concern that climate change is a serious problem with potentially large and damaging implications for their members.

By its very nature, sustainability requires foresight. If we are to avoid problems that could cause serious disruption not just for the United States, but also for the world's food production system, it is time to push efforts to understand and address emerging issues. However, there is precious little movement right now. Farmers, researchers, policy-makers, businesses, and consumers all have a role to play. It's time for all stakeholders—including the public—to reexamine their current efforts and find ways to make agriculture *truly* more sustainable.

REFERENCES

Adler, Robert W., Jessica C. Landman, and Diane M. Cameron. *The Clean Water Act 20 Years Later.* Natural Resources Defense Council. Washington, D.C.: Island Press, 1993.

Allen, Arthur W. *Conservation Reserve Program Contributions to Avian Habitat.* Colorado Wildlife Restoration Program, January 14, 1994.

Allen, Patricia and Carolyn Sachs. "Sustainable Agriculture in the United States: Engagements, Silences, and Possibilities for Transformation," In Patricia Allen, ed. *Food for the Future: Conditions and Contradictions of Sustainability.* New York: John Wiley and Sons, 1993.

Alt, Klaus. Personal communication. October 31, 1995.

Alt, K., et al. *Soil Erosion: What Effect on Agricultural Productivity?* Agriculture Information Bulletin no. 556, Washington, D.C.: Economic Research Service, United States Department of Agriculture, 1989.

American Farmland Trust. *Agricultural and Farmland Protection for New York.* Washington, D.C.: American Farmland Trust, 1993.

——. *Alternatives for Future Urban Growth in California's Central Valley: The Bottom Line for Agriculture and Taxpayers.* Washington, D.C.: American Farmland Trust, 1995.

Arbuckle, J. Gordon et al. *Environmental Law Handbook: Twelfth Edition.* Rockville, MD: Government Institutes, Inc., 1993.

Atwood, Jay Dee and Arne Hallam, "Farm Structure and Stewardship of the Environment," in *Determinants of Farm Size and Structure.* Ames, IA: Iowa State University, 1990.

Baker, B.B., et al. "The Potential Effects of Climate Change on Ecosystem Processes and Cattle Production on U.S. Rangelands," *Climatic Change,* Vol. 23 (October 1993): 97-117.

Ball, Eldon. Unpublished Statistics. Washington, D.C.: Economic Research Service, United States Department of Agriculture, 1995.

Biradur, D.P. and A.L. Rayburn. "Flow Cytogenic Analysis of Whole Cell Clastogenicity of Herbicides Found in Groundwater." *Archives of Environmental Toxicology,* Vol. 28 (1995):13-17.

Blair, Aaron and Sheila H. Zahm. "Cancer Among Farmers." *Occupational Medicine: State of the Art Reviews,* Vol. 6 (July-September 1991)3: 335-354.

Blisard, Noel and James R. Blaylock. *U.S. Demand for Food: Household Expenditures, Demographics, and Projections for 1990-2010.* Technical Bulletin no. 1818. Economic Research Service, December 1993.

Briggs, Shirley A. and Rachel Carson Council. *Basic Guide to Pesticides: Their Characteristics and Hazards.* Washington, D.C.: Hemisphere Publishing Corporation, 1992.

Brown, Lester R. and Hal Kane. *Full House: Reassessing the Earth's Population Carrying Capacity.* New York: W.W. Norton & Company, 1994.

Colborn, T., et. al. "Developmental Effects of Endocrine-Disrupting Chemicals in Wildlife and Humans," *Environmental Health Perspectives*, Vol. 101 (October 1993) no. 5: 378-384.

Consultative Group on International Agricultural Research. *Annual Report 1994-1995.* Washington, D.C: Consultative Group on International Agricultural Research, 1995.

Conway, Gordon R. and Jules N. Pretty. "Agriculture as a Global Polluter." Briefing papers on Key Sustainability Issues in Agricultural Development. Gatekeeper Series No. SA11/ IIED Sustainable Agriculture Programme, 1991.

Cook, Kenneth A. *So Long, CRP.* Washington, D.C.: The Environmental Working Group/The Tides Foundation, 1994.

Council for Agricultural Science and Technology. *Preparing U.S. Agriculture for Global Climate Change.* Summary 119. Washington D.C.: CAST, June 1992.

Crosson, Pierre. "Cropland and Soils: Past Performance and Policy Challenges." in, Kenneth D. Frederick and Roger A. Sedjo, eds. *America's Renewable Resources: Historical Trends and Current Challenges.* Washington, D.C.: Resources for the Future, 1991.

Daly, Herman E. "Towards an Environmental Macroeconomics," *Land Economics*, Vol. 67 (May 1991) no. 2:255-59.

Danish Environmental Protection Agency. *Male Reproductive Health and Environmental Chemicals with Estrogenic Effects.* 1995.

Darwin, Roy, et al. *World Agriculture and Climate Change: Economic Adaptations*, Washington, D.C.: Economic Research Service, United States Department of Agriculture, 1995.

Daugherty, Arthur B. *Major Land Uses of the United States.* AER-723. Washington, D.C: U.S. Department of Agriculture, Economic Research Service, 1995.

Davis, Devra Lee and H. Leon Bradlow. "Can Environmental Estrogens Cause Breast Cancer?" *Scientific American,* October 1995: 166-172.

Downing, Thomas E. *Climate Change and World Food Security.* NATO ASI Series, Series I: Global Environmental Change, Vol. 37. New York: Springer, 1996.

Edwards, E.O. and P.W. Bell. *The Theory and Measurement of Business Income.* Berkeley: University of California Press, 1961.

Faeth, Paul. *Growing Green: Enhancing the Economic and Environmental Performance of U.S. Agriculture.* Washington, D.C.: World Resources Institute, 1995.

Firor, John. Personal Communication. March 28, 1996.

Fowler, Cary and Pat Mooney. *Shattering: Food, Politics, and the Loss of Genetic Diversity.* Tucson, Arizona: The University of Arizona Press, 1990.

Frankel, O.H., and Michael E. Soulé. *Conservation and Evolution.* Cambridge, United Kingdom: Cambridge University Press, 1981.

Frazao, Elizabeth. "The American Diet: Health and Economic Consequences" Agriculture Information Bulletin No. 711. Washington, D.C.: Economic Research Service, February 1995.

Ghelfi, Linda M. "Rural Economic Disadvantage." in *Issues for the 1990s.* Agriculture Information Bulletin no. 664. Washington, D.C.: Economic Research Service, Sept. 1994: 121-122.

Gianessi, Leonard P. and James E. Anderson. *Pesticide Use in U.S. Crop Production.* Washington D.C.: National Center for Food and Agricultural Policy, February 1995.

Global Environmental Change Report. "Changes Observed in Greenland Sea." VIII (March 22, 1996)(6):5.

Hicks, John R. *Value and Capital: An Inquiry into Some Fundamental Principles of Economic Theory.* Oxford: Oxford University Press, 1946.

Hoag, Dana L. and Melvin D. Skold. "The Relationship between Conservation and Sustainability." *Journal of Soil and Water Conservation,* Vol. 51 (July-August 1996) no. 4: 292-295.

Hoover, Robert N. and Aaron Blair. "Pesticides and Cancer," *Cancer Prevention,* February 1991: 1-11.

Houghton, J.T., et al., eds. *Climate Change 1995: The Science of Climate Change.* Contribution of Working Group I to the Second Assessment Report of the Intergovernmental Panel on Climate Change. New York: Cambridge University Press, 1996.

Johnson, Richard and Earl Ekstrand. *Wildlife Economic Analysis of the Conservation Reserve Program.* Ongoing Report. Fort Collins, CO: National Biological Survey. February 10, 1994.

Kaiser, Harry M. and Thomas E. Drennen, eds. *Agricultural Dimensions of Global Climate Change.* Delray Beach, FL: St. Lucie Press, 1993.

Kantor, Linda Scott. Personal communication. 1995.

Karl, Thomas R., et al. "Trends in U.S. Climate During the Twentieth Century," *Consequences,* Vol. 1 (Spring 1995) no. 1: 2-12.

Katz, Richard W. "Use of Conditional Stochastic Models to Generate Climate Change Scenarios," *Climatic Change,* Vol. 32 (March 1996) 3:237-255.

Klinedinst, P.L., et al. "The Potential Effects of Climate Change on Summer Season Dairy Cattle Milk Production and Reproduction," *Climatic Change,* Vol. 23 (1993) no. 1: 21-36.

Kross, B.C., et al. *The Iowa State-Wide Rural Well-Water Survey*. Technical Information Series No. 19. Iowa City, IA: Iowa Department of Natural Resources, November 1990.

Lowengart, R., et al. "Childhood Leukemia and Parents' Occupational and Home Exposures," *Journal of the National Cancer Institute*, Vol. 79 (1987): 39.

Lowrance, Richard, et al. "A Hierarchical Approach to Sustainable Agriculture," *American Journal of Alternative Agriculture*, Vol. 1 (1986) no. 4: 170-173.

Majchrowicz, T. Alexander, et al. "Agriculture as a Rural Growth Strategy," in *Issues for the 1990s*. Agriculture Information Bulletin #664. September 1994: Economic Research Service, United States Department of Agriculture, 141-142.

McGranahan, David. "Population Loss in Remote Rural Areas," in *Issues for the 1990s*. Agriculture Information Bulletin no. 664. Sept. 1994: Economic Research Service, 153-154.

Mearns, Linda O., Cynthia Rosenzweig, and Richard Goldberg. "The Effect of Changes in Daily and Interannual Climatic Variability on Ceres-Wheat: A Sensitivity Study," *Climatic Change*, Vol. 32 (March 1996) no. 3: 257-292.

National Research Council, Committee on Characterization of Wetlands. *Wetlands: Characteristics and Boundaries*. Washington, D.C.: National Academy Press, 1995.

——. Committee on Managing Global Genetic Resources. *Managing Global Genetic Resources: Agricultural Crop Issues and Policies*. Washington, D.C.: National Academy Press, 1993a.

——. Committee on Long-Range Soil and Water Conservation. *Soil and Water Quality: An Agenda for Agriculture*. Washington, D.C.: National Academy Press, 1993b.

——. Committee on Scientific and Regulatory Issues Underlying Pesticide Use Patterns and Agricultural Innovation. *Regulating Pesticides in Food: The Delaney Paradox*. Washington D.C.: National Academy Press, 1987.

Nelson, Fred. Personal communication. May 31, 1994.

Oram, P.A. "Sensitivity of Agricultural Production to Climate Change, An Update." in International Rice Research Institute (IRRI), *Climate and Food Security: Papers presented at the International Symposium on Climate Variability and Food Security in Developing Countries, 5-9 February 1987, New Delhi, India*. Manila, The Philippines: IRRI, 1989.

Osborn, C. Tim. "Conservation and the 1996 Farm Act." *Agricultural Outlook*, November, 1996. Economic Research Service, United States Department of Agriculture.

——. "USDA Program Rules Finalize CRP Redesign," *Agricultural Outlook*, June 1991: 24.

Peart, B., et al. "Appendix C," in *The Potential Effects of Global Climate Change on the U.S.* J.B. Smith and D.A. Tirpak, eds. Washington D.C.: U.S. Environmental Protection Agency, 1989.

Pimentel, David, et al. "Environmental and Economic Impacts of Reducing U.S. Agricultural Pesticide Use." in David Pimentel, ed. *CRC Handbook of Pest Management in Agriculture, Volume 1.* Boca Raton: CRC Press, Inc., 1991.

Pimentel, David. Personal conversation. October 1995.

Raeburn, Paul. *The Last Harvest: The Genetic Gamble that Threatens to Destroy American Agriculture.* New York: Simon & Schuster, 1995.

Reid, Walter V. and Kenton R. Miller. *Keeping Options Alive: The Scientific Basis for Conserving Biodiversity.* Washington, D.C.: World Resources Institute, 1989.

Reilly, John M. and Margot Anderson, eds. *Economic Issues in Global Climate Change: Agriculture, Forestry, and Natural Resources.* Boulder, CO: Westview Press, 1992.

Reilly, John M., N. Hohmann, and S. Kane. "Climate Change and Agricultural Trade: Who Wins and Who Loses?" *Global Environmental Change,* Vol. 4 (1994) 1:24-36.

Reisner, Marc and Sarah Bates. *Overtapped Oasis: Reform or Revolution for Western Water.* Washington, D.C.: Island Press, 1990.

Repetto, Robert and Sanjay Baliga. *Pesticides and the Immune System: The Public Health Risks.* Washington, D.C.: World Resources Institute, 1996.

Ribaudo, Marc O. *Water Quality Benefits form the Conservation Reserve Program.* Agricultural Economic Report #606. Washington, D.C.: Resources and Technology Division, Economic Research Service, United States Department of Agriculture, 1989.

Richards, R. Peter, et al. "Pesticides in Rainwater in the Northeastern United States." *Nature,* Vol. 327 (1987):129-131.

Riha, Susan J., Daniel S. Wilks, and Patrick Simoens. "Impact of Temperature and Precipitation Variability on Crop Model Predictions," *Climatic Change,* Vol. 32 (March 1996) 3:293-311.

Robertson, T., et al. "Impacts of Climate Change in the Southern United States," in M. Meo, ed. *Proceedings of the Symposium on Climate Change in the Southern U.S.: Impacts and Present Policy Issues..* Norman, OK: Science and Public Policy Program, University of Oklahoma, 1987.

Rosegrant, Mark W., et al. *Global Food Projections to 2020: Implications for Investment.* Draft. Washington, D.C.: International Food Policy Research Institute, 1995.

Rosenzweig, C. and A. Iglesias, eds. *Implications of Climate Change for International Agriculture: Crop Modeling Study.* Washington, D.C.: United States Environmental Protection Agency, 1994.

Safe, Stephen. "Environmental and Dietary Estrogens and Human Health: Is There a Problem?" *Environmental Health Perspectives,* Vol. 103 (1995): 346-351.

Schneider, Stephen H. "The Future of Climate: Potential for Interaction and Surprises," in Thomas E. Downing, ed. *Climate Change and World Food Security.* NATO ASI Series, Series I: Global Environmental Change, Vol. 37. New York: Springer, 1996.

Sharpe, R. and N. Skakkebaek. "Are oestrogens involved in falling sperm counts and disorders of the male reproductive tract?" *Lancet,* Vol. 341:(1993) 1392–1395.

Smith, J.B. and D. Tirpak. *The Potential Effects of Global Climate Change on the United States:Draft Report to Congress.* U.S. Environmental Protection Agency, Office of Policy, Planning and Evaluation, Office of Research and Development. Washington, D.C.: Government Printing Office, 1988.

Solley, W.B., R.R. Pierce, and H.A. Perlman. *Estimated Use of Water in the United States in 1990.* AER-555. Washington, D.C.: U.S. Geological Survey, U.S. Department of Agriculture, Economic Research Service, 1993.

Terjung, W.H., et al. "Climate Change and Water Requirements for Grain Corn in the North American Plains," *Climatic Change,* Vol. 6 (1984): 193–220.

United States Department of Agriculture (USDA). World Agricultural Outlook Board. *Long-term Agricultural Baseline Projections, 1995-2005.* Staff Report WAOB-95-1. Washington, D.C.: United States Department of Agriculture, 1995a.

———. National Resource Conservation Service (NRCS). *Is There A Better Way? Final Report Data (Part II): Chief Paul Johnson's Reinvention Forums.* Washington, D.C.: United States Department of Agriculture, May 1995b.

———. NASS. *Agricultural Chemical Usage: 1993 Fruit Summary.* Washington, D.C.: Economic Research Service, United States Department of Agriculture, 1994a.

———. NASS. *Agricultural Chemical Usage: 1993 Field Crops Summary.* Washington, D.C.: Agricultural Statistics Board, United States Department of Agriculture, 1994b.

———. Economic Research Service (ERS). *Agricultural Resources and Environmental Indicators.* Agricultural Handbook no. 705. Washington, D.C.: United States Department of Agriculture, 1994c.

———. ERS. *Estimates of Producer and Consumer Subsidy Equivalents: Government Intervention in Agriculture, 1982-1992.* Statistical Bulletin no. 913. Washington, D.C.: United States Department of Agriculture, 1994d.

———. ERS. *Issues for the 1990s.* Agriculture Information Bulletin no. 664. Washington, D.C.: United States Department of Agriculture, 1994e.

———. NASS. *Agricultural Chemical Usage: 1992 Field Crops Summary.* Washington, D.C.: Economic Research Service, United States Department of Agriculture, 1993a.

———. NASS. *Agricultural Chemical Usage: 1992 Vegetables Summary.* Washington, D.C.: Economic Research Service, United States Department of Agriculture, 1993b.

———. NASS. *Agricultural Chemical Usage: 1991 Field Crops Summary.* Washington, D.C.: Economic Research Service, United States Department of Agriculture, 1992a.

———. NASS. *Agricultural Chemical Usage: 1991 Fruit* Summary. Washington, D.C.: Economic Research Service, United States Department of Agriculture, 1992b.

———. NASS. *Agricultural Chemical Usage: 1990 Field Crops Summary.* Washington,D.C.: Economic Research Service, United States Department of Agriculture, 1991a.

———. NASS. *Agricultural Chemical Usage: 1990 Vegetables Summary.* Washington,D.C.: Economic Research Service, United States Department of Agriculture, 1991b.

United States Department of Commerce, *Statistical Abstract of the United States 1993.* Washington, 113th Edition. Washington, D.C.: United States Department of Commerce, Economics, and Statistics Administration, Bureau of the Census, 1993.

United States Environmental Protection Agency. Proposed Environmental Goals for America with Milestones for 2005. In draft. As reported in the *New York Times,* September 24, 1994.

———. *1992 Needs Survey: Report to Congress.* EPA-832-R-93-002. Washington, D.C.: United States Environmental Protection Agency, Sept. 1993.

———. *The Potential Effects of Climate Change on the United States, Report to Congress.* EPA-230-89-050. Washington, D.C.: U.S. Government Printing Office, 1989.

United States General Accounting Office. *Conservation Reserve Program: Alternatives are Available for Managing Environmentally Sensitive Cropland.* Report to the Committee on Agriculture, Nutrition, and Forestry, U.S. Senate. Washington, D.C.: United States General Accounting Office, February 1995.

United States House of Representatives. *Food, Agriculture, Conservation, and Trade Act of 1990.* 101st Congress, 2nd sess., October 22, 1990: Conference Report to accompany S. 2830.

Ware, George W. *The Pesticide Book,* Fresno: Thomson Publications, 1989.

Warrick, R.A. "The Possible Impacts on Wheat Production of a Recurrence of the 1930's Drought in the Great Plains," *Climatic Change,* Vol. 6 (1984):5-26.

Weihe, W.H. and R. Mertens. "Human Well-being, Diseases and Climate." In J. Jager and H.L. Ferguson eds. *Climate Change: Science, Impacts and Policy, Proceedings of the Second World Climate Conference.* Cambridge: Cambridge University Press, 1991.

Wilkes, Garrison. Current Status of Crop Plant Germplasm. *CRC Critical Reviews in Plant Science,* Vol. 1, no. 2 (1983):133-181.

Wilks, Daniel S. and Susan J. Riha. "High-Frequency Climatic Variability and Crop Yields. A Guest Editorial." *Climatic Change,* Vol. 32 (March 1996)3: 231-235.

World Commission on Environment and Development. *Our Common Future.* New Delhi: Oxford University Press, 1987.

World Resources Institute, in collaboration with the United Nations Environment Programme, the United Nations Development Programme, and the World Bank. *The World Resources Report: A Guide to the Global Environment, 1996-97.* New York: Oxford University Press, 1996.

3.
DRIVING THE ROAD TO SUSTAINABLE GROUND TRANSPORTATION

James J. MacKenzie

Transportation plays a role in the functioning of a modern industrial economy nearly as critical as that of energy itself. Motor vehicles, planes, trains, and ships meet many essential needs, delivering goods and services and moving people about. Thus, any major interruption of transportation services—due to natural disaster, fuel shortages, or other causes—can lead quickly to economic and social disruption.

For all the benefits that it confers, though, the U.S. transport system—particularly the roadway system—cannot continue to function indefinitely as it is now structured. For reasons identified by the President's Council on Sustainable Development, reliance on today's system is unsustainable:[1]

◆ First, U.S. transportation is a major (30 percent) and growing source of atmospheric carbon dioxide (CO_2)—the principal "greenhouse" gas contributing to global warming.[2] By human time scales, global warming could change Earth's climate irreversibly. To mitigate this threat, atmospheric concentrations of greenhouse gases will soon have to be stabilized, and will eventually have to be reduced. In the case of CO_2, global atmospheric emissions would eventually have to be cut by as much as 60 to 80 percent below present levels. For transportation, a reduction of this magnitude means that new, nonfossil energy sources will be needed to power the planes, ships, and motor vehicles of the 21st century.

◆ Second, the transport system depends almost entirely on petroleum—a resource whose production continues to decline in the lower 48 states and in

Alaska. Some evidence also suggests that global production could begin to decline early in the next century because of resource constraints.[3] Clearly, new transportation energy sources are needed.

◆ Third, despite 25 years of engineering efforts and significant reductions in transportation-related emissions, motor vehicles are still important sources of carbon monoxide, organic compounds, nitrogen oxides, small particles, and other forms of air pollution—largely due to growth in the U.S. vehicle fleet. Without new approaches to this problem, domestic and global exposure to air pollution will persist, and in some areas, worsen, and ever-increasing numbers of people will be exposed.

◆ And, fourth, the nation's ground-transportation system is failing to provide citizens with convenient, affordable, and reliable access to services, goods, and job opportunities. As commuters know only too well, traffic and congestion in many metropolitan areas are growing worse each year. Even where congestion is not overwhelming, those too old, too young, or too physically challenged to drive an automobile are poorly served.

The first two factors—the threat of climate change from the build-up of greenhouse gases and the risks from depleting oil resources—represent major resource and ecological constraints. In this sense, they represent unsustainable trends and imply risks that most people consider unacceptably high. Air pollution and near total reliance on motor vehicles for ground transportation threaten sustainability less clearly. Arguably, they are less absolute and more tolerable—at least at levels found in the United States. But WRI's research indicates that vehicle air pollution is a threat to public health, particularly to urbanites. Further, near total reliance on automobiles, along with the congestion it brings, is making it increasingly hard to get to work and to shop for necessities.

How do we know that trends in transportation and population growth are fueling important national and international problems? First, more drivers in more cars and trucks inevitably mean more air pollution, more greenhouse gas releases, greater reliance on unstable sources of oil in the Persian Gulf, and more traffic. Left to themselves, matters will eventually become intolerable, which is what makes today's transportation enterprise unsustainable.

A Vision of the Future

Although the future cannot be predicted, it can be guided, and the grim consequences of continuing present practices are by no means inevitable.

A Vision of an Environmentally Sustainable United States

Nor are fuel constraints or growing traffic necessarily our destiny *if* a determined long-term national effort begins soon.

One of the most important conclusions to emerge from WRI's research is that population growth boosts both fuel use and traffic. Analysis of recent trends and projections of growth in traffic through around 2050 show that an expanding urban population and a commensurate increase in the number of licensed drivers is becoming the main force driving up the number of vehicle miles traveled (VMT) and making traffic worse. And, to many Americans, increasing traffic is *the* transportation problem.

Few options are available for dealing with growing vehicle use. We can build more urban freeways, endure the headaches and costs of thickening traffic, or view transportation differently. In fact, the third is the only one that makes sense in the long term, and to achieve it, major changes will be needed in vehicle technology, transportation planning, and Americans' attitudes toward population growth.

How much brighter might our transportation future be on this third path and what vision are we working toward? The easy part of the answer to this question centers on the use of sustainable fuels. If the nation were to shift to emissionless vehicles powered ultimately by electricity or hydrogen derived from photovoltaic cells, wind turbines, or other renewable technologies, today's atmospheric and resource risks from motor fuel use would ultimately disappear, no matter how large the vehicle fleet. Smog and carbon-monoxide concentrations stemming largely from vehicle emissions would be greatly reduced. Greenhouse gas emissions would too. The skies over our cities would begin to clear, and the threat of global warming (from U.S. emissions, anyway) would begin to recede.

Far more profound would be the changes in how Americans live and travel if we follow the prescripts to sustainability set out here. To begin with, thanks mainly to a national education effort, the U.S. population would stop growing by the mid-21st century, so the number of licensed drivers and, ultimately, the growth of VMT, would stabilize. In turn, pressures to build more freeways, more mega-malls, and more residential sprawl would wane.

With the right federal and state incentives, more resources would be available to make over major U.S. urban areas. Personal Rapid Transit (PRT) systems—a near-commercial technology based on small electric vehicles riding on dedicated guideways—would gradually connect urban and suburban residential developments, shops and shopping centers, universities, hospitals, train stations, airports, and business centers. With nonstop service between origin and destination, PRT systems would make people almost as mobile as they are with the personal motor vehicle, and for the elderly, the young, and the disabled, these systems would be far more usable and sensible.

Complementing the development of new forms of transit, land-use patterns would change. Today's retail strip developments would be replaced gradually with commercial and residential buildings, and retail shops would spring up around transportation nodes. PRT, bus, and other forms of public transit could then meet more of people's shopping and recreational needs. Bikeways and walking paths would line suburban areas, encouraging nonmotorized shopping and travel, as well as community development.

In this alternative future, U.S. cities would be far less subservient to cars. Streets would be quieter. The air would be cleaner. There would be more options for travel and less need to drive. People who want to live in single-family homes could, but would have less need to drive and more opportunities to bike, walk, take public transit, or telecommute. People who want to live in apartment houses closer to shops and transportation nodes could do so, too. Such a future—one of diversity and dynamic equilibrium—would accommodate various lifestyles and differing travel preferences. Best of all, it would be sustainable over the long haul.

Although the following discussion focuses on the U.S. transportation system's sustainability, many other nations face similar problems—pollution, resource depletion, climate change, congestion, and more equitable access to transport. For this reason, the analysis and recommendations put forth here are relevant to conditions in other countries.

OVERVIEW OF THE U.S. TRANSPORT SYSTEM

On almost any basis, the U.S. transport system is vast. More than 200 million cars, trucks, and buses roam nearly 4 million miles of roads and highways—about 750 vehicles for every 1,000 people. The motor vehicle fleet grew by an average 1.6 percent per year between 1985 and 1995. Although the United States accounts for only 4.5 percent of the world's population, it has more than 30 percent of all motor vehicles.

Other segments of the transportation system are equally impressive. The U.S. commercial airline fleet numbers over 5000 planes, while the general aviation fleet has more than 200,000 planes in service.[4] More than 600,000 buses traverse U.S. roads, and there are nearly this many passenger and freight railroad cars.

Transportation in the United States is big business. In 1994, expenditures for moving goods and passengers exceeded a trillion dollars.[5] Depending on one's definition, transportation and related businesses employ at least one-tenth of the U.S. workforce, and transportation-related activities account for nearly 20 percent of consumer spending.

Each year, Americans log almost 3.7 trillion (10^{12}) passenger-miles of travel and 3.4 trillion ton-miles of traffic. Nearly 90 percent of all passenger travel (measured in passenger-miles) is via motor vehicles (cars, trucks, buses, and motorcycles).[6] *(See Figure 3-1.)* Freight transportation is more evenly distributed by mode: 33 percent of ton-miles is by rail, 26 percent by water, 23 percent by truck, and 18 percent by pipeline.[7]

Given the dominance of the motor vehicle in U.S. transportation, roadway transport is the crux of both U.S. transport problems and their solutions.

UNSUSTAINABLE FUELS—A CLOSER LOOK

The U.S. transportation network is on a dead-end course for many reasons. Of the four factors governing ground transportation's sustainability, three—global warming, gradual resource depletion, and air pollution—relate primarily to the

FIGURE **3-1. Passenger Travel in the United States**

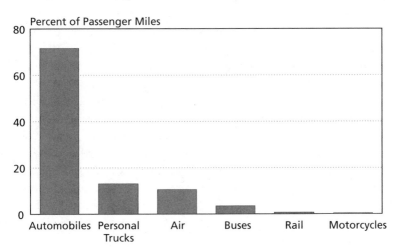

fuels used to power vehicles. The fourth (near exclusive reliance on automobiles and the congestion this gives rise to) is primarily a land-use issue stemming from the way that U.S. cities have evolved over the past 50 years. Conceptually, the fuel-related issues can be considered separately from the land-use problems: it is easy to imagine replacing oil-powered vehicles with emissionless ones without relieving congestion or enhancing personal mobility.

U.S. Motor Vehicles and Global Warming

In 1995, U.S. motor vehicles directly consumed about 142 billion gallons of fuels, principally gasoline and diesel fuel,[8] while the entire transportation sector (including motor vehicles, planes, ships, pipelines, and so forth) consumed 179 billion gallons.[9] If it is assumed that, on average, burning a gallon of fuel yields 5.5 pounds of carbon, in 1995 some 356 million (10^6) metric tonnes of carbon were directly released into the atmosphere from the U.S. motor vehicle fleet—about a quarter of national carbon emissions. (Worldwide, motor vehicles account for about 15 percent of fuel-related carbon emissions.[10])

Figure 3-2 shows the trends in oil consumption and carbon dioxide emissions for the entire U.S. transportation sector, in which oil accounts for about 97 percent of energy supply. The temporary reductions in oil use in the early and late 1980s and the early 1990s stemmed at least partly from fall-offs in national economic activity in the earlier cases associated with oil disruptions.

Will these trends change? Is the major reduction in emissions needed to control atmospheric concentrations in the offing? Or are emissions likely to continue growing, compounding the difficulties of mitigating global climate change? These questions have no certain answers. In principle, vehicle-related CO_2 emissions could fall if fuel efficiency in the fleet improved, but on this point the history of the past two decades is not encouraging. Between 1970 and 1995, the amount of fuel annually consumed per motor vehicle declined on average by about 1.05 percent per year. (In 1995, actual consumption was about 17 percent less than in 1970.) This slow, downward trend was the end result of two opposite patterns: a significant improvement among passenger cars (33 percent reduction in fuel use per vehicle) and an equally significant deterioration in performance among medium and heavy-duty trucks (a 30 to 50 percent increase in fuel consumption per vehicle). More recently, as real gasoline prices have continued to drop, Americans began purchasing more trucks, minivans, and sport utility vehicles (SUV). Minivans and SUVs (classified as trucks) have much poorer fuel economies, typically 20 miles per gallon (mpg), compared with 28 for new

FIGURE **3-2. Direct Oil Consumption and Carbon Emissions by the U.S. Transportation Sector**

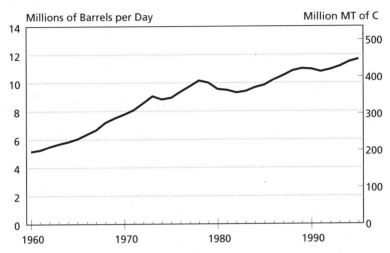

cars.[11] Constituting about 40 percent of new light-duty vehicle sales, trucks, minivans, and SUVs are slowing improvements in overall fleet fuel efficiency, currently around 22 mpg.

Figure 3-3 shows how these trends affected overall motor vehicle fuel consumption. Compared to 1970, the number of motor vehicles increased by 85 percent by 1995.[12] This enormous growth in fleet size overwhelmed the relatively modest reduction in per-vehicle fuel use (17 percent). As a result, motor vehicle fuel use increased by more than 50 percent.

In turn, growth in the vehicle fleet can be attributed to both an increase in per capita ownership—largely the result of growing individual wealth and the availability of better, less expensive cars—and to a growing population: from 1970 through 1995, per capita registration of vehicles increased by 44 percent while the overall population grew by 28 percent. For perspective, consider that had the U.S. population remained at its 1970 level, all other things being equal, fuel use in 1995 would have increased by only 18 percent over 1970 levels, not 55 percent.

Largely because the motor vehicle fleet is growing, total motor vehicle fuel consumption—as well as carbon dioxide emissions—has been rising steadily. Will such growth continue? Using current trends, it's possible to make a rough estimate as follows. First, assume that the decline in fuel use per vehicle (1.05

FIGURE 3-3. **Fuel Trends Among U.S. Motor Vehicles**

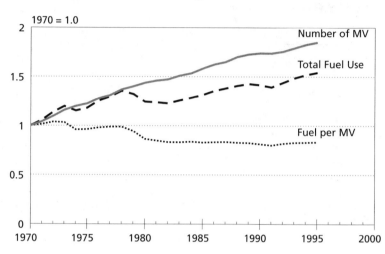

percent per year) continues along the trend that it has in the past. To estimate the number of vehicles in 2010, consider first that the trend in per-capita motor vehicle registrations suggests there will be about 0.812 motor vehicles per person by the year 2010. Using the U.S. Census Bureau's mid-range estimates for the size of the U.S. population in 2010 (298 million), some 242 million motor vehicles will be on the road. Thus, if present trends continue, fuel consumption and CO_2 emissions from the U.S. vehicle fleet will be about the same in 2010 as they were in 1993: net reductions in fuel use per vehicle would be just about offset by increases in the motor vehicle fleet. The Department of Energy's (DOE) computer model (the National Energy Modeling System) forecasts a 15-27 percent increase in motor vehicle oil use by 2010.[13] Hence, in the best of circumstances, motor vehicle oil use and CO_2 emissions could stay even; under conditions of high economic growth, they could swell by more than 25 percent.

Oil: A Finite Resource

Oil is a fossil fuel formed in significant amounts only over millions of years. For time periods of human interest, it is a nonrenewable resource. Oil represents almost 40 percent of both U.S. and world commercial energy supply (the fuels that are bought and sold in commercial trade including oil, natural gas, coal, and electric power).

Worldwide, motor vehicles account for about one-third of world oil consumption. In the industrialized (OECD) countries, motor vehicles represent over 40 percent of oil demand, and in the United States, over half.[14] Transportation activities as a whole make up two-thirds of U.S. demand for oil.

U.S. crude oil production in the lower 48 states has been declining for more than 26 years. Production in Alaska peaked in 1988 and had fallen by 26 percent by the end of 1995. *(See Figure 3-4.)* In its 1996 *Annual Energy Outlook*, the Department of Energy (DOE) projected that by 2015 U.S. crude oil production will fall an additional 20 percent below 1992 levels[15] and that U.S. demand for oil products in 2015 will be about 25 percent higher than in 1992. To satisfy the nation's appetite for oil, imports are projected to double by 2015, when they will account for over 55 percent of supply. Under DOE's forecast, real world oil prices will rise steadily over this period.

While U.S. production declines and imports increase, worldwide production will shift ever more to the Organization of Petroleum Exporting Countries (OPEC) producers. In 1996, OPEC accounted for about 41 percent of world oil supply.[16] By 2010, this could rise to 53 percent, essentially the same share that OPEC controlled in 1973 when the Arab oil embargo hit.[17] As reliance on Persian Gulf production grows, so will the economic and security risks associated with this region's volatile and unstable oil production. However difficult to quantify, such risks are real to those who depend heavily on Middle East oil.

Today, with world oil prices very low, interest in estimating when world crude oil production might peak and start declining is minimal. Yet, compelling evidence suggests that if global consumption continues rising moderately (say, 1.5 to 2.0 percent per year), conventional crude oil production may be only a decade or two away from peaking and beginning its inevitable long-term, resource-driven decline.

If it proves true that a decline in global crude oil production is this near, the U.S. transportation system as we know it would be even more unsustainable than generally acknowledged and major efforts to develop replacement transportation energy sources would be needed.

Understanding why unsustainability is a distinct possibility requires a little knowledge of geological change and economics. The production of a nonrenewable resource such as oil begins with the discovery of oil fields. Production rises as oil fields are found and developed (usually the largest first) and eventually declines as they are exhausted. In this general pattern lies the history of crude oil production in both the lower 48 states and in Alaska. *(See Figure 3-4.)* The

FIGURE 3-4. **Trends in U.S. Crude Oil Production**

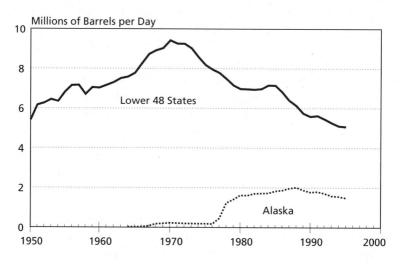

area under the curve of annual production (the sum of annual production from its beginning up to a given time) equals cumulative production up to that time. Cumulative production over the entire lifetime of a well or field is the total amount of oil ever to be produced, an amount termed Estimated Ultimately Recoverable (EUR) oil. In the case of the United States, EUR crude oil can be estimated from Figure 3-4 to be about 189 billion (10^9) barrels for the lower 48 states. Of this, 159 billion barrels—about 84 percent—had already been produced by the end of 1995.

Like U.S. production, world production can be expected to generally follow the behavior of Figure 3-4. As world production declines, oil prices can be expected to rise, new technologies and processes for extracting the remaining recoverable oil will come on line, and, eventually, production (and energy) costs will become too high and new energy sources will be introduced.

How much oil is expected to be recovered globally over all time? Many estimates have been made of global EUR oil over the years. For the past 20 years, these estimates have ranged from 1,800 to 2,200 billion barrels, and most authorities believe EUR resources are at the lower end of this range.

Figure 3-5 shows five possible scenarios for world crude oil production assuming EUR resources of 1,800; 2,000; 2,200; 2,400; and 2,600 billion barrels (Gb), respectively. All five projections match both 1994 actual world oil produc-

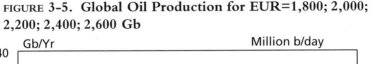

FIGURE 3-5. Global Oil Production for EUR=1,800; 2,000; 2,200; 2,400; 2,600 Gb

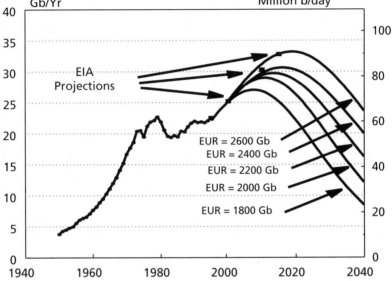

tion and DOE's forecast for the year 2000. For EUR oil equal to 1,800 Gb, peaking occurs around 2007; for 2,000 Gb, peaking occurs in 2011; for 2,200 Gb, 2013; for 2,400 Gb, 2016; and for 2,600 Gb, 2019. Remarkably, a spread of 800 billion barrels between the lowest EUR (1,800 billion barrels) and the highest (2,600)—a quantity greater than all past oil production—delays peaking by only 11 years. From a long-term perspective, it hardly matters whether the global EUR oil is 1,800 Gb or 2,600 Gb: the decline is imminent.

No one knows for sure what the actual trends will be in the years ahead for world oil prices or global demand. Certainly, the curves in Figure 3-5 only illustrate what might transpire. Still, production is certain to peak early in the next century if estimates of EUR oil prove anywhere near right and if projected growth trends prove correct. Conversely, if consumption stayed relatively flat—because oil prices rose sharply or alternative energy sources were introduced—a production decline could be delayed by several decades more. The key point is that the time when world crude oil production will begin to decline is measured not in centuries but in years, and that is how much time is available to develop and introduce alternative energy sources.

Many knowledgeable analysts have come to similar conclusions. Jean Laherrere of Petroconsultants in Geneva estimates ultimate global resources at 1750 billion barrels and calculates that world production will peak around the year 2000 at about 66 million barrels per day, followed by a decline of 2.7 percent per year.[18] Colin Campbell, also of Petroconsultants, assuming 1,800 billion barrels, also projects the peaking will occur around the year 2000. John F. Bookout, retired president and CEO of Shell Oil Company, accepts the USGS estimate of EUR oil as consistent with that of Shell and agrees that world oil production will peak around the year 2010.[19]

This analysis does not suggest that the world will soon "run out" of oil or hydrocarbon fuels. Oil production will continue—though at a declining rate—for many decades once production peaks.[*] And enormous amounts of coal, heavy oils, tar sands, and bitumen could be used to produce liquid or gaseous substitutes for crude oil. But the facilities to make such synthetic fuels are costly to build, environmentally damaging to operate, and—because of the need to upgrade the heavy oil—have higher carbon dioxide emissions than those used to process conventional crude oil.

The transportation sector is likely to be particularly hard hit by rising prices and a decline in oil availability. Motor vehicles alone account for about 53 percent of U.S. oil consumption, and few viable short-run alternatives to oil are available.

Motor Vehicles and Air Pollution

Motor vehicles have long been known as major sources of air pollution, including carbon monoxide (CO), volatile organic compounds (VOCs), nitrogen oxides (NO_x), toxic compounds, and particulates. Gasoline-fueled vehicles are the principal sources of carbon monoxide and nitrogen oxides, while diesel engines are major contributors of particulates and toxic chemicals. Some of these pollutants (CO, VOCs, NO_x) contribute directly to ozone formation and acid deposition and, indirectly, to global warming.

Recent studies of mortality in polluted cities suggest that small-particle pollution, much of it from motor vehicles, is associated with higher rates of lung and heart

---

[*]   Analysts sometimes cite the ratio of proven reserves to annual production (R/P), measured in years, as an indication of how much oil is left to be recovered. This ratio *cannot* be used as an indicator of how long oil production can continue at present levels. The issue of paramount interest is when oil production peaks and starts to decline; it is at this time that new energy sources will be needed.

diseases and premature death.[20] Studies in California conclude that nonsmokers living where air pollution is heavy lose almost as much lung function as cigarette smokers do. Recent British studies suggest that episodes of high levels of urban pollution, much of it from vehicles, can increase mortality from respiratory and cardiovascular diseases by 10 percent.[21] And a study of more than 200 American metropolitan areas concludes that about 64,000 people may die prematurely from heart and lung disease each year due to particulate air pollution.[22] On balance, many people's lives are being shortened by one to two years.

American consumers have spent billions of dollars on emission-control devices for their cars and trucks, and estimated motor vehicle emissions in the United States have declined significantly. According to EPA, despite growth in the nation's fleet, national carbon-monoxide emissions from vehicles decreased by about 18 percent between 1984 and 1993, while NO_x emissions dropped by 6 percent.[23] Emissions of VOCs from vehicles declined by 27 percent over this same period. Yet, vehicles remain significant sources of air pollution.[24] By EPA estimates, motor vehicles still account for 78 percent of all U.S. carbon monoxide emissions, 45 percent of NO_x emissions, and 37 percent of VOC emissions.[25]

Despite dramatic reductions in new-vehicle emissions, some 80 million Americans lived in areas where air quality standards were violated in 1995.[26] Fifty million live where the ozone standard is breached. For several reasons, efforts to reduce vehicle air pollution have met with only partial success. One is that emission controls on oil-powered vehicles deteriorate over time,[27] so on average, motor vehicles in the United States emit hydrocarbons at much higher rates than the rates at which they were certified. Another is that some drivers modify or remove their pollution devices. EPA found that almost one-fourth of cars and light trucks that were not covered by inspection and maintenance programs had been tampered with.[28] (If the catalytic converter is missing or damaged, VOC and CO emissions can increase by more than 400 percent.)

Finally, the number of motor vehicles and the number of persons exposed to pollution, especially in urban areas, are growing relentlessly. In the United States, the number of motor vehicles grew from 108 million (10^6) in 1970 to 193 million in 1993 with a commensurate growth in travel and fuel use. If present trends continue, there could be well over 240 million motor vehicles on U.S. roads by the year 2010. Continued growth in the number of sources—combined with less-than-expected reductions from emission controls—has overwhelmed much of the technological advance of the past 25 years in reducing motor vehicle air pollution.

In sum, despite heroic engineering efforts over the past several decades, vehicles remain important sources of air pollution. Growth in the number of vehicles, growth in the number of people exposed to pollution, and the deterioration of and removal of pollution-control equipment on vehicles have combined to make progress in protecting public health and welfare less than experts once anticipated.

THE SEARCH FOR SUSTAINABLE TRANSPORTATION FUELS

A more sustainable transportation system would alleviate or solve the most pressing problems of nonsustainability posed by today's technology, patterns of land use, and social institutions. Specifically, it would cut emissions of greenhouse gases (especially CO_2), reduce transportation oil consumption, increase reliance on sustainable energy sources, lower air pollution emissions, and reduce congestion and auto dependency by improving mobility for all.

Since motor vehicles today depend totally on petroleum, the pollution, security, climate, and resource problems that vehicles pose probably can't be solved until economically and technologically attractive vehicles that run on alternative fuels become widely available. Consumers the world over will have little alternative but to continue buying oil-powered cars and trucks. As a result, air pollution will affect more and more people and oil consumption will rise, as will carbon dioxide emissions.

The world desperately needs alternatives to oil-powered cars and trucks. And the burden for developing the technological alternatives rests squarely on the industrial countries that make the world's motor vehicles—80 percent are made in the United States, Japan, and Europe—since no one else has the industrial muscle to do it.

Of the various legacies of motor vehicle use, air pollution has received by far the most political attention. Since the health and other effects of air pollution are more or less immediate and visible, most countries have clean air legislation. At least until recently, proposals to cut carbon monoxide and ozone concentrations in the United States have emphasized the introduction of so-called "alternative fuels" for vehicles—methanol (wood alcohol), compressed natural gas, ethanol (grain alcohol), electricity, and hydrogen. But how sustainable is their use?

Methanol

Methanol is currently made almost exclusively from natural gas. In the United States, methanol would be introduced in "flexible-fuel" vehicles that can burn methanol-gasoline blends ranging from pure gasoline (M0, for zero percent methanol) to a mixture of 85 percent methanol and 15 percent gasoline (M85). According to a 1991 National Research Council (NRC) report on the American ozone problem and the potential benefits of methanol and other fuels in reducing ozone exposures,[29] any ozone reductions from switching to M85 would be modest and would be quite sensitive to the percentages of methanol and gasoline in the blend. The overall air quality impacts of using methanol blends and methanol oxygenates (MTBE) were also assessed by a group of university-based engineers, who reported in the July 2, 1993 issue of *Science* magazine that methanol and ethanol "especially when blended with gasoline and used in flexible fuel vehicles, provide little or no air quality advantages beyond the reduction of CO" and that "the addition of oxygenates, such as methyl tertiary butyl ether (MTBE) and ethanol, has been found to have little or no effect on the problems of atmospheric reactivity and ozone formation."[30]

As for its global warming impacts, methanol made from fossil fuels offers cold comfort in the battle against greenhouse warming. According to a Congressional Office of Technology Assessment (OTA) report, methanol "...use is expected to provide, at best, only a small greenhouse gas benefit over gasoline, and then only if the vehicles are significantly more efficient than gasoline vehicles."[31] Mark DeLuchi, a researcher at the University of California at Davis, confirms this result: light-duty vehicles running on pure methanol (made from natural gas) have about the same greenhouse gas emissions as gasoline vehicles.[32] Heavy-duty vehicles running on pure methanol (made from natural gas) would emit about 20 percent more greenhouse gases. If the methanol were made from coal, the greenhouse gas emissions would increase (relative to gasoline and diesel fuel) by 70 to 100 percent.

Compressed Natural Gas

Compressed natural gas (CNG) is another potential option for reducing air pollution and reliance on imported oil. CNG vehicles emit comparatively little carbon monoxide, reactive hydrocarbons, and particulates. According to the U.S. Office of Technology Assessment and the National Research Council, switching to CNG vehicles would definitely reduce ozone concentrations, at least in regions where reactive-hydrocarbon reductions are required.[33]

Per vehicle mile traveled, CNG cars emit less carbon dioxide than gasoline-pow-ered cars. But methane—the principal component of natural gas—is itself a potent greenhouse gas. On balance, using CNG vehicles could reduce overall greenhouse gas emissions for light-duty vehicles by 15 percent and increase those of heavy-duty diesel-powered vehicles by about 5 percent.[34] Moreover, because CNG is a depletable fossil fuel that releases carbon dioxide when burned, CNG vehicles aren't the long-term answer to fuel depletion and climate problems. Still, their limited use would reduce pollution emissions (especially from such heavy vehicles as buses and trucks) and dependence on imported oil during a switch to other fuels. Also, experience with CNG vehicles might make it easier to introduce hydrogen-powered vehicles later since the technologies for natural gas and hydrogen are similar.

Ethanol

Another alcohol fuel, ethanol is commonly made from biomass feedstock, such as corn. In a study of ethanol's impacts on air quality, Sierra Research—a California-based consulting firm—concluded that gasohol (a mixture of 10 percent ethanol and 90 percent gasoline) would on average reduce CO concentrations by 25 percent, but would increase NO_x emissions by 8 to 15 percent and evaporative emissions of VOCs by 50 percent.[35] Switching to gasohol, the researchers estimated, would *increase* ozone concentrations by at least 6 percent. The academic study already cited concluded that the "addition of ethanol to gasoline is generally counterproductive with respect to ozone formation."[36]

Estimates of the net greenhouse impacts of ethanol are subject to great uncertainty. All depend on which feedstock is used, which fuels (coal, gas, wood) are used in distillation, how up-to-date the equipment is, and how much energy in ethanol production is credited to the manufacture of by-products (such as animal feeds, in the case of corn). DeLuchi concludes that switching to motor vehicles running on pure ethanol (E100) made from corn (as it is made in the United States) could increase greenhouse gas emissions by 20 percent (for light-duty vehicles) to 50 percent (for heavy-duty vehicles).[37] According to his study, both the combustion of coal at the production facility and the use of fertilizers in corn farming yield large greenhouse gas emissions.[38] Fertilizers can lead to emissions of both nitrous oxide (N_2O) and nitrogen oxides (NO_x). Other studies conclude that if the most modern technology were used, ethanol production would lead to a net energy benefit and, hence, a reduction in carbon dioxide emissions relative to those associated with gasoline use.

Alcohol Fuels from Other Biomass Sources

Ethanol and methanol could be produced from a variety of biomass sources, including paper, municipal solid waste, agricultural and forestry residues and wastes, or low-grade "junk wood" stands.[39] But, in some settings, removing forest and agricultural residues could reduce soil humus and increase erosion, water runoff, and nutrient loss.[40]

By far the largest potential source of alcohol fuels would be energy plantations, where crops or trees would be grown as a feedstock.[41] DeLuchi estimates that under favorable conditions—very energy-efficient processes using little or no fertilizer—total emissions of all greenhouse gases from wood-based ethanol could decline by as much as 75 percent compared with those from gasoline.[42] He also points out, though, that under unfavorable conditions, biofuel cycles will provide little or no reductions in greenhouse gas emissions.[43]

Growing energy crops on a large scale could trigger significant environmental problems.[44] For example, growing short-rotation trees can lead to a loss of topsoil and nutrients and to soil compaction and increased water runoff. It can also require heavy applications of fertilizer and herbicides. Soil erosion where trees are harvested every five years has been estimated at 2 metric tons per hectare per year, while forest soils are formed at only 0.3 metric tons per hectare per year.[45] Detailed analyses of erosion rates for energy crops have yet to be made, and most data come from small field trials collected over short periods.[46]

Viewed solely as a way to cut air pollution, each of these energy sources discussed has some merit, though sometimes it is small. All three fuels would reduce carbon monoxide emissions relative to those from burning conventional gasoline. A switch to CNG would also reduce ozone concentrations. But when their impacts on global warming are also taken into account, these fuels look less attractive overall than so-called zero emission vehicles (ZEVs).

Zero Emission Vehicles: Electric, Hydrogen, and Hybrid Vehicles

Could electric vehicles (EVs)—cars and trucks powered with electric drive trains by on-board batteries, flywheels, or hydrogen fuel cells—curb air pollution, reduce oil consumption, and cut greenhouse gas emissions?[47] And what role could be played by hybrid vehicles—electric-drive cars and trucks with some form of electrical storage device and an on-board generator powered by a fuel such as gasoline, CNG, or hydrogen?*

---

* The generator in a hybrid vehicle can be used to charge the battery or to run the vehicle directly. In some hybrids, all the electricity consumed by the car is generated from the on-board engine. Alternatively, the engine can serve as a range extender, to be used only when the battery of the car is nearly depleted. In a range-extender application, the engine might be used very little in urban driving where trips are short and the vehicle is kept charged using the wall outlet.

Air-Pollution Impacts

The overall impacts of EVs on air quality, global warming, and oil use would be determined by the emissions from the power plants used to charge the EVs. Where modern power plants supply the electricity, EVs would have much lower carbon monoxide or hydrocarbons emissions than even the cleanest gasoline vehicles—97 percent lower emissions of carbon monoxide and reactive organic compounds according to the California Air Resources Board. Nitrogen oxide emissions would be reduced 24 to 47 percent.[48] In one study of the pollution impacts of EVs, carbon monoxide and organic compound emissions were estimated to fall by 99 percent and nitrogen oxide emissions by 70 percent.[49] In short, if the power plants are reasonably clean, introducing EVs would greatly reduce carbon monoxide and ozone problems. Toxic emissions (such as benzene) would also be drastically cut. Moreover, pollution emissions would not increase much, if at all, as EVs aged. Offsetting these benefits, emissions of sulfur dioxide and, hence, acid precipitation could increase in areas where high-sulfur emissions resulted from the generation of the electricity used to power the EVs—not a problem in the United States since total national sulfur emissions are capped by the 1990 amendments to the Clean Air Act.

If EVs were recharged at night—as they ought to be initially—no new power plants would have to be built and ozone formation (which needs sunlight) would fall even more. If batteries were charged by renewable energy sources, pollution and greenhouse gas emissions would be minuscule.

EVs and Global Warming

How significantly battery-powered EVs can cut carbon dioxide emissions depends mostly on two factors: the electric efficiency of the vehicles and the emissions from the power plants that produce the electricity used to charge them.* If EVs are charged by electricity from steam power plants fired by natural gas, the carbon-dioxide emissions would be about 50 percent less than those from similar gasoline-powered vehicles. But charging electric vehicles with electricity made from coal would cut greenhouse gas emissions by only about 15 to 20 percent. In California, which derives much of its electricity from natural gas, hydro plants, wind machines, and geothermal energy, CO_2 emissions would

---

*   In the following comparisons, emissions are compared between gasoline vehicles and their electric drive conversions.

BOX 3-1. Zero-Emission Vehicles (ZEVs) from California

In 1990, California's Air Resources Board (CARB) adopted regulations to help improve state air quality. These regulations required that two percent of new cars and light trucks offered for sale in 1998 have zero emissions (in effect, that they be electric vehicles). In 2001, five percent of new light vehicles would be electric and by 2003, ten percent. In December 1995—after reviewing current electric-vehicle technology, especially batteries—CARB proposed substituting a voluntary "market-based ZEV launch" of up to 3,750 EVs in total between 1998 and 2000. (The ten percent requirement in the year 2003 would remain intact.) The vehicles would be introduced through the "Cal/Big 7 Technology Development Partnership." (The "Big 7" include GM, Ford, Chrysler, Toyota, Honda, Nissan, and Mazda.)

The details of the change were developed by CARB staff and adopted by CARB in March 1996. Most of the electric vehicles are expected to be sold in Los Angeles, Sacramento, and the San Francisco Bay area. The changes in California's regulations are expected to influence similar programs in New York, Massachusetts, and other states.

---

be cut by about 70 percent. In the northeastern United States, where more coal is burned, these emissions would be cut by about 30 percent. In the longer term, emissions could be eliminated entirely by using either renewable or other nonfossil energy sources to charge the batteries.

Reductions in Oil Use

No matter how the batteries are recharged, switching to battery-powered EVs will reduce oil consumption. The largest drop would occur where the electricity used for recharging the EVs is generated without burning any oil. But even where oil is used, the reduction would matter. If refining (oil) and transmission (electricity) losses are taken into account, an EV recharged with electricity from an oil-fired steam plant would consume 25 to 30 percent less oil than a comparable gasoline-powered vehicle.

Hydrogen-Powered Vehicles

Interest in hydrogen-powered vehicles has increased over the past decade, especially in Japan, Europe, and Canada. Hydrogen vehicles can use either variants of conventional internal combustion engines, or hydrogen fuel cells (truly emissionless technologies that give off only electricity, water, and waste heat).

The most likely long-term source of hydrogen is electrolysis of water, a CO_2-free process with energy supplied by renewable energy sources.

Although hydrogen vehicles offer the potential for long range and fast refueling, their widespread commercial use awaits creation of a supporting infrastructure—such as hydrogen pipelines, storage facilities, and refueling stations. Still, like battery-powered electric vehicles, they would form a natural link in developing a sustainable energy system, and their use would reduce oil imports, alleviate trade deficits, and cut air pollution and greenhouse gas emissions.

As a result of regulations in California (which represents about 10 percent of the U.S. vehicle market), automakers around the world are developing electric- and hydrogen-powered vehicles. While electric vehicle prices still range higher than those of conventional vehicles, as technology improves—better batteries, more efficient heating and cooling systems for EVs, and so forth—prices will drop and a global transition will begin. Already, electric vehicle fleet tests have been conducted in France, Germany, Japan, and the United States. This vehicular revolution will dramatically cut street-level pollution and—as power plants become more efficient—carbon dioxide emissions. Refueled eventually by emissionless power sources, these vehicles will serve as one leg of a sustainable transportation system.

Although electric-drive vehicles are far more sustainable than those powered by oil, important technological, economic, and institutional obstacles to their widespread introduction remain in place. The driving range of EVs is still less than that of comparable oil-powered vehicles, and refueling takes longer. Battery and fuel-cell costs—though dropping—are still high, and the need for refueling stations and other infrastructure remains. But these technological and economic shortcomings are being addressed through both federal and private research, here and abroad. At the moment, EVs are most attractive as leased commuter/urban vehicles in multiple-car families (so a shorter range is not a major problem) and in buses and trucks in use in cities (so refueling and maintenance can be centralized).

Hybrid Vehicles

Low pollution emissions, low greenhouse gas emissions, and a driving range at least as great as that of present gasoline vehicles all recommend hybrid vehicles. Still, their actual pollution and greenhouse benefits remain uncertain. Air pollution emissions would obviously depend on the engine's emissions and how much it is used. Performance deteriorates over time and if engines are started frequently—emissions are greatest when the engine is cold—their pollution benefits could be only marginal.

The relative carbon dioxide emissions of gasoline vehicles, EVs, and hybrids can be estimated by comparing the fuel efficiency of gasoline vehicles with that of similar EVs. Consider a Geo Metro, a car also sold as an EV—the Solectria Force. As a conventional gasoline vehicle, the car has a measured efficiency of about 39 miles per gallon in the city.[50] As an EV, its efficiency is about 4.8 miles per kilowatt-hour (electricity measured at the wall outlet). If the electricity comes from an oil-fired power plant, and transmission losses are taken into account, the EV would have an effective efficiency of about 47 mpg. If all the electricity for the EV were supplied by an on-board gasoline generator (with a 25 percent oil-to-electricity efficiency), the resulting hybrid would have an efficiency of about 41 mpg. In short, for a given driving distance, the hybrid would emit about 5 percent less CO_2 than the comparable gasoline version, but the EV would emit about 15 percent less than the hybrid (assuming the electricity for the EV is generated at an oil-fired plant). Of course, if the electricity came from a renewable energy source, such as solar-hydrogen, wind, or photovoltaic cells, the pollution and climate benefits would be enormous.

The performance of hybrids (or EVs or gas-powered vehicles, for that matter) could be improved by better use of aerodynamic designs, low-resistance tires, and lightweight materials. But right now, hybrids hold no clear-cut environmental benefits over battery-powered or hydrogen fuel-cell electric vehicles.

URBAN CONGESTION: THE FACTORS BEHIND THE TRAFFIC JAMS

Congestion ranks high among drivers' complaints and high on the agendas of urban transportation planners. Although nearly everyone intuitively recognizes highway congestion by the obvious symptoms—slow or stop-and-go traffic, crowded lanes, gridlock—a technical definition of the condition is surprisingly hard to pin down. The Institute of Transportation Engineers describes "congestion" as what happens when the number of vehicles attempting to use a roadway at a given time exceeds the roadway's ability to carry the load at generally acceptable service levels.[51]

Nationally, congestion is projected to grow. By 2005, delays have been projected to increase between 300 to 500 percent over 1985 levels.[52] In Los Angeles, congestion has already reduced average freeway speeds to less than 31 miles per hour (mph). According to the Federal Highway Administration

FIGURE 3-6. **Trends in Roadway Congestion in Some Major U.S. Cities**

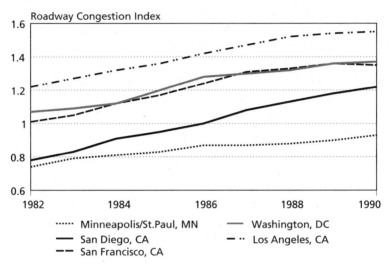

Roadway Congestion Index

..... Minneapolis/St.Paul, MN   —— Washington, DC
—— San Diego, CA   — ·· Los Angeles, CA
— — San Francisco, CA

(FHWA), congestion is serious and rapidly worsening elsewhere too. On interstates and other major roads, congestion caused an estimated 8 billion hours of delay in 1989, lowering productivity and raising the costs of shipping freight by truck.[53] By 1989, almost 70 percent of daily peak-hour travel on the urban interstate system occurred under near stop-and-go conditions, a 30 percent increase since 1983.[54] "Congestion now affects more areas, more often, for longer periods, and with more impacts on highway users and the economy than at any time in the nation's history," according to the FHWA.[55]

Congestion intensifies environmental problems, increases commuting times, raises vehicle operating costs (through wasted fuel, excess wear on parts, etc.), lowers worker productivity, boosts insurance costs by increasing the risk of accidents, and slows the delivery of business products. In 1991, the cost of those problems totaled an estimated $340 per capita in the United States.[56] Annual costs for delay and wasted fuel average $780 million for large urban areas, and far more in Los Angeles ($7.7 billion), New York City ($6.6 billion), and Washington, D.C. ($2.4 billion).[57] The congestion cost per vehicle in the Washington, D.C., area is the country's highest, some $1,400 per year (1990).[58] Although difficult to estimate, the toll of congestion on drivers' health and mental

FIGURE 3-7. **Trends in Vehicle Miles Traveled and Highway Construction**

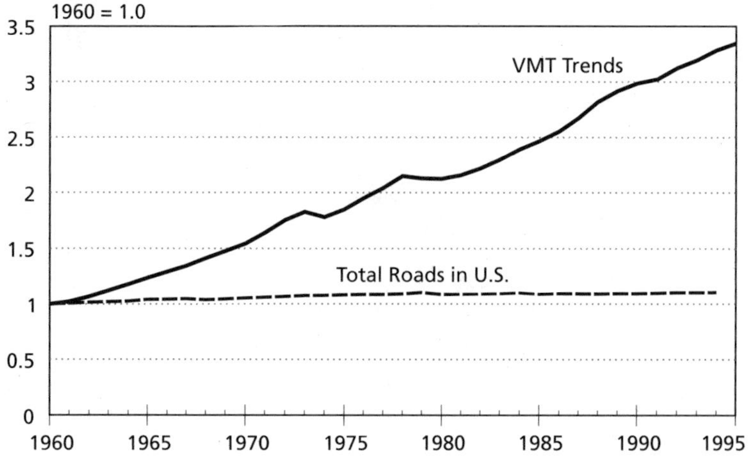

well-being is also real. Congestion is believed to increase blood pressure, frustration, and aggressive driving habits, even as it saps drivers' patience.[59]

Partly because of an illusive definition, congestion trends are hard to quantify. One possible characterization of congestion is revealed in Figure 3-6, which shows the growth in the Roadway Congestion Index (RCI) for a number of major U.S. urban areas.[60] (An RCI value greater than one indicates congestion, and the greater the RCI, the greater the congestion.) The data show that congestion is rapidly getting worse in many regions, and in some, hundreds of lane-miles of new highways would be needed just to keep congestion from getting worse.[61]

As Figure 3-7 shows, for more than 30 years the number of vehicle miles traveled (demand) has grown far faster than the road system (supply). During the oil price shocks of 1974 and 1979, vehicle miles traveled (VMT) declined for one to three years. But when gasoline prices began falling in 1976 and 1982, VMT started climbing again.

Understanding Growth in Traffic

Dealing with increasing congestion means grappling with two underlying issues. One is when and why people are driving. The question here is whether trips can

FIGURE **3-8. Total VMT by Trip Purpose Including Commercial Travel**

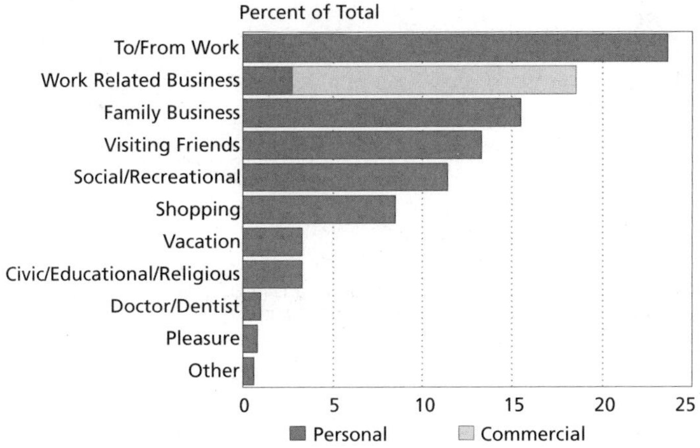

Percent of Total

be shifted to other times or days or reduced to cut congestion. Second is what factors drive traffic growth. Ultimately, the need here is to identify public policies that can slow and then stop congestion.

Two kinds of driving contribute to VMT: personal and commercial. Commercial driving includes driving a car, bus, or truck for hire to deliver goods or passengers, or working at a job that involves too much driving to report trip by trip (e.g., as a police officer in a patrol car).[62] In 1990, about 14 percent of all drivers (19 percent of all workers) fell into this second category.[63] Commercial driving constitutes about 16 percent of all VMT, much of it involving large trucks.[64] The remaining 84 percent of VMT is personal driving.

Figure 3-8 shows the breakdown of total VMT, by trip purpose, averaged over the entire week.[65] Perhaps surprisingly, only 24 percent of all VMT involve commuting to and from work. Another 3 percent is work-related driving and, as indicated above, another 16 percent is commercial driving. Almost half of all personal travel is to carry out family business (16 percent), visit friends (13 percent), enjoy social and recreational opportunities (11 percent), and shop (9 percent). But though most personal driving is not related to work, a significant fraction of these personal trips (dropping off the kids at day care, stopping at the cleaner's, etc.) are taken on the way to or from work. About one-third of commuting trips to work involve one or more intermediate stops (transportation

FIGURE 3-9. **Vehicle Miles Traveled by Day of Week and Purpose**

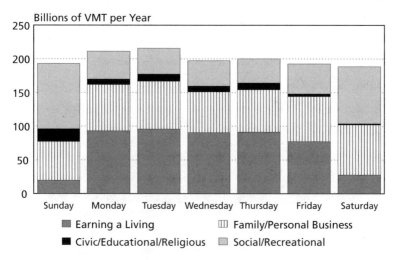

Billions of VMT per Year

Earning a Living □ Family/Personal Business

Civic/Educational/Religious □ Social/Recreational

planners call this "trip chaining"), with many more such stops on the way home than on the way to work.[66] Trip chaining also occurs in about 30 percent of non-work-related travel.

In principle, combining trips reduces overall driving, fuel use, and pollution. But almost 90 percent of the trip chaining that occurs during commuting and 60 percent of that during personal trips take place during peak hours, thus adding to congestion and excess fuel use.[67]

Could some of these personal trips during rush hours be moved to another time of day or to the weekend to reduce weekday congestion peaks? As Figure 3-9 shows, work-related travel on weekdays accounts for nearly half of daily travel, while commuting on weekends drops to very low levels, even though total daily VMT remain about the same. In short, daily VMT is nearly constant seven days a week. As a result, shifting a large number of trips from weekdays to weekends may prove difficult, if not counterproductive.

Perhaps shifting trips to less busy hours would be possible. Travel data show that, on average, the term "rush hour" is no longer really appropriate: rush hours have evolved into "rush days," characterized by high travel rates, beginning at 6 A.M. and extending to 7 P.M. The much disparaged morning rush-hour period of 6 to 9 A.M. actually accommodates no more travel (VMT per hour) than the

FIGURE **3-10.** **Total Annual Weekend Travel By Time of Day and Trip Purpose**

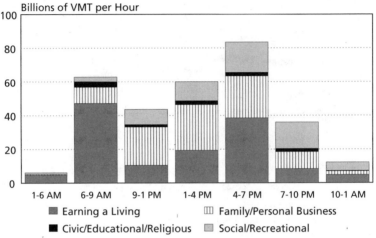

1 to 4 P.M. period and considerably less than the 4 to 7 P.M. period. Commuting to work and work-related driving (except for commercial driving) constitutes about 75 percent of the travel between 6 and 9 A.M., but only 45 percent of the travel between 4 and 7 P.M. Driving for family and personal business and for social reasons constitutes only 15 percent of the 6 to 9 A.M. travel and about 30 percent of the 4 to 7 P.M. travel. During midday (9 A.M. to 4 P.M.), overall travel (VMT per hour) is somewhat less than it is from 6 to 9 A.M. About 70 percent of the travel during this period is for family and personal business and for social and recreational purposes. Still, the weekend data make clear that travel is somewhat lower from 9 A.M. to 1 P.M. and from 7 to 10 P.M., so with appropriate incentives, some of the travel in the 4 to 7 P.M. crunch could be shifted to these other, less congested periods.

As Figure 3-10 shows, work-related travel constitutes a tiny part of weekend travel, most of which is focussed on shopping for food and other goods, getting haircuts, picking up dry cleaning, going to religious activities, visiting friends, attending movies, and so forth. Unlike on weekdays, people sleep in on the weekends and trip-taking from 6 to 9 A.M. is less frequent than during the week or during other weekend periods. From 9 A.M. to 10 P.M., travel rates are fairly comparable.

FIGURE **3-11. VMT Growth (By Year and Trip Purpose)**

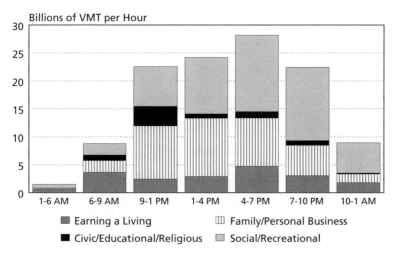

Billions of VMT per Hour

Earning a Living   Family/Personal Business
Civic/Educational/Religious   Social/Recreational

The Week in Review

Clearly, drivers have distributed their trips fairly evenly over the week. Work travel, largely a weekday phenomenon, is greatest in the morning (6 to 9 A.M.) and evening (4 to 7 P.M.). Personal trips fill out the day, while the 7 to 10 P.M. slot could be better utilized. Driving from 4 to 7 P.M. (the busiest period) could be reduced by shifting some of the commuting or personal trips by an hour or more. Weekend trip rates tend to be uniform over most of the day; weekend congestion might be reduced by persuading drivers to take some trips during the underutilized "shoulders" of 6 to 9 A.M. and 7 to 10 P.M.

Factors Fueling VMT Growth

What factors affect growth in the number of VMT, especially in urban areas? A tentative answer can help policy-makers deal better with the problem's underlying causes.

Personal driving grew by more than 80 percent between 1969 and 1990. *(See Figure 3-11.)* Over these two decades, commuting and work-related driving increased by about 50 percent. Yet, driving for other purposes grew so much that work-related travel dropped from about 42 percent of total personal VMT to about a third. Driving for social and recreational purposes increased by 48

FIGURE 3-12. **Factors Contributing to Increase in VMT (Noncommercial Travel)**

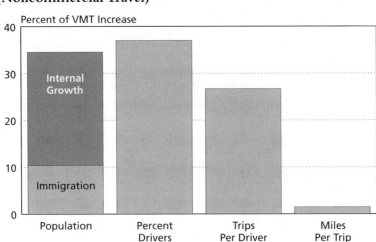

percent; for civic, educational, and religious purposes, by 64 percent; and for family and personal business, by 208 percent. This rapid growth in part reflects the declining costs of owning and operating cars: between 1969 and 1995, these costs declined, in real terms, by almost 20 percent.[68]

For a given year, VMT can be written as the product of several contributing factors.

$$\text{VMT} = \text{POP} \times \frac{\text{DRI}}{\text{POP}} \times \frac{\text{TRIPS}}{\text{DRI}} \times \frac{\text{VMT}}{\text{TRIP}}$$

The first factor on the right is the total U.S. population; all things being equal, VMT would increase with overall population growth. The second is the fraction of the population licensed to drive; the more drivers on the road, all things being equal, the more VMT. The third is the average annual number of trips (for all noncommercial purposes) per driver. And the fourth is the average number of miles traveled per (vehicle) trip. The trends in these four factors show that the average length of trips (miles/trip) fell by about 10 percent until the early 1980s, when it again reached its approximate 1969 value. The number of trips per driver (trips/driver) held steady until the early 1980s, when it began to grow; by 1990, it had increased by about 16 percent.

By far the most important factor in VMT growth has been the increase in the number of drivers. Between 1969 and 1990, the number of licensed drivers

FIGURE **3-13.** **Trends in VMT, Actual and With No Increase in the Number of Drivers**

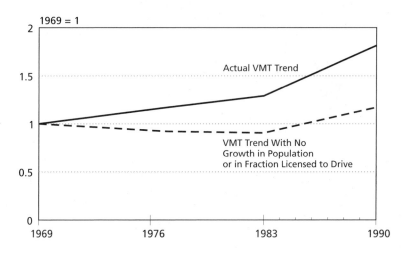

increased by 54 percent, partly because the number of women licensed to drive rose so fast—by 35 million (a 77 percent increase), compared with an increase of only 23 million (a 41 percent increase) among men.

The percentage contribution of each of these four factors to the overall growth rate of VMT from 1969 to 1990 is shown in Figure 3-12. Continued growth in population by itself leads to more drivers. Population growth (including immigration) contributed about 35 percent to the increase in VMT while an increase in the fraction of the population licensed to drive accounted for another 37 percent. In short, the boom in the number of drivers accounts for over 70 percent of the increase in VMT. Had the number of drivers held steady over this period—and all other things remained equal—total VMT would have increased by only 19 percent, rather than 84 percent. *(See Figure 3-13.)*

Trends in VMT Growth

With no new policy intervention or major technological or demographic changes, will VMT growth slow down and eventually stop? A look at the growth factors and at simple projections of current trends give us some clues.

FIGURE 3-14.   **Trends in U.S. Population Growth**

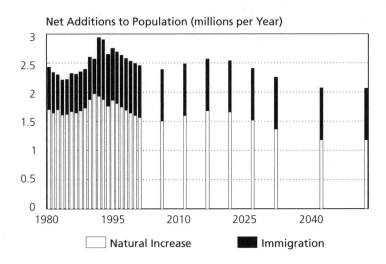

Net Additions to Population (millions per Year)

Natural Increase   Immigration

Population

Figure 3-14 shows past and projected annual population growth for the United States, broken down by rural and urban growth. If the U.S. Bureau of the Census's middle forecast materializes, the U.S. population will continue to grow, expanding by almost 2.5 million persons per year for the next half century and, probably, well beyond. By the year 2010, there would be 300 million Americans; by 2050, nearly 400 million. The 50 percent increase in the U.S. population (2050) would reflect an average annual growth rate of 0.7 percent from 1995 to 2050. Unless transportation technology or travel habits or both change dramatically, population growth will probably eventually lead to a corresponding increase in the number of licensed drivers on the road and in their number of VMT—a factor virtually absent from the debate over transportation trends.

Licensed Drivers

The fraction of the population licensed to drive continues to increase. In 1960, slightly less than half (48 percent) of the population had driver's licenses; by 1992, about 67 percent did. *(See Figure 3-15.)* It appears that the percentage of the population licensed to drive will level off at about 72 percent early in the next

BOX 3-2. Going to the City

For the past century, the U.S. population has become increasingly urbanized. The fraction of the U.S. population living in urban areas rose from 40 percent in 1900, to 64 percent in 1950, to 75 percent in 1990. Similar trends are evident worldwide; in the industrialized countries, 70 to 80 percent of the population lives in cities. With the world's urban population expected to double from 2.6 billion in 1995 to 5.3 billion in 2025, by then 60 percent of all people could live in cities compared with only 40 percent in 1990. These global trends imply that the same kinds of transportation problems affecting the United States—especially those related to air pollution and congestion—are likely to emerge worldwide.

Source: "Towards an Urban World," Highlights of the OECD Future Studies Information Base, May 1995.

century, and increases in this fraction are unlikely to affect VMT growth much after the turn of the century.

Trips per Driver

The number of trips per driver remained more or less constant from 1969 through 1983. But between 1983 and 1990, the number of trips drivers took grew by about 15 percent. By far the largest contributing factor was a two-thirds increase in the number of trips (per driver) for family and personal business. In part, such trips became more numerous because personal income grew and average household size fell from 3.15 persons in 1969 to 2.56 in 1990. Today, there are 4 million more households than the same number of people would have formed in 1983.[69] Meanwhile, vehicle occupancy has declined from 1.9 in 1977 to 1.6 in 1990, so more vehicle trips are needed to achieve the same trip goals.

Length of Trips

The average length of a trip declined between 1969 and 1983. Between 1983 and 1990, the trend reversed and the trip length grew by about 12 percent; by 1990, it was about the same as in 1969. Those trips that grew the most in number (family and personal business) barely grew in length. Conversely, the trips that lengthened (those related to earning a living, civic/educational/religious, social and recreational) did not become much more numerous.

FIGURE **3-15. Fraction of U.S. Population Licensed to Drive**

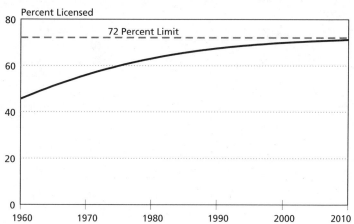

The Nationwide Personal Transportation Survey (NPTS) suggests that several factors undergird these trends.[70] First, people are moving to large metropolitan regions. In 1990, more than 75 million people lived in areas of 3 million or more, compared to 60 million in 1980. Work trips in small towns are much shorter than in metropolitan areas. Suburban trip length growth also suggests that householders are locating farther out from the city center to find affordable housing and are commuting longer distances to central city or suburban job destinations. As a result of suburban job growth, someone moving from a central city to its suburb might well have to travel 50 percent farther to work. At the same time, as jobs move to the suburbs, central city residents commute longer distances.

The increase in the number of women with driver's licenses also bears on trip length. Women who get licenses typically travel 50 percent farther to work, perhaps going the distance to find better jobs.[71] Finally, other factors, such as rising incomes—needed to buy more or better cars—encourage longer trips of all types.

Projection of VMT Growth

Using the U.S. Bureau of the Census population estimate and the growth curve in Figure 3-15, one can estimate the number of U.S. drivers there will be in the future. One can also estimate how many miles they will drive. The pattern over

FIGURE **3-16. Projected Growth in U.S. VMT**

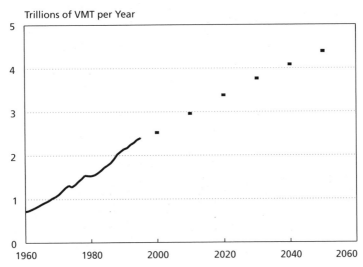

Trillions of VMT per Year

the past 40 years suggests that annual VMT will grow from the present 13,000 miles per driver to nearly 16,000. By combining these two trends—the expected number of drivers and the vehicle-miles-driven-per-driver—one can project future VMT in a "business as usual" future. As Figure 3-16 suggests, unless present trends change course, VMT in the United States will double by the middle of the next century, and still be growing at about 0.7 percent per year.

Figure 3-17 shows the relative contribution of the three factors to overall VMT growth between 1990 and 2050. If only the projected 3,000 miles per year growth in individual driving occurred over the coming decades, total VMT would grow by about 23 percent by 2050. If, in addition, the fraction of the population licensed to drive increased modestly, as expected, total VMT would increase by about a third over 1990. Finally, if the expected increase in population is added, total VMT would increase by over 100 percent. In short, by the year 2050, VMT growth would be propelled largely by population growth, with the two other factors contributing only modestly.

OPTIONS TO REDUCE TRAFFIC

Compared with reducing vehicle emissions, controlling traffic over the coming decades will be far more difficult: congestion can only get worse as the

FIGURE **3-17. Dependence of VMT (for the year 2050) on Population, the Fraction Licensed to Drive, and Per Capita Driving**

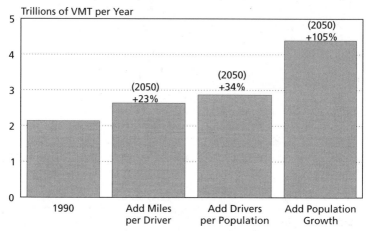

ever-increasing use of motor vehicles bumps up against a virtually fixed supply of roadways to handle traffic. Moreover, the growth in vehicle use is driven by population growth and long-term demographic changes, including greater individual wealth, smaller households, and urbanization. Obviously, something must give if Americans are to avoid being increasingly trapped in gridlock.

The traditional response to congestion would be to build more roads, and some additional lanes will no doubt be added, particularly to some highly clogged major urban freeways. Yet, the prospect is slim that new road building can keep pace if urban VMT continues to grow, perhaps doubling over the next half century. Then too, as experience attests, expanded roadways soon attract more cars and trucks, making much of the anticipated relief from congestion temporary. According to University of California professor Mark Hansen, a 1.0 percent increase in lane miles induces a 0.9 percent increase in VMT within five years.[72]

What, then, are the prospects for halting or at least slowing VMT growth? Several of the trends that have contributed to recent VMT growth are unlikely to persist long. The trend in licensed drivers, for example, suggests that by the year 2010 most people eligible to have driver's licenses will have them; so after that point the number of licensed drivers will probably grow no faster than the population. If the Census Bureau's projections that the U.S. population will grow well into the next century at 2 to 2.5 million per year prove right,[73] the number of drivers will continue to grow at 1.4 to 1.8 million per year on average, at least through 2050 or so.

Many possible approaches can be taken to reduce traffic congestion. Some involve changing driving habits and patterns of trip making. At least in principle, these could be initiated fairly soon. A second approach involves adopting new technologies to increase the carrying capacity of existing roads. A third approach involves alternatives to the car as the sole option for meeting daily needs. What may work best is some combination of these three approaches.

Changing Driving Patterns

The primary goal of changing driving patterns would be to reduce the number of vehicles on the road during peak periods. Effective measures would cause drivers to reschedule their trips from peak to off-peak periods, to combine trips, to ride-share, to change their mode of travel, or even to forego some trips. Such behavioral changes are best sparked by such economic measures as increasing fuel prices (while perhaps reducing income or social security taxes), changing parking fees, or adopting congestion pricing.

Increasing Road Capacity

ITS Programs

Congestion can be temporarily relieved by introducing new technology. The Intelligent Transportation Society of America (ITS America), a Congressionally mandated public-private organization that coordinates the development of new transportation technology, contends that "A broad range of diverse technologies, known collectively as intelligent transportation systems (ITS), holds the answer to many of our transportation problems" and "will save lives, save time, and save money."

The Intermodal Surface Transportation Efficiency Act (ISTEA) authorized $600 million for ITS development over six years. Important systems under development include Advanced Traffic Management Systems (including freeway ramp metering, arterial signal controls, and other technologies that collect real-time data on congestion), Advanced Traveler Information Systems, Advanced Vehicle Control Systems, Commercial Vehicle Operations, and Advanced Public Transportation Systems.

Although ATMS are the basic building blocks of ITS, Advanced Traveler Information Systems (ATIS) also provide travelers with information on how to reach their destination via private vehicle or public transportation, or a combination of both. On-board navigation systems are key. Cars equipped with computers can transmit their exact location to global positioning satellites; the location then appears on a map displayed on a dashboard computer. ATIS alerts drivers to accidents and tie-ups, weather and road conditions, optimal routes, recommended speeds, and lane restrictions.

Advanced Vehicle Control Systems enhance the driver's control of the vehicle to make travel safer and more efficient. Intended to help prevent accidents, these systems include collision-warning systems. In advanced automated roadway systems, vehicles would automatically brake or steer away from a collision.

Commercial Vehicle Operations are already widely used in commercial fleets. Examples include automated location systems and two-way radios to link drivers with their dispatch centers and automated vehicle-identification systems (AVI) to speed toll collection.

Advanced Public Transportation Systems will inform users of mass transit and ride-sharing options of schedules and costs and provide real-time ride-matching for car pools. One option, the Smart Card, will enable consumers to board transit vehicles and pay fares and parking fees, all without carrying cash.

ITS technologies are in various stages of development. In October 1994, the Department of Transportation awarded General Motors a $160-million contract to develop an Automated Highway System. The daunting goal is fully automated hands-off driving in dedicated lanes by linking the steering, brakes, and throttles of specially equipped vehicles through sensor and communications devices to the highway. According to proponents, cars would be bumper to bumper, all moving safely at approved highway speeds.[74]

Oldsmobile's Guidestar system is a computer-based route-guidance system that gives drivers turn-by-turn directions to selected destinations. Drivers get detailed guidance on a small video screen (mounted near the car's instrument panel) and simple voice instructions. Priced at about $2,000, Guidestar became available nationwide in 1996. Other examples of vehicle-installed technology include obstacle detection and collision-avoidance systems for cars, trucks, and buses that can be integrated into cruise-control systems.

In the San Francisco area, a new system dubbed TravInfo was launched in 1996. TravInfo gathers, organizes, and disseminates information about traffic and road conditions, public transit routes and schedules, ride-sharing, park-and-ride

> **BOX 3-3.**
>
> We cannot build our way out of congestion, and providing better information to drivers so that they can potentially alter their travel patterns—time, mode, et cetera—will help transportation professionals better manage the system and help lessen the effects of congestion—such as bad air quality, accidents, and short tempers.
>
> ...Rodney Slater, Secretary of Transportation

and bikeway facilities, and van and taxi services for disabled travelers. All TravInfo information is free to the public.[75] Like other advanced traffic management systems, this one collects, uses, and disseminates traffic information from sensor readings, police reports, roadside mounted video cameras, and even call-ins.

The safety benefits of some advanced ITS systems are uncertain. With one system, involving electronic traffic controls at intersections, FAST-TRAC, the accident rate went up by about 21 percent. According to Brian O'Neill, president of the Insurance Institute for Highway Safety, collision-avoidance detectors and navigation systems "offer further potential for distraction from driving."[76] Clearly, the overall safety impacts of ITS will depend on how and where they are used.

Some elements of ITS make good sense. Some simple systems that provide real-time information to travelers can increase transportation productivity and alleviate congestion at reasonable costs. The net benefits of other applications—such as the automated highway systems—are less clear. These could prove very costly to build, requiring the installation of guideways in major roads. Nor do they get to the heart of congestion. Increasing the density of cars on the road would increase traffic on side streets near highway exits, for instance. More important, systems would have to be very reliable. The failure of an electronic component—for whatever reason—that guides scores of speeding cars packed closely together could bring havoc and death. Finally, even if these automated roads worked, they would buy only a few years of breathing room until they too were congested.

BOX 3-4. State Land-Use and Transportation Initiatives

Taking the Portland Trolley

Portland Oregon's light-rail system is considered a success story for land-use and transportation planning. Operating since 1986, the system has revitalized downtown Portland and is expected to handle the predicted 50 percent population growth through the next 20 years. Portland adopted several simple principles to guide urban growth:

- High-density growth, spurred by new re-zoning laws, would be nearest the transit stations
- Parking near the transit system would be limited
- No new road construction would be allowed in downtown

Local governments played an important role in encouraging high-density development by

- Reclassifying zones to permit only mixed-use development
- Limiting parking spaces
- Improving pedestrian mobility
- Adopting a traffic impact tax

In return, the light-rail system would accommodate growth. The results include a 50 percent increase in jobs in downtown since 1975, improved air quality, and control of traffic congestion. The trolley system is a success: 40 percent of downtown Portland's labor force travels to work on it and 66 percent of the businesses located near stations believe they benefitted from their proximity to it. Real estate values around transit stations have increased much faster than the nationwide rate. Portland voters have approved expanding the project from 15 to 58 miles.

The Pittsburgh Express

Pittsburgh's approach to reducing congestion is dedicated freeway lanes for buses. The results? Lower commuting costs, shorter commuting times, and less stress.

Providing Alternatives to Driving

Avoiding the Commute, Sometimes

Telecommuting is working at home or very near home.[77] Most telecommuters typically work at home one or two days per week. Benefits include reduced travel and energy consumption, less stress, higher productivity, improved management skills, higher office morale, and greater job access for disabled persons. An

BOX 3-4 continued.

The idea has caught on: along one corridor, ridership has grown by 50 percent—from 20,000 to 30,000—since the lanes opened in 1983. Construction of a new busway connecting the airport with downtown is under way and a plan is in the works to expand the East Busway to further increase ridership by 2005.

Riding the Bus in Aspen

Seeing its air quality deteriorate, Aspen responded by developing a multi-faceted public transport program. The Aspen plan consists of restricted parking downtown, double bus service, free parking and shuttle service at an airport satellite parking lot, and a carpool program to service areas not within range of transit. In only ten months transit ridership increased 30 percent. Now tourists find it easier to park and walk and enjoy the city's village atmosphere.

Parking at Work in California

In 1992, the legislature passed a bill affecting workplace parking in California. Under the new law, eligible employees could receive tax-exempt workplace parking benefits of up to $155 per month. The goal was to provide a cash benefit for those who forego their parking benefits, thereby rewarding carpooling or public transportation use. Workers can choose between a taxable cash allowance and a taxable parking space. Researchers predicted that 20 percent of commuters who drive alone would opt for alternative transportation. As a side benefit, the U.S. Treasury would gain $1.2 billion in tax revenue. Yet, so far the parking law is rarely enforced because both employers and employees would have to pay taxes on previously tax-exempt parking spaces. Because of the air pollution benefits, the California Air Resources Board is currently encouraging its enforcement.

Source: Various issues of *Progress,* Surface Transportation Policy Project, Washington, D.C.

estimated 4 to 6 million people telecommute—between 3 and 5 percent of the U.S. adult workforce, or about the same number of Americans as take mass transit.[78] The great bulk of telecommuters work in small companies (fewer than 100 employees).[79] The Department of Transportation projects that as many as 15 million workers could telecommute by 2002—about 10 percent of the workforce or 17 percent of all so-called information workers. Up to 30 million could telecommute by 2010.[80] If this scenario materialized, upwards of 4 percent of commuting VMT could be eliminated with an annual savings of about 4 million metric tons of carbon.[81] Right now the number of telecommuters is growing at about 20 percent per year.

The short-term impacts of telecommuting on vehicle travel appear positive. According to a study of a Californian pilot project, total VMT by telecommuters (both work and non-work-related) can decline by 75 percent on telecommuting days, while peak-period trips declined by 60 percent. Freeway travel by telecommuters declined by 90 percent.[82] Telecommuters make fewer peak-hour trips and stick closer to home when they do travel but continue their peak-hour nonwork trips. The long-term impacts of telecommuting on where people live and work are uncertain.

The potential benefits of telecommuting on congestion obviously depend on how many workers seize the option and how other drivers react to the decrease in traffic. The direct benefits of less commuting, like those of HOV lanes, are somewhat offset by so-called latent demand—new vehicles take their place. Telecommuting may also allow people to live farther from work, inviting sprawl, more driving, and higher fuel consumption.[83] These offsetting effects could cancel half of the VMT reduction from telecommuting. According to researchers at Oak Ridge, even with 30 million telecommuters in 2010, reductions in delay, energy consumption, and air pollution emissions are likely to be no more than a few percentage points of the total.[84]

Many workers do not have the freedom to telecommute. Their jobs may require them to be on the spot, management may not approve, or technological limitations may stand in the way. According to the Office of Technology Assessment, providing modest tax incentives to businesses and telecommuters and making local zoning changes to cut parking space in new buildings would encourage more telecommuting.

Reducing Driving Through Urban Design

Some planners believe that land use, zoning, and transportation policies that encourage low-density residential development and that separate residential from commercial and retail development contribute to metropolitan traffic problems. To be sure, policies that are focused almost exclusively on motor vehicles for travel make nonmotorized travel difficult, if not hazardous.

Can strip shopping centers and other developments along many of our suburban arterial highways be modified to relieve these symptoms? Yes, says Rich Untermann, professor of urban design and planning at the University of

Washington.[85] Looking for ways to gradually redevelop today's commercial developments into more attractive mixed-use centers, Untermann proposes concentrating most retail uses along highways into "centers" located about a mile apart. These nodes would be "denser, mixed-use areas with shopping, work and housing all intertwined." Over time, the land along the highway between the centers would be converted from strip retail uses to apartments, condominiums, and office buildings, which generate less traffic than retail uses. Higher density housing in and around the centers and new pedestrian and bicycle paths would permit residents living in the surrounding lower density neighborhoods to reach transit, shopping, work, and recreation without having to drive.

Besides shops, centers might contain offices, a library, child-care facilities, schools, meeting rooms, and a public transportation or transit stop. Untermann believes that the elderly and handicapped, as well as many nontraditional families would feel at home in these higher density developments. Also, both buses and van pools make more sense where population is dense. Clearly, restructuring our metropolitan areas to encourage more walking and bicycling and less motor travel, as Untermann urges, would take time and strong leadership at all levels of government.

New development patterns could help metropolitan areas avoid many of the transportation problems they now face. According to community planners Andres Duany and Elizabeth Plater-Zyberk, entirely new towns and villages could be laid out on grids the way that older communities were built, rather than in the contemporary hierarchial system of feeder streets and major arterial highways.[86] These "neo-traditional" town centers would have wide sidewalks and on-street parking as a buffer between pedestrians and traffic. Buildings would abut the streets with parking behind them, if needed. These centers could also serve as public transportation nodes, further reducing the need to drive. As in older traditional towns, residential and commercial buildings—of comparable height and size—would be intermixed.

The concepts underlying neo-traditional towns can also be applied to redeveloping major urban areas, including old abandoned shopping malls.[87] Untermann and others argue that—though the institutional barriers are formidable—priority should be given to redeveloping urban areas, increasing centers' density rather than developing new lands. Careful planning of new centers will be required to take present housing and travel patterns, transit possibilities, and feasible walking distances into account.

BOX 3-5.   Some Ingredients for a Pedestrian-Friendly Environment

- Wide sidewalks to protect pedestrians from traffic and provide access to activities.
- Shortcuts through long blocks, diagonals through parks or shopping areas, and steps for steep grades.
- Crosswalks that are well lit and marked with stripes and signs; raised medians for streets with more than two lanes.
- Trees, benches, and bus shelters to make sidewalks inviting.
- Walking access to such goods and services as grocery stores, hardware stores, drug stores, banks, and restaurants.
- Buffering of pedestrians by on-street parking, wide sidewalks, and planter strips.

Source: Rich Untermann, "Reshaping our Suburbs, Linking Land Use and Transportation to Better Serve Pedestrians, Bicycles and Transit," May 1995.

The neo-traditional planning concepts espoused by Duany, Plater-Zyberk, Peter Calthorpe, and other town planners reflect far more than transportation considerations. Imbedded in their approach is a belief that neighborhoods should be socially alive and active, "user friendly" for children and the elderly, and amenable to walking and bicycling as well as driving. In their visions, the automobile is still around but plays a decidedly less dominant role than in most contemporary American developments. In short, there should be a sense of community that is often absent from much of today's suburban sprawl.

Not all planners agree that higher density neo-classical towns are really what homeowners want. New York planner Oscar Newman describes neo-classical villages as "retrogressive sentimentality."[88] And economist Peter Gordon of the University of Southern California believes people are more or less happy living in subdivisions: "If you look at how people really want to live in this country, suburbanization is not the problem, it is the solution."[89]

Given these starkly differing assessments, perhaps the best that can be said is that neo-classical developments offer one more way to lessen dependence on the automobile and break up congestion. In any case, the concept is applicable primarily in new developments, so it offers little relief for the bulk of urban America.

Mass Transit, Paratransit, and Personal Rapid Transit

By shifting riders from cars into trains, buses, and other vehicles, public transport can reduce vehicle congestion. There are nearly 7,000 public transit systems operating in the United States providing passenger services on 65,000 diesel and electric buses, 11,000 heavy-and light-rail vehicles, 4,500 commuter train cars, and 100 ferry boats.[90] Together, these vehicles account for over 39 billion passenger miles traveled each year, but only 2.5 percent of all passenger miles of travel in 1990.[91]

Why do people take public transit? And what are the trends in transit use? Commuting to work is the largest single use (about 41 percent), followed by school and church (mostly school) attendance (22 percent). Family, personal, social, and recreational trips together make up the remaining 37 percent.[92] The percentage of public transit trips in all of these categories continues to decline. According to University of California (Berkeley) professor Melvin Webber,

> *"Since 1964 the federal government has spent more than $100 billion to improve and expand transit service, and yet trips to and from work in the urbanized areas, the ones widely believed to be most amenable to transit, have been falling even more dramatically: from 25 percent of work trips in 1960, to 14 percent in 1970, to 10 percent in 1980, down to perhaps 5 percent today. In the suburbs, transit use is down to about 2.5 percent of trips to work. Nationwide, people use transit for only 2 percent of their urban trips."[93]*

These declines have occurred in cities and rural areas, and especially in suburbs. Riders from all user groups have jumped ship: women, the very young and old, the rich and the poor, and the in-between. Reductions have been greatest in the Northeast, where transit use is greatest.

The decline undoubtedly stems from many factors. These include the time saved using a personal vehicle rather than transit, flexibility in allowing side trips while commuting, the perceived personal safety advantages of traveling in a locked car, and the conveniences of door-to-door service without waiting for scheduled service or putting up with crowds. Additional factors are a decline in real fuel prices and an increase in the quality and longevity of used cars, which make reliable, affordable auto transportation available to lower income families.

A host of subsidies enjoyed by car owners also encourage driving. A World Resources Institute report estimates the costs that cars and trucks impose on society—but that drivers don't directly pay—at upward of $300 billion per year (an average of more than $2 per gallon of gasoline and diesel fuel consumed by the nation's cars and trucks).[94] Such enormous subsidies make driving seem cheaper than it really is—to the detriment of public transportation, bicycling, and walking—and invite congestion, road deterioration, smog, and dependency on foreign oil. Examples include:

◆ *Road Construction and Repair.* According to the WRI report, only 60 percent of the $74 billion collected in 1989 came from user fees, such as gasoline taxes, registration fees, and truck charges. The remainder (over $29 billion) was raised from property taxes, general funds, and other sources.

◆ *Free Commuter Parking.* About 86 percent of U.S. workers drive to their jobs and more than 90 percent of these commuters enjoy free parking. This benefit makes solo commuting almost irresistible.

◆ *Air Pollution.* Despite major reductions in pollution from new motor vehicles over the past 20 years, cars and trucks still contribute heavily to smog and acid rain, which cost Americans tens of billions of dollars per year in health and materials costs.

◆ *The Military Costs of Protecting the Persian Gulf.* According to a recent analysis, the U.S. Defense Department spends about $50 billion per year[95]—apart from the costs of actually waging a war—maintaining a military presence in the Middle East primarily to ensure access to Persian Gulf oil. Oil consumers don't pay those costs.

With this array of subsidies for auto driving, the construction of many new heavy- or light-rail systems seems unlikely. Expensive to build and operate, they may not be flexible enough to meet present consumer transportation needs either. Subway systems are best suited to carrying passengers to central downtown locations—while current development trends across the country are toward decentralized commercial and retail construction in the suburbs.

Even so, today's mass transit systems could help revitalize older American cities. With creative design and planning, neighborhoods around the stations of older transit systems could be redeveloped to provide attractive mixed development, including retail stores, commercial buildings, and high-density housing.

Public transport more flexible than rail could cut congestion down to size. Vans that can carry half a dozen people could serve designated areas and carry passengers from their homes to near-common destinations. Computer dispatched in response to a phone call, the vans could provide reasonably flexible service for less than taxis cost, reducing vehicle use and providing viable transport for the elderly, the young, the handicapped, and other nondrivers. Such van service would require only a modest capital investment to initiate. Unfortunately, though, vans can get caught in traffic. Another form of transit soon to become commercially available, personal rapid transit (PRT), could help us largely avoid this problem. A PRT system consists of small, dedicated, computer-guided electric vehicles operating on their own electrified guideways, either elevated or underground. Capable of carrying four seated passengers, the vehicles—spaced about 100 feet from each other—will travel at about 30 mph. A single guideway can move more than 10,000 passengers per hour. PRT vehicles are small, so support structures and tunnels for the guideways are too. Consequently, PRT systems can be installed much more quickly, cheaply, and less disruptively than mostly underground systems can. They are also less prone to vandalism and present no barriers to neighborhood movement. On balance, for a comparable investment, many more PRT stations and guideways than subway systems can be constructed so service can be extended to a wider area. Indeed, no user should have to walk more than about ten minutes to a PRT station.

Important progress in PRT systems-development has occurred over the past 20 years. Detailed engineering and construction plans have been completed by the Massachusetts-based Raytheon Company, which already has a small experimental PRT system running in Marlborough, Massachusetts. Assuming no major technical glitches, the first commercial system will be constructed in Rosemont, Illinois, near O'Hare International airport to serve local hotels, a convention center, businesses, and the local train station.

Unlike trips on traditional rail-based transit systems, those on PRT vehicles will be nonstop from origin to destination. There will be no waiting time for a vehicle off peak; at rush hour, passengers should not have to wait more than three minutes. PRT eliminates the need to change from one transit vehicle to another—a major disadvantage of conventional systems.

A PRT system could displace a large volume of automobile traffic. The total capacity of a single lane of PRT varies directly with the average vehicle speed and inversely with the distance between the vehicles. In other words, the higher the vehicle speed and the shorter the spacing between vehicles, the more vehicles will pass by in a given period. At initial design capacity, a single PRT guideway could carry about 1440 vehicles per hour.

Raytheon's PRT system will have redundant control, communications, and drive systems. Instead of tickets, passengers will use magnetically encoded cards.

Upon entering the station, passengers insert cards into a ticket machine that encodes them with the selected destination and subtracts the fare. An empty car would then arrive and each passenger (or self-selected group of four or fewer) would enter and be taken without stopping to the final station. No turnstiles will be needed in the stations so people in wheel chairs can easily use the system.

One of PRT's most attractive features is its expected favorable economics: fare box revenues should cover all operating and maintenance costs. If so, this could mark an end to transit subsidies, which typically amount to 50 percent or more of operating costs.

Confronting Population Growth

The pressure of population growth helps fuel virtually all problems related to sustainability. Yet, largely because the issue is politically thorny, there is little public discussion of population in the debate over resource consumption and environmental deterioration. However complex the issue, a few words on its pivotal role in transportation are in order.

As an underlying driver of consumer demand, population growth plays a major role in sustainable transportation. Had the United States reduced population growth, the nation would have consumed less fuel, contended with less air pollution, depended less on imported oil, battled less congestion, and enjoyed a longer breathing period to develop alternatives to oil-powered vehicles.

As population increases, solving many fundamental problems related to improving the quality of life becomes more difficult.[96] The crunch comes because pressures to reduce per capita resource consumption to satisfy the requirements of more people utilizing limited or declining resources are inevitable. All resources, even renewable ones, are finite with respect to the rate at which they can be used, and maintaining adequate supplies of clean air, clean water, farm lands, parks, open spaces, fuels, and material goods becomes more challenging, technologically and financially, as the pressures of growing consumption mount. In addition, growth in human numbers has enormous momentum. After all, except for relatively brief periods of global wars and pandemics, the world's human population has been steadily increasing for thousands of years. Stabilizing or reducing population is clearly a long-term effort measured in decades if not centuries.

POLICY RECOMMENDATIONS

Continuing the United States' near-exclusive reliance on oil-powered vehicles as the backbone of U.S. ground transportation is not sustainable. If Americans maintain today's total commitment to oil, they cannot address the threats to climate and national security from reliance on imports (increasingly from the Persian Gulf), further global warming as CO_2 concentrations grow, and the deterioration of urban air pollution from carbon monoxide, ozone, and acid precipitation. Similarly, staying wedded almost exclusively to motor vehicles to meet ground transportation needs will condemn the nation's metropolitan areas to deepening traffic congestion for those who drive and to increasing immobility for those who don't.

Clearly, various policy approaches can be used to stimulate technological development, change driving behavior, make a transition to more sustainable land-use patterns, introduce zero-emission vehicles, and reduce U.S. population growth. These include:

◆   A market approach based primarily on "getting the prices right," which means reducing or eliminating subsidies that distort the true market value of goods. If transportation-related prices were revamped to more accurately reflect the total social costs of providing goods and services, the transportation system would run more efficiently. Consumers would purchase cleaner, more efficient vehicles and would drive them more efficiently. And they would make more economically rational decisions about public transit, ride sharing, bicycling, and telecommuting, as well as where to live and work. In practice, this market oriented approach might take the form of congestion, pollution, or carbon taxes. Accurate pricing is necessary but by itself may not be enough to solve our transportation problems.

◆   A "command-and-control" approach based mainly on government regulation. The cornerstone of much environmental protection, this approach yields fairly predictable results. Examples include pollutant-emission standards for mobile sources (motor vehicles) and stationary sources (power plants and factories) and CAFE (Corporate Average Fuel Economy) fuel-economy standards. California took this approach in its original ZEV (Zero Emission Vehicle) mandate.

◆   "Subsidies" to encourage certain behavior or the use of certain technologies or products. Subsidies appeal to U.S. consumers because they feel they ought to be rewarded for doing the "right thing," rather than punished by higher prices or taxes for behavior once considered acceptable. Examples of subsidies include reduced federal motor fuel taxes on ethanol and compressed natural gas when used as motor fuels, and a 10 percent federal tax credit on EV purchases. Helping to create markets through government purchase programs—sometimes called "market pulls"—is another form of subsidy. Pulling the market brings down prices by achieving economies of scale in production.

◆ "Technology Research and Development" with support from federal funding that spurs innovation. Federal co-funding for developing new automotive technology and for advanced batteries for electric vehicles exemplifies this approach.

◆ "Public education." This approach—sometimes called social marketing—relies primarily on advertising and communication techniques to convince consumers to change their behavior. Examples of effective programs include those advocating reduced drinking before driving and greater use of safety belts.

A fully integrated campaign to encourage a more sustainable transportation system might rely on a combination of these policy tools. But since the public strongly opposes new taxes, federal and local policies will most likely rely more on financial subsidies and regulations. While this approach may improve the mix of vehicles we drive, it also subsidizes excessive driving and passes some of those costs to innocent parties.

Policies to Encourage Sustainable Fuels

Policies to make motor vehicles more efficient and to encourage the introduction of zero-emission vehicles (ZEVs) would have both short- and long-term elements. For the next decade or so, relatively few ZEVs can be put on the road, so U.S. efforts should emphasize improving new vehicle fuel efficiency, reducing emissions from the current fleet, and cutting the growth in vehicle miles traveled.

In an ideal world, gasoline consumption could be reduced by some combination of the policies described below. No doubt gasoline and diesel fuel demand could be forced down along with emissions of associated air pollution and greenhouse gases. With less driving, congestion would also ease. But these goals will elude us unless fuel prices are adjusted to more fully reflect unpaid social, environmental, and security costs.

Introducing cleaner (reformulated) gasoline to cut pollution emissions has helped, as will continuing inspection and maintenance programs, supplemented with the use of roadside instruments that can identify high-emitting vehicles.[97] But, unless the American public understands and accepts the necessity for these changes—a herculean education task—they are unlikely to be adopted.

The transition to ZEVs will take three to four decades, so it should begin as soon as possible. Measurable benefits could be realized within 15 years if electric-drive vehicles were phased in aggressively. If the original California ZEV mandate had been implemented in both California and the northeastern United States—as once seemed possible—upwards of 15 million light-duty electric-drive vehicles could have been on the road by 2010 (though they would still account for less than 1 percent of the total.)[98] This many ZEVs would displace around 500,000 barrels per day of imported oil and reduce national CO_2 emissions by more than 8 million metric tons of carbon per year. If current goals for battery performance and cost can be met, these benefits could be realized for about what the United States is willing to pay now to achieve the same benefits using alternative means.[99]

Politics permitting, the policies summarized below would help pave the way for electric-drive vehicles.

Increase the Price of Motor Fuels

Higher gasoline and diesel fuel prices could be achieved by increasing fuel taxes. The justification would be to make prices more fully reflect the full costs of driving. The taxes need not raise the nation's overall tax burden: additional taxes on fuels could be used to offset, for example, social security or healthcare taxes, or to repair or pay for new transportation infrastructure. A common objection to higher fuel taxes is that they hurt low-income families, who spend proportionately more of their income on energy. But recent analyses suggest that, for gasoline at least, the tax is not as regressive as many think. MIT economist James Poterba argues that while the poor do spend a sizable fraction of their income on gasoline, overall household expenditures are a better index than income for measuring the regressiveness of higher gasoline taxes. According to Poterba, such benefits as food stamps, Medicaid, and other programs give low-income families more purchasing muscle than their total income would suggest. If this hidden power is taken into account, he found, the poorest 10 percent of households spend less than 4 percent of their total outlays on gasoline—less than any other income bracket except for the very wealthy.[100] As it turns out, the greatest burden

FIGURE **3-18. Relation Between Gasoline Price and Annual Auto VMT (1970-1992)**

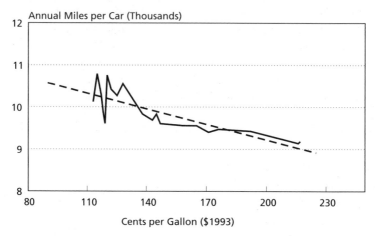

falls on middle-income households, where 4 to 6 percent of spending is for gasoline, making the tax politically hard to sell.

Higher gasoline and diesel-fuel prices would raise driving costs and help reduce traffic growth. If prices got steep enough, some unimportant trips would be skipped or combined into a single outing. Drivers would become more sensitive to traffic conditions and congestion and schedule their trips to reduce fuel costs. They might speed less, thereby improving fuel economy and reducing gas consumption and accident rates. Overall, the road system would probably be used more efficiently and even more safely. Higher gasoline prices would also make the use of alternatively powered vehicles, such as ZEVs, more attractive, thus greatly reducing pollution and CO_2 emissions.

As Figure 3-18 shows, in years when fuel prices were about $1.10 per gallon (all prices in constant 1993 dollars), motorists averaged around 10,300 miles per year. In years when fuel prices were higher, near $2 per gallon, for instance, driving dropped around 10 percent to about 9,200 miles. The data shows that a 1 percent increase in the real price of gasoline (currently about $1.25) would almost immediately reduce the fleet's annual VMT by 0.1 percent. Similarly, a $0.50 per gallon increase in the price of gasoline might cut annual mileage by about 6 percent.

FIGURE **3-19. International Comparison of Gasoline Prices and Annual Auto VMT (1991)**

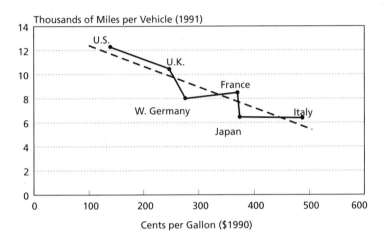

The longer term effects of higher fuel prices on driving are harder to ascertain. Initially, drivers would probably drive less. As their cars age, drivers would most likely buy more efficient vehicles, reducing driving costs and offsetting somewhat the impacts of the higher fuel tax. With correspondingly lower driving costs, drivers would be tempted to drive more than if they still owned their inefficient cars. In Europe and Japan, where fuel prices have been high for years, gasoline prices range from $3.28 per gallon in Luxembourg to $4.66 in the Netherlands.[101] *(See Figure 3-19.)* Average annual mileage (per car) in the United Kingdom (UK) is about 15 percent less than in the United States; in Italy, where gas costs about $4.39 per gallon, annual mileage is about half what it is here.

Just how applicable is overseas experience to the United States? Certainly, gasoline prices are by no means the only difference between the transportation systems in the United States and other countries. The widespread availability of good public transport, more pedestrian- and bicycle-friendly roads, higher population densities, and mixed residential and commercial zoning all contribute to a reduced need for driving in many countries. Even so, over the long term, an additional $0.50-per-gallon gasoline tax could reduce annual VMT per vehicle by 7 percent, and a $1-per-gallon additional tax would cut VMT per vehicle by about 15 percent. While higher fuel taxes would reduce overall driving modestly, these reductions would not necessarily occur at peak driving periods.

Many benefits of higher fuel taxes could be captured by "pay-at-the-pump" insurance. Under these proposals, vehicle liability insurance would be paid for through an extra premium levied at the gas pump. The idea is to turn a fixed cost of driving (insurance is usually bought annually) into a variable cost that would increase or decrease in direct proportion to how much a person drives—one component of the actual risk of accidents.

By raising fuel prices, pay-at-the-pump insurance would encourage people to purchase more efficient vehicles, discourage unnecessary driving, reduce the overall cost of automobile liability insurance, and guarantee that everyone on the road has such insurance. Unlike a simple increase in the fuel tax, Pay-at-the-pump, Private, No-fault (PPN) auto insurance should be financially attractive to drivers since it is expected to substantially cut their annual insurance costs. Under one such scheme, the premium on gasoline would be about $0.40 per gallon.[102] The money would go into a fund from which claims would be made on a no-fault basis.

Creating a Market for Electric-Drive Vehicles Through Fleet-Purchase Programs

The purchase or lease of EVs for utility and government fleets would definitely help establish ZEV markets. Right now, governments and utilities operate 23 percent of all fleet vehicles. Cars in these fleets average 55-58 miles per day, and light trucks about 56-70 miles per day.[103] Of the approximately 3.2 million government vehicles in federal, state, and local fleets,[104] 1.1 million are passenger vehicles (cars, station wagons, and jeep-like vehicles), while well over 800,000 are utility fleet vehicles. If 10 percent of those vehicles were replaced each year (assuming each vehicle has a 10-year life) the potential market for electric-drive vehicles would be 400,000 new vehicles annually—far greater than present ZEV production capacity. A market this big gives great leeway in designing a government/utility purchasing program.

Government fleets are especially promising candidates for electric-drive vehicles, and several federal laws support fleet purchases. Under the Clean Air Act Amendments of 1990, some 150,000 alternative-fuel vehicles (AFV) must be sold in the pilot state of California beginning in 1996 and 300,000 per year by 1999. The law also requires other states that don't meet clean air standards to introduce AFVs. Under the executive order that implements the Energy Policy Act of 1992, one-third of new vehicles purchased by federal agencies in fiscal year 1997 must be AFVs. In 1998, half must be AFVs, and in 1999, three-fourths. In meeting these goals, agencies can count an electric vehicle as two AFVs.

Almost 90 percent of government gasoline-fueled fleets and 80 percent of utility fleets are refueled on-site—a plus for ZEVs. Given the national importance of introducing emissionless vehicles, utilities and governments should explore the use of EVs in their car and light-duty truck fleets. Because of the questionable long-term benefits offered by vehicles burning methanol, ethanol, and compressed natural gas, government fleet purchase programs should focus mainly on ZEVs—battery and hydrogen fuel-cell electric cars, trucks, and buses.

BOX 3-6. What Other Countries Are Doing

Costa Rica. Placing in operation 15 electric buses in San José, along with 50 electric cars and trucks and 50 electric scooters.

France. Offering one thousand $3,300 government payments to local authorities that purchase EVs. Operating 300 EVs and 130 recharging spots in Paris. Planning more EVs and charging facilities.

Germany. Offering a five-year EV tax exemption to EV purchasers.

Japan. Offering reduced value-added, purchase, and annual taxes for EV purchasers. (The Japan Electric Vehicle Association has an EV-leasing program financed by public subsidies.)

Mexico. Planning to purchase EVs to help combat air pollution.

The Netherlands. Providing an accelerated-depreciation tax break for EVs (one year for EVs versus five years for gasoline cars). Listing EVs as environmentally friendly products that qualify for tax breaks.

Sweden. Providing a subsidy of $1,900 for a car or estate wagon and $3,800 for a minibus or panel van for the first 1,000 EV purchasers. Powering Swedish Postal Service postal van in Stockholm with a replaceable zinc-air battery. In Gothenburg, offering parking permits to EV drivers for $675 for three years versus $810 per year for drivers of conventional cars and free on-street parking for EVs (normally $2 per hour).

Switzerland. Co-funding vehicle purchases, infrastructure development, and preferred parking in seven cities.

The United Kingdom. Exempting EVs from the yearly road tax (average of $154).

Source: Electric Transportation Coalition, Washington, D.C.

Provide Subsidies to Purchasers of New EVs

At the moment, new electric vehicles—which are mostly conversions of gasoline-powered vehicles—are considerably more expensive than conventional vehicles. The batteries cost a lot and the companies performing the conversions are "mom and pop" by nature. These costs are partly offset by federal and state tax incentives to EV buyers. (The Energy Policy Act of 1992 offers both purchase tax credits and deductions for the purchase of electric vehicles and the construction of recharging stations. The federal 10 percent tax credit—scheduled to be phased out by the year 2004—is worth up to $4,000 on the purchase or lease of a new EV.

Still untried, a so-called "feebate" program that taxed new relatively inefficient gasoline vehicles and used the money raised to subsidize EV purchases would also reduce the purchase price of electric vehicles. Alternatively, state and local vehicle fees and taxes (annual registration fees, sales and personal-property taxes, and fuel taxes) could be waived for a five- to ten-year honeymoon.

Providing temporary tax subsidies, feebates, and other economic instruments unquestionably makes EVs more attractive to price-sensitive consumers. How much is hard to say, though, since some number of consumers who care about environmental issues or who like to own cutting-edge technology would probably have acted even without them.

Other Incentives

Other low-cost measures would also encourage the introduction of zero-emission vehicles:

◆ Allow EVs to use HOV lanes regardless of how many passengers are on board.

◆ Reduce or waive fees for EVs on urban toll roads.

◆ Give EVs access to preferential parking spaces equipped with electric recharging facilities.

◆ Provide federal grants to municipalities that install charging meters or photovoltaic-charging arrays in public facilities.

◆ Showcase companies that use electric-drive vehicles in their fleets.

◆ State public utility commissions (PUCs) could require utilities to adopt time-of-day rates for recharging EVs and include in these rates the costs of installing EV recharging equipment at homes and offices.

◆ PUCs might permit utilities to own batteries and lease them to consumers—a way to keep consumers' costs down and to ensure that batteries get recycled.

The costs of preferential parking, access to HOV lanes, reduced tolls, and recognition programs are likely to be modest. Politically, such programs should be fairly easy to adopt. So that "EV congestion" never becomes a problem, transportation measures such as these should have a clearly stated "sunset" provision—say, ten to fifteen years.

A remaining need is technological: standardized driving cycles should be adopted and fuel-efficiency labels provided for EVs so that consumers can gauge their efficiency more accurately under realistic driving conditions. Much like the federal driving cycle used to measure gasoline-powered vehicle fuel efficiency, this would allow potential buyers to easily compare various vehicles' relative performance (including effective range and acceleration).

Policies to Reduce Congestion

There are several policy options available to curb urban congestion. Some are economic measures affecting the prices of fuels, parking, and road use. Others emphasize new technologies to increase road capacity. Finally, some encourage substitutes for driving: telecommuting, land-use changes, and public transit. Determining which combination will work best in a given region requires analyzing local land-use patterns, transportation infrastructure, and driving trends.

As for "getting the prices right," only correct economic signals will prompt drivers (and society as a whole) to find truly viable, long-term alternatives to the exclusive use of the car for transportation.

Reducing Subsidies to Motor Vehicle Use

Higher fuel prices would encourage a switch to electric-drive vehicles. Along with pay-at-the-pump insurance, they would also reduce growth in VMT and help relieve congestion. By increasing the variable cost of motor vehicle driving, higher fuel prices would encourage ride sharing, trip-chaining, the use of public transport, skipping of unneeded trips, telecommuting, and even moving closer to work and shops. Conversely, as long as the cost of driving is kept artificially low, as it surely is today, congestion will worsen, no matter how valiant the attempts to increase road capacity.

BOX 3-7. Examples of What States Are Doing for EVs

Arizona. Clean Air Fund provides $1,000 tax credit for individuals or businesses that install charging stations. EVs are exempt from special alternative-fuel license plate fee ($100 to $400). EVs and other AFVs can use HOV lanes.

Arkansas. Provides nearly $17 million to convert vehicles to alternative fuels and to provide infrastructure.

California. Adopted original ZEV mandate (later amended) to require sale of EVs beginning in 1998. From 1996 to 1998, the first 1,200 buyers of EVs receive a $5,000 price reduction. From 1991 to 1996, buyers received a tax credit equal to 55 percent of the value of converting a vehicle to an EV. Special license plates allow EVs special parking privileges in public parking lots.

Colorado. Gives a rebate for purchase or conversion to AFV. Rebates range from $1,500 for cars to $6,000 for heavy trucks. From 1994 to 1998, a 5 percent tax credit on purchase of EVs, limited to half the cost of the EV conversion (expires 7/1/98).

Connecticut. Exempts EVs and EV conversions from sales tax ($1,550 limit for purchase, $900 for a conversion). Corporations get tax credit of 50 percent of total conversion cost. And utilities, railroads, and carriers are eligible for tax credit of 10 percent of differential or conversion cost (valued at $693 for differential, $1,500 for conversion).

Florida. Exempts new EVs from state use and sales tax for five years.

Louisiana. Provides a 20 percent credit of conversion cost to EV.

Maryland. Exempts machinery to convert cars to EVs from sales and use taxes.

Massachusetts. ZEV sales goals similar to those of California. The largest auto manufacturers must sell at least 3,750 EVs between 1998 and 2000.

New York. Conversion costs to EV are exempt from sales tax.

Eliminating Parking Subsidies

Most Americans—95 percent—enjoy free parking at work, a major incentive to commute in cars and trucks.[105] Such parking subsidies total many tens of billions of dollars per year, and reducing this perk could have a large impact on solo commuting. Requiring employers to simply provide taxable parking allotments to their workers, and then having workers, in turn, pay for their parking, would give ride-sharing and other alternative forms of commuting a huge boost.

BOX 3-7 continued.

North Dakota. Provides a 10 percent tax credit for EV conversions, with a limit of $200 for under 10,000 pounds and $500 for above. Expires 12/97.

Oklahoma. An income tax credit of 50 percent of the conversion cost for EV or for charging facilities. The credit drops to 20 percent over the period 1/1/1997 to 1/1/2002. $1.5 million available to help defray costs of converting public vehicles (maximum of $3,500 per conversion).

Oregon. Through December 1995 gave a 35 percent business tax credit for purchase of or conversion of EVs.

Pennsylvania. Exempts the extra cost of an EV from retail sales tax. No annual registration fee for EVs. Provides $3.5 million in grants to school districts, municipalities, and corporations to convert or buy EVs.

Utah. Provides a 20 percent tax credit for purchase (maximum of $500) or conversion (maximum $400). Provides $330,000 for purchase or conversion of government vehicles to EVs.

Virginia. Provides a tax credit equal to 10 percent of federal tax credit to individuals and businesses. From January 1, 1996, sales tax on EVs cut from 3 percent to 1.5 percent. From July 1, 1994, EV drivers are exempt from motor vehicle license fees.

Washington. Eliminated license fees for EV taxicabs and for-hire vehicles. (King County)

West Virginia. Established minimum fleet purchases for the state and political subdivisions for 1995 to 1997.

Wisconsin. Grants up to $30,000 to municipalities for EV conversions, a maximum of $2,000 per vehicle.

Source: The Electric Transportation Coalition, Washington, D.C.

Free commuter parking is worth considerably more than the gasoline needed to drive to work. In downtown Los Angeles, the free parking enjoyed by 50,000 solo commuters is equivalent to a subsidy of 11 cents per commuting mile.[106] Per mile driven, this subsidy is 16 times greater than the federal gas tax (about $0.007 per mile) and about 50 percent greater than the cost of the fuel consumed in commuting. The end result is dramatic. According to UCLA'S Donald Shoup, employer-paid parking shifts 27 percent of all commuters into solo driving from other modes, and among solo commuters 41 percent drive to work alone *only* because their employers pay for parking.[107]

Employers offer free parking to workers partly because this fringe benefit is not taxed federally. Consider the following example. In the Washington, D.C., area, an employer can provide a parking space as a fringe benefit for an employee for about $8 per day—about $2,000 per year—and the recipient pays no federal tax on the benefit. To provide the same employee with an extra $2,000 of take-home salary, an employer would have to spend about $4,400 per year (including federal, state, and local taxes, pension contributions, and other benefits).

According to Shoup and Willson, simply ending employer-paid parking would reduce the number of solo commuters between 18 and 81 percent, depending on local circumstances and transportation alternatives, and it would cut the number of cars driven to work by 15 to 28 percent.[108] But ending free parking abruptly would be politically tough. A more palatable alternative, Shoup contends, is requiring employers who offer free parking to give employees the option of a taxable payment equal to the value of the (tax-free) parking subsidy or a ride-sharing or public transit subsidy.[109] Such a program would add little to employer costs, benefit low-income and disabled employees, make downtown jobs more attractive, and, where workers choose the taxable payment, increase government revenues. Under this proposal, Shoup estimates, solo commuting in downtown Los Angeles would decline by 20 percent.

Congestion Pricing

Highway capacity is now rationed by congestion during rush hours. At some point, conditions get so bad that additional drivers simply refuse to get onto crowded freeways. Under proposals to reduce congestion through road-pricing, economic signals would be used to discourage driving during peak periods. Introducing tolls on clogged urban roads is another way to reduce peak-hour traffic simply, economically, and efficiently. If tolls are pegged to traffic conditions, drivers have an incentive to ride-share, travel when traffic is light, or find new ways to get around or to avoid having to get around. The political prospects for congestion pricing are not bright for existing freeways. But time-of-use road pricing might be viable on current toll roads or new highways.

Varying tolls to reduce peak demands is not a new idea. Telephone companies have long employed time-of-use rates, and some electric utilities have experimented with them. Some public transportation systems, including Washington

D.C.'s Metrorail and Bus, employ time-of-day rates. Variable charges are also routine in the movie theater and airline industries. With new technology, tolls can be charged instantly to moving vehicles. Such electronic tolls (though not time-of-day pricing) have been fully demonstrated in Hong Kong and on some U.S. toll roads.

According to Kenneth Small, congestion pricing must have two characteristics if it is to cut peak-time traffic. First, the road tolls must vary significantly during the day. Second, peak prices must be high enough to deter or shift travel.[110]

Congestion pricing has a number of advantages over such technological fixes as HOV lanes or car pooling. Such fixes initially reduce the number of vehicles on the road but give drivers no incentive to travel at other times or to find alternative means of travel. As a result, drivers otherwise kept off the road by heavy traffic quickly notice the traffic reduction and take to their cars during the peak periods. The end results? More cars on the road and, often, no less congestion. Congestion pricing, on the other hand, sends a signal to all drivers, both those using the road and those who are not, and gives them choices: they can evaluate all the rush-hour alternatives before chancing traffic.

With no strong constituency, congestion pricing faces major political obstacles.[111] Most American drivers complain when freeways are converted to toll roads. Drivers also object to paying for access to roads that they feel they have paid for through fuel taxes. And, last, congestion pricing may be regressive, hurting low-income drivers disproportionately.[112] But public resistance to congestion pricing might be reduced if revenues were dedicated to highway repair, public transport, or reductions in income taxes or sales taxes. If it happens at all, congestion pricing will most likely be introduced on new highways or on today's toll roads and bridges.

Increasing Road Capacity

Intelligent Transportation Systems can be used to increase highway capacity and reduce congestion. While not attacking the causes of congestion, technologies for information processing, communications, vehicle control, freeway ramp metering, arterial signal controls, and on-board navigation systems can temporarily reduce the symptoms. ITS systems (aside from automated highways) represent effective improvements in increasing highway capacity, and they are also politically popular and easy to implement.

Alternatives to Driving

Some ways to reduce the need for driving—including telecommuting, changes in urban development patterns, and the introduction of more flexible forms of public transit—go to the heart of how American life and the American landscape are organized.

Telecommuting

Telecommuting can reduce travel and, hence, congestion. While telecommuting reduces the amount of work-related driving by the participant, its impacts on overall VMT and congestion remain uncertain. Up to 30 million workers could be telecommuting by 2010, though even this number would represent only a 4 percent reduction in commuting VMT and a smaller percentage reduction in overall VMT. The concept meets with little or no resistance where it is compatible with job requirements, and federal information programs could encourage growth in telecommuting.

Reducing Driving Through Urban Design

Today's land use, zoning, and transportation policies encourage excessive driving and congestion by promoting suburban low-density residential developments separate from commercial establishments and inaccessible to public transportation. Land-use proposals for reducing urban traffic abound, and many advocate changing zoning restrictions and development practices to encourage higher density developments that commingle residential units with public transportation facilities and commercial buildings. These changes in development patterns—subsidized in some communities—would reduce the need to use autos to meet many daily needs.

Some urban planners are making similar proposals for new communities. New developments would have town centers comprised of shops, restaurants, and other commercial buildings of roughly comparable size and scale that people could easily reach on foot or by bike. But changing urban design to reduce auto dependency takes time, and since it is also controversial it's unlikely to affect congestion seriously for decades.

Mass Transit, Personal Rapid Transit (PRT)

A promising new approach to reducing congestion, PRT is a public-transit technology that could overcome many of the problems that beset public transport systems today. The precommercial PRT system that the Raytheon Corporation is field testing in Massachusetts may represent a partial technical fix for much of our urban congestion. Yet, PRT in general is also a decades-long solution, not a quick fix.

Stabilizing the U.S. Population

Population growth plays an important role in the unsustainability of the U.S. transportation system. With reduced or zero population growth, the nation would consume less fuel, reduce air pollution, depend less on imported oil, and spend less time stuck in traffic. It would also have more time to develop alternatives to oil-powered vehicles.

Since population growth makes solving many fundamental quality of life problems more difficult,[113] transportation issues should be addressed as just one part of an educational campaign stressing the long-term benefits of a stable population. In the United States, such an effort might best be spearheaded by charitable foundations and other nongovernmental organizations since the issue is a political hot potato in Washington.

To help begin stabilizing the population, the President's Council on Sustainable Development (PCSD) recommended that all citizens have access to basic reproductive health services, such as family planning, education, and pre- and postnatal care.[114] The PCSD also recommended that partnerships be created to enhance opportunities for women, giving special attention to the socioeconomic factors behind the higher levels of unintended and teen pregnancies among the disadvantaged.

Of course, our looming transportation problems aren't the only impetus for facing the ethical and potentially divisive issues surrounding population growth and immigration. But planners in particular and all Americans in general must realize the extent to which population growth works against efforts to solve transportation problems. To give our proposals and those aimed at other pressing social problems a chance of success, and to mark a starting point for a national debate on the issue, we recommend that continued immigration be permitted only to offset a declining resident U.S. population.

CONCLUSIONS

The unsustainability of the U.S. motor vehicle transportation system reflects several of the same sustainability issues affecting other sectors of the economy. First, climate change plays a pivotal role in the problem. Cars, trucks, and buses account for almost a third of U.S. carbon dioxide emissions, and these emissions are growing along with the number of vehicles, drivers, and miles driven. All other issues aside, the need to greatly reduce these emissions—in an absolute sense—almost certainly dictates a fundamental change in our choice of energy sources to power ground transportation. Fossil fuel burning will have to be largely phased out in the 21st century if we are to stabilize the earth's climate. Making the transition to electric-drive vehicles powered by electrical (battery) or chemical (hydrogen) storage will greatly reduce climate risks, provided that the ultimate energy sources are nonfossil. Ancillary benefits of this transition include deep reductions in urban air pollution and reduced security risks from reliance on imported oil from politically unstable regions. Policies are needed that will encourage the introduction of electric-drive vehicles with the electricity and hydrogen ultimately supplied by renewable energy technologies.

The second factor undermining our ground-transport system can be captured in one word: growth. In the case of motor vehicles, growth in driving—primarily within major urban areas—is the primary cause of increasing congestion. Were the number of vehicle miles traveled not increasing, measures such as parking reform, congestion pricing, and increased fuel taxes could keep congestion at a tolerable level. But as long as the U.S. population continues to grow—and everyone can afford a car—there will be more drivers in more vehicles trying to drive on a virtually fixed supply of roadways. Increased gridlock is bound to be the outcome.

There is a trade-off, then, between continued population growth—there could well be 400 million Americans by the year 2050—and the suburban, car-dominated lifestyle many seem to want. It seems clear that if we accept continued urban population growth, we won't necessarily be able to drive any time and any place we want to.

In any case, different models of urban development are possible. Higher density mixed development centered around transit stations (either traditional or personal rapid transit) could greatly reduce the need to drive so much. Unfortunately, current trends augur ill: low-density sprawl is still the development norm, and transit ridership continues downward while subsidies rise—partly the result of immense hidden subsidies to cars. The changes needed to deal with long-term congestion could be stimulated by reforming land-use planning, supporting innovative forms of public transportation, "un-subsidizing" the use of motor vehicles, and rapidly stabilizing population.

NOTES

1. President's Council on Sustainable Development, *Sustainable America: A New Consensus for Prosperity, Opportunity, and a Healthy Environment for the Future* (Washington, D.C.: U.S. Government Printing Office, February 1996), 54.

2. Intergovernmental Panel on Climate Change, *Climate Change 1995: The Science of Climate Change* (New York: Cambridge University Press, 1996).

3. James J. MacKenzie, "Oil as a Finite Resource: When is Global Production Likely to Peak?" (Washington, D.C.: World Resources Institute, March 1996); James J. MacKenzie, "Heading Off the Permanent Oil Crisis," *Issues in Science and Technology,* Summer 1996, Vol. 12, No. 4: 48-54.

4. U.S. Bureau of Transportation Statistics, *Transportation Statistics in Brief* (Washington, D.C.: U.S. Department of Transportation, January 1996).

5. U.S. Bureau of Transportation Statistics, *Transportation Statistics in Brief* (Washington, D.C.: U.S. Department of Transportation, June 1995), 21.

6. Oak Ridge National Laboratory, *Transportation Energy Data Book* (Oak Ridge, TN: U.S. Department of Transportation, Federal Highway Administration, 1993), Vol. 13, xxiii.

7. Ibid.

8. American Automobile Manufacturers Association (AAMA), *Motor Vehicle Facts and Figures, 1996* (Washington, D.C.: AAMA, 1996), 66.

9. Energy Information Administration, *Annual Energy Review, 1995,* DOE/EIA-0384(95) (Washington, D.C.: U.S. Department of Energy, 1995), Table 5.12b, 163.

10. James J. MacKenzie and Michael P. Walsh, *Driving Forces: Motor Vehicle Trends and Their Implications for Global Warming, Energy Strategies, and Transportation Planning* (Washington, D.C.: World Resources Institute, 1990).

11. Pietro Nivola and Robert Crandall, *The Extra Mile* (Washington, D.C.: The Brookings Institution, 1995).

12. American Automobile Manufacturers Association (AAMA), *Motor Vehicle Facts and Figures, 1996* (Washington, D.C.: AAMA, 1996), 32.

13. Energy Information Administration, *Annual Energy Outlook 1996, With Projections to 2015,* DOE/EIA-0383(96) (Washington, D.C.: U.S. Department of Energy, 1996), Tables B7 and C7, 148, 205.

14. James J. MacKenzie and Michael P. Walsh, *Driving Forces: Motor Vehicle Trends and Their Implications for Global Warming, Energy Strategies, and Transportation Planning* (Washington, D.C.: World Resources Institute, 1990), 1.

15. Energy Information Administration, *Annual Energy Outlook 1996, With Projections to 2015* (Washington, D.C.: U.S. Department of Energy, 1996), 215.

16. "Oil and Gas Reserves, Oil Output Rise in 1996," *Oil & Gas Journal Special,* December 30, 1996, Vol. 94, No. 53: 37-38.

17. "Steady Rise in Oil, Gas Demand Ahead," *Oil & Gas Journal*, June 6, 1994, Vol. 92, No. 23: 34-36; *BP Statistical Review of World Energy* (Washington, D.C.: 1994), 4.

18. Jean Laherrere, "World Oil Reserves—Which Number to Believe?," *OPEC Bulletin*, February 1995, Vol. 26, No. 2: 12.

19. John F. Bookout, "Two Centuries of Fossil Fuel Energy," *Episodes*, December 1989, Vol. 12, No. 4: 257-262.

20. William Brown, "Dying From Too Much Dust," *New Scientist*, March 12, 1994, Vol. 141, No. 1916: 12-13; Deborah Sheiman Shprentz, *Breath Taking, Premature Mortality Due to Particulate Air Pollution in 239 American Cities* (Washington, D.C.: Natural Resources Defense Council, May 1996).

21. William Brown, "Deaths Linked to London Smog," *New Scientist*, June 25, 1994, Vol. 142, No. 1931: 4.

22. Deborah Sheiman Shprentz, *Breath Taking, Premature Mortality Due to Particulate Air Pollution in 239 American Cities* (Washington, D.C.: Natural Resources Defense Council, May 1996).

23. U.S. Environmental Protection Agency, *National Air Quality and Emissions Trends Report, 1993*, EPA 454/R-94-026 (Washington, D.C.: October 1994).

24. National Research Council, *Rethinking the Ozone Problem in Urban and Regional Air Pollution* (Washington, D.C.: National Academy Press, 1992).

25. U.S. Environmental Protection Agency, *National Air Pollutant Emission Trends: 1900-1994*, EPA 454/R-95-0116 (Washington, D.C.: 1995).

26. U.S. Environmental Protection Agency, *National Air Quality and Emissions Trends Report: 1995*, EPA 454/R-96-005 (Washington, D.C.: October 1996).

27. S.P. Beaton, et al., "On-Road Vehicle Emissions: Regulations, Costs, and Benefits," *Science*, May 19, 1995: Vol. 268, 991-993.

28. U.S. Environmental Protection Agency, *Motor Vehicle Tampering Survey: 1988* (Washington, D.C.: Office of Air and Radiation, Office of Mobile Sources, May 1989).

29. National Research Council, *Rethinking the Ozone Problem in Urban and Regional Air Pollution* (Washington, D.C.: National Academy Press, 1992).

30. Ibid., 42.

31. U.S. Congress, Office of Technology Assessment, *Replacing Gasoline, Alternative Fuels for Light-Duty Vehicles*, OTA-E-364 (Washington, D.C.: U.S. Government Printing Office, September 1990), 71.

32. Mark A. DeLuchi, *Emissions of Greenhouse Gases from the Use of Transportation Fuels and Electricity*, ANL/ESD/TM-22 (Argonne, IL: Center for Transportation Research, Energy Systems Division, Argonne National Laboratory, November 1991), Vol. 1, 69-72.

33. U.S. Congress, Office of Technology Assessment, *Replacing Gasoline, Alternative Fuels for Light-Duty Vehicles*, OTA-E-364 (Washington, D.C.: U.S. Government Printing Office, September 1990), 98; National Research Council, *Rethinking the Ozone Problem in Urban and Regional Air Pollution* (Washington, D.C.: National Academy Press, 1992), 402.

34. Mark A. DeLuchi, *Emissions of Greenhouse Gases from the Use of Transportation Fuels and Electricity*, ANL/ESD/TM-22 (Argonne, IL: Center for Transportation Research, Energy Systems Division, Argonne National Laboratory, November 1991), Vol.1, 70.

35. Sierra Research, Inc., *The Air Pollution Consequences of Using Ethanol-Gasoline Blends in Ozone Non-Attainment Areas* (Sacramento, CA: May 1990).

36. J.G. Calvert et al., "Achieving Acceptable Air Quality: Some Reflections on Controlling Vehicle Emissions," *Science*, July 2, 1993, Vol. 261, 42.

37. Mark A. DeLuchi, *Emissions of Greenhouse Gases from the Use of Transportation Fuels and Electricity*, ANL/ESD/TM-22 (Argonne, IL: Center for Transportation Research, Energy Systems Division, Argonne National Laboratory, November 1991), Vol.1, 59.

38. Ibid., 72.

39. C.E. Wyman et al., "Ethanol and Methanol from Cellulosic Biomass," *Renewable Energy, Sources of Fuels and Electricity*, T.B. Johansson et al., eds. (Washington, D.C.: Island Press, 1993); "Meeting Transportation Needs Through Biofuels," *Energy*, February 1993: 23-24; L.R. Lynd et al., "Fuel Ethanol from Cellulosic Biomass," *Science*, Vol. 251, March 15, 1991: 1318-1323.

40. U.S. Congress, Office of Technology Assessment, "Potential Environmental Impacts of Bioenergy Crop Production," Background Paper, OTA-BP-E-118 (Washington, D.C.: U.S. Government Printing Office, 1993), 34.

41. National Renewable Energy Laboratory, *The Potential of Renewable Energy: An Interlaboratory White Paper*, Interlaboratory Report (Golden, CO: 1990), B-5.

42. Mark A. DeLuchi, *Emissions of Greenhouse Gases from the Use of Transportation Fuels and Electricity*, ANL/ESD/TM-22 (Argonne, IL: Center for Transportation Research, Energy Systems Division, Argonne National Laboratory, November 1991), Vol.1, 126.

43. Ibid.

44. J.M. Ogden and M.A. DeLuchi, "Solar Hydrogen Transportation Fuels" (presented to the Conference on Transportation and Global Climate Change: Long Run Options, Asilomar, CA, August 1991); J.H. Cook et al., "Potential Impacts of Biomass Production in the United States on Biological Diversity," *Annual Reviews of the Energy and the Environment*, 1991: 401-31; D. Hall et al., "Biomass for Energy: Supply Prospects," in *Renewable Energy, Sources of Fuels and Electricity*, T.B. Johansson et al., eds., (Washington, D.C.: Island Press, 1993); National Renewable Energy Laboratory, *The Potential of Renewable Energy: An Interlaboratory White Paper*, Interlaboratory Report (Golden, CO: 1990).

45. David Pimentel and John Krummel, "Biomass Energy and Soil Erosion: Assessment of Resource Costs," *Biomass*, 1987, Vol. 14: 15-38.

46. National Renewable Energy Laboratory, *The Potential of Renewable Energy: An Interlaboratory White Paper*, Interlaboratory Report (Golden, CO: 1990), 23.

47. James J. MacKenzie, *The Keys to the Car: Electric and Hydrogen Vehicles for the 21st Century* (Washington, D.C.: World Resources Institute, 1994).

48. California Air Resources Board, *Technical Support Document, Zero-Emission Vehicle Update* (Sacramento, CA: 1994).

49. Northeast States for Coordinated Air Use Management (NESCAUM), *Impact of Battery-Powered Electric Vehicles on Air Quality in the Northeast States, Final Report* (Boston, MA: 1992).

50. Southern California Edison, Electric Transportation Division, "Electric Vehicle Efficiency Test," April 1992.

51. For an excellent summary of various ways of measuring congestion, see "Traffic Congestion: Trends, Measures, and Effects," GAO/PEMD-90-1 (Washington, D.C.: U.S. Government Accounting Office, November 1989), 35-49.

52. Shawn M. Turner, "An Examination of the Indicators of Congestion Level," No. 920729 (presented to the Transportation Research Board 71st Annual Meeting, Washington D.C., January 12-16, 1992).

53. U.S. Federal Highway Administration, *The 1991 Status of the Nation's Highways and Bridges: Conditions, Performance, and Capital Investment Requirements* (Washington, D.C.: July 2, 1991) 5. The 8 billion hours is the difference between congestion at D and F levels of service and is based on FHWA models of travel speed decline for given vehicle densities and travel conditions (Harry Caldwell, FHWA, phone interview, August 12, 1991).

54. U.S. House of Representatives Committee Print, Committee on Public Works and Transportation, *The Status of the Nation's Highways and Bridges: Conditions and Performance* (Washington, D.C.: U.S. Government Printing Office, September 1991), 21.

55. U.S. Federal Highway Administration, *The 1991 Status of the Nation's Highways and Bridges: Conditions, Performance, and Capital Investment Requirements,* (Washington, D.C.: July 2, 1991) 24.

56. U.S. Department of Commerce, *Statistical Abstract of the United States: 1995* (Washington, D.C.: 1995), 115th edition, Table 1032, 637.

57. Federal Highway Administration, *Estimates of Urban Roadway Congestion: 1990* DOT-T-94-01 (Washington, D.C.: U.S. Department of Transportation, Federal Highway Administration, 1990), x.

58. Ibid., xi.

59. California psychologist Raymond Novaco, as quoted in *Time*, September 12, 1988, Vol. 132, No. 11:55.

60. Shawn M. Turner, "An Examination of the Indicators of Congestion Level," No. 920729 (presented to the Transportation Research Board 71st Annual Meeting, Washington D.C., January 12-16, 1992).

61. Federal Highway Administration, *Estimates of Urban Roadway Congestion: 1990* DOT-T-94-01 (Washington, D.C.: U.S. Department of Transportation, Federal Highway Administration, 1990), x.

62. Oak Ridge National Laboratory, *1990 NPTS Databook, Volume 1* (Oak Ridge, TN: U.S. Department of Transportation, Federal Highway Administration, November 1993), 1-10.

63. Oak Ridge National Laboratory, *1990 NPTS Databook, Volume 2* (Oak Ridge, TN: U.S. Department of Transportation, Federal Highway Administration, October 1994), 9-11.

64. Oak Ridge National Laboratory, *1990 NPTS Databook, Volume 1* (Oak Ridge, TN: U.S. Department of Transportation, Federal Highway Administration, November 1993), 1-11.

65. Oak Ridge National Laboratory, *1990 NPTS Databook, Volume 1* (Oak Ridge, TN: U.S. Department of Transportation, Federal Highway Administration, November 1993), Table 2.7, 2-15

66. Nationwide Personal Transportation Survey, "Implications of Emerging Travel Trends, Conference Proceedings," July 1994, p. 16; Nationwide Personal Transportation Survey, "Special Reports on Trip and Vehicle Attributes," Chapter 1: Understanding Trip Chaining, Draft (Washington, D.C., April 1994).

67. Nationwide Personal Transportation Survey, "Special Reports on Trip and Vehicle Attributes," Chapter 1: Understanding Trip Chaining, Draft (Washington, D.C., April 1994) 29.

68. American Automobile Manufacturers Association, "Motor Vehicle Facts and Figures, 1995," (Washington, D.C.: U.S. Department of Transportation, 1995), 58.

69. Alan E. Pisarski, "1990 Nationwide Personal Transportation Survey: Travel Behavior Issues in the 90s," FHWA-PL-93-012 (Washington, D.C.: U.S. Department of Transportation, Federal Highway Administration, July 1992), 7.

70. Ibid.

71. Ibid.

72. Mark Hansen, "Do New Highways Generate Traffic?" *Access,* University of California Transportation Center, Fall 1995, No. 7: 16-22.

73. U.S. Bureau of the Census, *Statistical Abstract of the United States: 1994* (Washington, D.C.: 1995), 114th edition.

74. "GM-led Consortium Awarded Automated Highway Contract," *ITS America,* November/December 1994, Vol. IV, No. 10: 1.

75. "Bay Area's TravInfo Up and Running," *ITS World,* November/December 1996, Vol. 1, No. 6: 10.

76. Jill Vitiello, "Traffic 2000," *In Sync,* Erie Insurance Group, Summer 1995: 8-12.

77. Patricia L. Mokhtarian, "Telecommuting: What's the Payoff?," *Access,* University of California Transportation Cenber, Spring 1993, No. 2: 25-28.

78. U.S. Congress, Office of Technology Assessment, *Replacing Gasoline, Alternative Fuels for Light-Duty Vehicles,* OTA-E-364 (Washington, D.C.: U.S. Government Printing Office, September 1990), 239.

79. Patricia L. Mokhtarian, "Telecommuting: What's the Payoff?," *Access,* University of California Transportation Center, Spring 1993, No. 2: 25-28.

80. Oak Ridge National Laboratory, Center for Transportation Analysis, *Intersections,* Vol. 3, No. 5, March 1995.

81. U.S. Congress, Office of Technology Assessment, *Saving Energy in U.S. Transportation,* OTA-ETI-589 (Washington, D.C.: U.S. Government Printing Office, July 8, 1994), 240. Carbon reduction estimate based on projected fuel savings of 1.679 billion gallons of gasoline.

82. Ibid.

83. Oak Ridge National Laboratory, Center for Transportation Analysis, *Intersections,* Vol. 3, No. 5, March 1995.

84. Ibid.

85. Rich Untermann, *Reshaping our Suburbs, Linking Land Use and Transportation to Better Serve Pedestrians, Bicycles and Transit* (Seattle, WA: Bullitt Foundation, May 1995).

86. Andres Duany and Elizabeth Plater-Zyberk, "The Second Coming of the American Small Town," *Wilson Quarterly,* Winter 1992, Vol. 16, No. 1: 49-50.

87. Jerry Adler, "15 Ways to Fix the Suburbs," *Newsweek,* May 15, 1995, Vol. 125, No. 20: 50.

88. Ibid., 51.

89. Jerry Adler, "Bye-Bye, Suburban Dream," *Newsweek,* May 15, 1995, Vol. 125, No. 20: 45.

90. U.S. Bureau of Transportation Statistics, *Transportation Statistics in Brief* (Washington, D.C.: U.S. Department of Transportation, January 1996).

91. Alan E. Pisarski, "1990 Nationwide Personal Transportation Survey: Travel Behavior Issues in the 90's", FHWA-PL-93-012 (Washington, D.C.: U.S. Department of Transportation, Federal Highway Administration, July 1992), 17.

92. Ibid.

93. Melvin M. Webber, "The Marriage of Autos and Transit, How to Make Transit Popular Again," *Access*, University of California Transportation Center, Fall 1994, No. 5.

94. James J. MacKenzie, Roger C. Dower, and Donald D.T. Chen, *The Going Rate: What it Really Costs to Drive* (Washington, D.C.: World Resources Institute, 1992).

95. Earl C. Ravenal, *Designing Defense For a New World Order* (Washington, D.C.: The Cato Institute, 1991), 46.

96. President's Council on Sustainable Development, *Sustainable America: A New Consensus for Prosperity, Opportunity, and a Healthy Environment for the Future* (Washington, D.C.: U.S. Government Printing Office, February 1996), 143.

97. Charles Lave, "Clean for a Day, California Versus the EPA's Smog Check Mandates," *Access*, University of California Transportation Center, Fall 1993, No. 3.

98. James J. MacKenzie, *The Keys to the Car: Electric and Hydrogen Vehicles for the 21st Century* (Washington, D.C.: World Resources Institute, 1994).

99. Ibid.

100. James Poterba, "Is the Gasoline Tax Regressive?" *Tax Policy and the Economy*, Vol. 5, 1991: 145-164.

101. Michael Drew and Dita Smith, "Paying Through the Hose," *The Washington Post*, May 11, 1996: A21.

102. Andrew Tobias, *Auto Insurance Alert*, A Fireside Book (New York, NY: Simon and Schuster, 1993).

103. S.P. Miaoi et al., *Fleet Vehicles in the United States: Composition, Operating Characteristics, and Fueling Practices*, ORNL-6717 (Oak Ridge, TN: Oak Ridge National Laboratory), 38.

104. Ibid., 18.

105. Donald C. Shoup, *Curbing Gridlock, Peak-Period Fees to Relieve Traffic Congestion*, Special Report 242 (Washington, D.C.: National Research Council, National Academy Press, 1994); Donald C. Shoup, "An Opportunity to Reduce Minimum Parking Requirements," *Journal of the American Planning Association*, Winter 1995, Vol. 61, No. 1: 14-28.

106. Donald C. Shoup, *Curbing Gridlock, Peak-Period Fees to Relieve Traffic Congestion*, Special Report 242 (Washington, D.C.: National Research Council, National Academy Press, 1994).

107. Donald C. Shoup, "Cashing Out Employer-Paid Parking," *Access*, University of California Transportation Center, Spring 1993, No. 2.

108. Donald C. Shoup and Richard W. Willson, "Employer-Paid Parking: The Problem and Proposed Solutions," Draft, July 1991.

109. Donald C. Shoup, "Cashing Out Employer-Paid Parking," *Access*, University of California Transportation Center, Spring 1993, No. 2.

110. Kenneth A. Small, "Congestion Pricing: New Life for an Old Idea," *Access*, University of California Transportation Center, Spring 1993, No. 2.

111. Martin Wachs, "Will Congestion Pricing Ever Be Adopted?" *Access*, University of California Transportation Center, Spring 1993, No. 4.

112. Ibid.

113. President's Council on Sustainable Development, *Sustainable America: A New Consensus for Prosperity, Opportunity, and a Healthy Environment for the Future* (Washington, D.C.: U.S. Government Printing Office, February 1996).

114. Ibid.

4.
CHALLENGES TO SUSTAINABILITY IN THE U.S. FOREST SECTOR

Nels Johnson and Daryl Ditz

Today, the U.S. forest sector ranks among the most productive, diverse, and well managed in the world. Forest area is stable, productivity has outpaced consumption for decades, nearly two million Americans have jobs in the sector, and a wide range of laws, regulations, and policies intended to sustain the diverse benefits America's forests provide are in place. Still, conflicts over forest-sector activities have grown highly visible and divisive in recent decades. Logging of old-growth forests in the Pacific Northwest and the release of persistent pollutants in pulp and paper manufacturing have triggered high stakes public struggles over policy and management practices. Disputes over what is not being sustained—whether timber industry jobs, recreational opportunities, or natural salmon stocks in the Pacific Northwest—suggests that at least some forest goods and services are reaching finite limits. Meanwhile, some forest resources could be overwhelmed in coming decades as domestic and international demand swells, unless policies and practices change.

The U.S. forest sector does face serious sustainability challenges. But it also has the luxury of time, a history of innovation, and recourse to market and policy processes that work better than those of many countries.

Why should Americans—most of whom live in large metropolitan areas increasingly isolated from natural resource economies—care about the future of forests? First, the U.S. forest sector generates an enormous volume and range of products and services that Americans use every day—from lumber, paper, and engineered wood products to wildlife and fisheries, watershed protection, and energy. Second, the forest sector provides jobs, recreation, educational opportunities, and aesthetic benefits. Third, this sector provides valuable unsung assets including growing stores of carbon that offset greenhouse gas emissions and reservoirs of biodiversity that support healthy ecosystems and fuel advances in biotechnology.

SCOPE AND APPROACH

This chapter addresses three overarching questions. First, what are the most important sustainability challenges facing the U.S. forest sector in coming decades? Second, how can it meet sustainability challenges as demands for all types of forest products and services grow? Third, how can it make the United States as a whole more sustainable?

Broadly speaking, the U.S. forest sector comprises the *ecosystems* that constitute forest areas (*see Figure 4-1*), the broad array of *raw materials and environmental services* that forests supply, the *industrial processes* that transform forest resources into manufactured products, the *economic relationships* among forest resources, industries, and consumers, and the *social processes* that mediate economic demands and impacts on forest ecosystems and the environment. All these dimensions must be addressed to put sustainability into perspective.

FIGURE **4-1.** **Forest Cover in the United States, 1993**

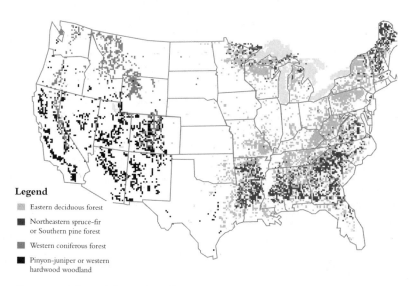

Legend
- Eastern deciduous forest
- Northeastern spruce-fir or Southern pine forest
- Western coniferous forest
- Pinyon-juniper or western hardwood woodland

Source: Adapted from EPA forest cover maps based on Powell et al., 1993.

Sustainability is about the future. This study is cast in terms of the next few decades, to the year 2020 or so—beyond today's controversies and far enough away to permit turnover in industrial technology and a shift in the biological capital and yet still relevant vis-à-vis strategic public and private sector decision-making. Anticipating problems over this time scale preserves options and flexibility, lowers risk, and leaves time to define acceptable costs and desired benefits.

Key Challenges to a More Sustainable Forest Sector

In many respects, the U.S. forest sector's environmental performance has improved significantly over the last century. Overall forest area has remained basically stable since 1920, even though the country's population has grown by 150 million. The release of conventional water pollutants by pulp and paper mills declined from 210 pounds per ton of production in 1945 to less than 10 pounds per ton in 1990 (Miner and Unwin, 1991). Tree planting after harvest increased from less than 400,000 acres annually in 1950 to approximately 3 million acres in 1990 (MacCleery, 1992). The net annual growth of timber species increased by 62 percent between 1952 and 1987, while timber inventories increased by 27 percent (Haynes et al., 1995). Today, recycled paper accounts for more than one-third of the fiber used in the pulp and paper industry, compared to less than 8 percent in 1962 (Ince, 1994). Meanwhile, the area of legally protected forest has increased from less than 20 million acres in 1960 to 47 million acres in 1992, or about 6 percent of the total U.S. forest area (Powell et al., 1993).

Nevertheless, steadily rising demands for both timber and nontimber forest products and benefits, will force Americans to make choices about how to use their forests. For example, the average American's appetite for paper products has nearly tripled in three decades to 700 pounds annually. Meanwhile, net annual timber growth is stagnating, and a growing number of forest-dwelling species are showing evidence of population declines due to the cumulative effects of logging, water pollution, and forest fragmentation as housing, infrastructure, and commercial development encroach. If current trends in wood-products consumption and net annual timber growth continue over the next decade, the United States will consume more wood fiber than it grows in its timber base.[*]

---

[*]   This does not include forest areas that are legally protected or woodland areas which have such low productivity they are not included in the timber base.

If this happens nationally, it would be the first time since the 1920s that the country has not managed its timberlands on a sustained-timber yield basis.

Four main issues will determine sustainability in the U.S. forest sector in coming decades: *maintaining forest health and productivity, safeguarding biological diversity, adapting to climate change,* and *preventing industrial pollution.* If these challenges are met, the forests of the 21st century will continue to provide Americans with a wide range of products and essential environmental services, enhance their quality of life, and help the country adapt to changes in the environment and in the global marketplace.

The first challenge is important because timber productivity has stagnated in many parts of the country after decades of steady growth. The many possible explanations—from a lack of investment to extended regional droughts—vary from place to place. But evidence also points to increased tree mortality, which may signal widespread or systemic problems in the health of forest ecosystems in the United States caused by insect pests, diseases, or drought, which in turn could be exacerbated by air pollution, forest practices, and even climate change. If forest health and productivity do not improve, the only U.S. options will be consuming fewer wood products, importing more wood fiber, establishing tree plantations on lands used for agriculture or other purposes, harvesting timber-lands unsustainably, or opening legally protected forest areas to logging. Several of these options (especially the last two) fly in the face of sustainability.

The second challenge—protecting the variety and variability of living organ-isms and the ecosystems in which they live—has rapidly become one of the most visible and complex issues in forest management. Biodiversity measured at all levels—ecosystem, species, and genetic—shows significant signs of decline in U.S. forest areas. As examples, "old-growth" forest stands have virtually disappeared from most types of forest, and more than half of all endangered or threatened species in the United States live in forest ecosystems. Biodiversity is likely to become more valuable—as a raw material to fuel biotechnology development and as a factor that allows forest ecosystems to adapt to environmental change. Species extinction and the loss of genetic diversity within a species are irre-versible, but even experts can't predict the consequences of losing a particular species on ecosystem productivity, the survival of other species, or biotechnology development. For these reasons, biodiversity conservation is a strategic sustain-ability issue in the U.S. forest sector.

Should climate warming occur—and most scientists believe that it has already begun—U.S. forests are likely to figure prominently in future climate change policies. Climate warming could well increase the prevalence of insect pests and diseases in forest ecosystems, and it could also open the door to severe fire in drier areas that are already marginal for forests. At the same time, trees and forests can help offset greenhouse gas emissions and could play an important role in the drive to replace fossil fuels with renewable energy sources.

Conventional air- and water-pollution from the forest sector are generally on the decline, but clean production remains a strategic sustainability issue. To date, the forest products industry has tried to control pollution, rather than prevent it. While largely successful, this tack increases industry's capital investment and operating costs mightily and imposes huge regulatory costs on government to monitor and enforce compliance. Furthermore, recent ecological and epidemiological research is raising new concerns about how chlorinated compounds and other pollutants generated in pulp and paper manufacture harm human and ecosystem health. While the forest sector is only one of several major industries that release such compounds and the effects of low-level exposures are very uncertain, the effects of accumulating persistent toxics in the environment are hard to undo.

Envisioning a More Sustainable Forest Sector

Since forest health and productivity, biodiversity, climate change, and industrial pollution are all driven by growing demands on essentially fixed resources, finding ways to derive more from less is fundamentally important. Eco-efficiency—defined by the World Business Council for Sustainable Development as producing competitively priced goods and services to satisfy human needs and raise the quality of life while reducing overall ecological impacts (BCSD, 1994)—is the key: A sustainable U.S. forest sector will ultimately depend on lower per capita consumption of virgin wood fiber, the judicious use of nonwood fibers, the outright substitution of electronic media or other new technology for paper, and the adjustment of collective demand to the capacity of a sustainable forest sector.

Our vision of a sustainable forest sector integrates responses to the four sustainability challenges. In it, Americans derive more value from both forest products and environmental services. For example, pulp and wood-chip production will be concentrated on a smaller but more productive land base, leaving most timberland areas to produce longer rotation saw timber and nontimber forest benefits. The forest sector will support more diverse rural economies, supply more educational and recreational opportunities (especially for urban residents), and make communities, neighborhoods, and workplaces more attractive and livable. There will be fewer logging and sawmill jobs, but growing markets for specialty wood products, recreation, and nontimber forest products; and such environmental services as clean water and carbon sequestration will provide new opportunities to make a decent living and stay on the land.

Steps Toward Sustainability

A more integrated approach to using and managing forest resources is central to increasing the U.S. forest sector's sustainability. This means changing how we allocate land uses, rewarding sustainable business practices, and enabling consumers and citizens to make informed choices about the products they buy and the policies they support. It also means producing more fiber, conserving biodiversity more effectively, protecting vital ecological functions, and satisfying social demands for recreation, jobs, and other benefits. Meeting these goals will become ever harder unless we encourage more diverse, sustainably managed uses on all forested landscapes. One example is an ambitious program—using land swaps, targeted tax benefits, and conservation easements—to protect biodiversity and increase recreational opportunities, especially in the eastern and southern United States.

Forest industries must be encouraged—through policy changes, competitive pressures, and greater public accountability—to make business decisions that reflect the fundamentals of sustainability. Some companies will have to expand their concept of forest products to include biodiversity and climate protection as well as fiber. Others might develop substitutes, possibly from agricultural fibers. Future investments in manufacturing technology and new product development should provide a hedge against uncertainties. Greater public disclosure of corporate environmental performance in land management, manufacturing, and products will also help to make sustainability a competitive advantage.

These shifts in land management and business practice will be reinforced by informed consumers. If forest practices are rated by an independent group, the marketplace can school private decision-makers and reward firms committed to sustainable practices. Meanwhile, more open and participatory processes—for example, in land-use planning and the development of tax and policy incentives—will allow citizens and consumers to see the consequences of their choices. Finally, given the large U.S. stake in the fate of forests worldwide, the United States needs to show more ambitious international leadership on technology, market development, and the implementation of international agreements related to forests (such as those on climate change and biodiversity).

THE U.S. FOREST SECTOR'S BOUNDARIES

To assess sustainability, what is the best way to draw the boundaries of the U.S. forest sector? Clearly, the *natural resources* that constitute forests—the forest flora and fauna, soil, water, and nutrients that underlie sustainable ecosystems—help to define the physical boundaries, and such potentially competing land uses as agriculture, commercial, and residential development must be taken into account. It is also possible to examine the sector in terms of the *forest products* that forests supply—timber harvest, boardfeet of lumber, tons of pulp, and so on. But the forest sector is also defined by economic relations among consumers, processors, manufacturers, distributors, retailers, and the communities with historical ties to forests—all concerned with jobs, skills, and regional economic stability. Finally, *national borders* and the goods and services that cross them also come into play: the United States has interests in other countries' forest resources and in world markets for U.S. forest products.

A Brief History of Forest Change

When the Pilgrims first set foot on the shores of North America in 1620, forests blanketed more than one billion acres, roughly half of what is now the continental United States. Today, forests cover about 737 million acres, close to one-third of total U.S. land area (Powell et al., 1993). Over the first 150 years, deforestation proceeded slowly as the small colonial population—fewer than four million during the American Revolution—slowly chipped away at forests to make way for agriculture and to get wood for energy, housing, furniture, and masts for the British king's fleet.

As the frontier crossed the Appalachians after independence, the conversion of forests to agriculture and degraded lands accelerated. In many cases, lumber companies leveled old-growth timber, such as the valuable wood in eastern white pine forests. Then farmers stepped in. Forests yielded the fuel and timber needed to build the railroads, and wood charcoal stoked the iron mills at the beginning of the Industrial Revolution. By the 1880s, the timber frontier had crossed the continental divide, and lumber camps sprouted in the towering coniferous forests of California and the Pacific Northwest before the century's end. In the last two decades of the 19th century, tens of millions of acres of logged-over forests strewn with mounds of wasted wood burned every summer from West Virginia's highlands to New England's mountains and across to the lake country of northern Minnesota. At least 80 million acres were degraded and abandoned, and three million more were being cleared each year for agriculture. Timber harvests nationally exceeded forest growth, wood waste was rampant, and reforestation nearly unheard of (MacCleery, 1992). By 1920, U.S. forest cover had been whittled down to 732 million acres (Powell et al., 1993).

What drove much of the concern over forests a century ago was what happened after the forest was logged, not how it was logged. When the loggers picked up and moved on, marginal agriculture or outright land abandonment usually followed. This pattern fueled fears that the country was permanently losing its forest estate—much like international concerns over tropical deforestation today. In many areas east of the Mississippi River, forest cover during the 1800s had fallen from 70 percent or more to less than 20 percent (MacCleery, 1992). To save the West from a similar fate, in the 1890s Congress set aside timber reserves on unclaimed federal lands that eventually became the backbone of today's National Forests.

Gifford Pinchot, a close confidant of Theodore Roosevelt, convinced the President to create the U.S. Forest Service in 1905. The policies started by Roosevelt, together with improvements in agricultural productivity and rural-urban migration, stemmed the net loss of forest land and averted the "timber famine" Pinchot feared. For the next several decades, conservation meant protecting forests from waste and destruction, but not from logging as long as it ensured undiminished future harvests—that is, "sustained timber yield."

In the 1920s and 1930s, the federal government bought up abandoned or degraded farmlands in Appalachia and the northern Great Lakes states to reestablish forests, protect watersheds, and supply local timber industries. At the same time, Pinchot and others fought in vain to regulate forest practices on private lands (Clary, 1986). Over the intervening decades, U.S. forest area recovered modestly, reaching 762 million acres in 1963. In the last 30 years, forest cover shrank about 3 percent to 737 million acres (Powell et al., 1993). During these decades, the reclamation of agricultural lands by forests along the east coast slowed and forest land was cleared to make way for suburban development and transportation infrastructure. By comparison, forest loss in tropical countries averaged nearly 25 percent over these years (WRI, 1996).

For the first four decades after their creation, the National Forests were managed as forest reserves—storehouses of timber on the stump to fall back on when private forest lands were exhausted. The Organic Act of 1897 and the Transfer Act of 1905 (which transferred federal forest reserves from the Department of Interior to the Department of Agriculture) charged the U.S. Forest Service with improving and protecting the forests, managing them for water supplies, and furnishing a continuous supply of timber. Mainly custodial, the agency protected the forests from fires, pests, and timber poachers.

As demands for wood products rose sharply during the Second World War and the postwar housing boom, and as supplies of old-growth and mature timber on private lands dwindled, production was stepped up in the National Forests nationwide; from less than 1.5 billion boardfeet in 1941, timber production grew to 12.8 billion boardfeet by 1968 (Sample and LeMaster, 1992; O'Toole, 1988). By the 1950s, it became clear that logging on National Forests could threaten water supplies, fisheries, wildlife, soil, and recreation (Lyden et al., 1990). In 1960, Congress passed the Multiple-Use Sustained-Yield Act (MUSY), which expanded the U.S. Forest Service's management mandate to include outdoor recreation, range, fish and wildlife, and wilderness, in addition to watersheds and timber. The Act also defined "sustained-yield" to cover not just timber but also other renewable forest resources—the so-called multiple-use model. Meanwhile, the forest industry began to invest heavily in replanting trees, improving timber yields, and buying additional timberlands. Between 1952 and 1976, the industry more than tripled the annual acreage planted or replanted, saw timber productivity increase by approximately 50 percent, and expanded its timberland holdings by nearly 20 percent (Powell et al., 1993; Haynes et al., 1995). In other words, while the Forest Service began to slowly expand its management mission to nontimber forest resources, the industry plunged heavily into making wood fiber production profitable and renewable on its own lands.

U.S. Forests at a Crossroads

Controversies over extensive clearcutting in Montana, West Virginia, and elsewhere during the early 1970s convinced many conservationists that multiple-use management either wasn't being applied or didn't work. Congress eventually agreed, and first passed the Forest and Rangelands Resource Planning Act (RPA)[*]

---

[*] The Forest and Rangelands Planning Act (RPA) requires the Forest Service to develop a long-range national plan to assure the long-term sustainable management of all renewable resources in the National Forest system. Updated every five years, this plan requires the agency to inventory and set clear goals and objectives for a range of resources from timber to wildlife, wilderness, and recreation. Finally, on an annual basis, the Forest Service is to provide Congress with an annual budget that reflects the priorities of the program and a policy statement and an annual report on progress implementing the "program."

BOX 4-1. Regional Dimensions of the U.S. Forest Sector

	Pacific	Rocky Mountains	North Central
Timberland (million acres)	69.8	59.1	81.9
Ownership:			
Public land	55%	70%	28%
Forest industry	18%	5%	5%
Private, nonind.	28%	25%	67%
Roundwood Harvest (billion cubic feet)	4.2	0.9	1.9
Major Forest Types	Douglas-fir, Hemlock-Sitka spruce, Ponderosa pine, Redwood	Ponderosa pine, Douglas-fir, Fir-spruce, Lodgepole pine, Western White pine	Oak-hickory, Maple-beech, birch, Aspen-birch, Spruce-fir, White-red-jack pine
Recent Points of Conflict	Endangered species, timber shortages, old-growth forest protection	Forest health (fire), public land management, below-cost timber sales, endangered species	Impact of mills on water quality, wildlife management

in 1974 and then the National Forest Management Act (NFMA)[*] in 1976 to address MUSY's weaknesses. By the 1980s, controversies over logging practices, especially on public lands, focused increasingly on real or potential impacts on wildlife, fisheries, and endangered species. While the definition of "natural" forests is complicated due to the influence of Native Americans on forest ecosystems (e.g., MacCleery, 1992), few deny that the diverse mature forest ecosystems that early colonists found are now virtually gone. Outside of Alaska, 95 percent of America's forest area has been logged or cleared at least once. Only a few spots in the Pacific Northwest or the higher elevations in the intermountain West have

[*] The National Forest Management Act (NFMA) essentially amended the RPA, extending long-range planning to the local level. For each National Forest, the law requires a detailed, ten-year plan on land and resource management. The Act promotes sustainable forest management by restricting timber operations on lands physically and economically unfit for such use. Under NFMA-mandated planning, several management alternatives are produced, typically ranging from preserving nontimber forest resources to the intensive management of timber and other commodities. The idea is to integrate national "top-down" RPA planning with "bottom-up" perspectives from individual national forests.

			BOX 4-1 continued.
	Northeast	**Southeast**	**South Central**
Timberland (million acres)	79.5	84.8	114.5
Ownership:			
Public land	11%	11%	10%
Forest industry	15%	19%	20%
Private, nonind.	74%	70%	70%
Roundwood Harvest (billion cubic feet)	2.2	3.8	4.8
Major Forest Types	Maple-beech-birch, Oak-hickory, Spruce-fir, White-red-jack pine	Oak-hickory-Loblolly-shortleaf pine, longleaf-slash pine, Oak-gum-cypress	Oak-hickory, Loblolly-shortleaf pine, Oak-pine, Oak-gum-cypress
Recent Points of Conflict	Forest health (air pollution), forest fragmentation, watershed management	Forest health (insects, air pollution), endangered species, conversion to pine plantations	Endangered species, forest fragmentation, conversion to pine plantations

never-logged forests left (Bryant et al., 1997). Today's forests still represent a rich and diverse mosaic of forest ecosystems, but compared to the forests the colonists encountered, they house fewer species, contain fewer large old trees, and yield diminished ecological services.

As the U.S. population continues to rise—by an estimated 70 million over the next 30 years (WRI, 1996)—the pressures on the U.S. forest sector will intensify. Within this essentially fixed area, Americans and others around the world will demand more of virtually everything they now get from forests—from 2x4s, toilet paper, and plywood, to clean water, recreation, residential subdivisions, wildlife, fisheries, and biodiversity. Conflicts over how to manage and use forest resources are already strong, most visibly in the Pacific Northwest. (*See Box 4-1.*) The relatively fixed resource base and a growing population make further conflicts certain, but better use of scientific, social, and economic knowledge coupled with more participatory approaches to planning and decision-making could make them less frequent and less intense.

Toward Ecosystem Approaches to Forest Management

The cumulative impact of numerous local management decisions has led many scientists and resource managers to conclude that biodiversity, water quality, and other forest resources can be conserved only through cooperative efforts across large landscapes controlled, in most cases, by many owners (e.g., PCSD, 1996; Aplet et al., 1993). For the public and a growing number of forest managers, timber is no longer the only forest management goal, and sustained timber yield is no longer the only measure of success. Increasingly, the emphasis is on integrating ecological knowledge into natural resources management—or "ecosystem management."

What distinguishes ecosystem management from earlier approaches? First, decisions can be made at multiple geographic scales, and the best available scientific, social, and economic information (PCSD, 1996) is used to identify which ecosystem processes are essential to the productivity of a wide variety of natural resources and to determine which political and economic strategies will best meet public demands and landowners' objectives. Second, many such efforts are voluntary and cooperative—essential where property lines are crossed (Keystone Center, 1996).[*]

In 1992, the U.S. Forest Service and the Bureau of Land Management formally adopted policies for using ecosystem management on over 300 million acres of federal lands. More recently, one University of Michigan study identified more than 100 projects around the country that claim to be using ecosystem approaches to natural resources management (Yaffee et al., 1996), and another effort inventoried over 100 more (Keystone Center, 1996)—many initiated by conservation groups, state and county governments, private landowners, and forest-products companies.

Still, the ecosystem approach faces stiff challenges. Most efforts started after 1990, so it's hard to assess their success. Also, they require interagency and public/private collaboration, better integration of research and management activities, and the maintenance of biodiversity and ecosystem processes—none of which was understood when the policies and institutions that still govern forest management were created decades ago. Finally, this approach to forest

---

[*]   The Keystone Dialogue group defined ecosystem management as "A collaborative process that strives to reconcile the promotion of economic opportunities and liveable communities with the conservation of ecological integrity and biological diversity."

management may require new scientific, social, and economic knowledge and management skills.

Demands for U.S. Forest Resources

Along with the history of U.S. forest management, understanding current and future trends in the use of forest resources is vital. Trends in demand for forest products—very broadly defined—tell the story.

Forest Products

Over the past century, the United States has become the world's largest producer of forest products. As Table 4-1 shows, the sheer productive capacity of the U.S. forest sector dwarfs that of most industrialized and developing regions. The one exception is in energy: the United States converts relatively little of its forest wealth to meet basic energy needs, unlike such forest giants as China, India, and Brazil. Per capita, Canada, Sweden, Finland, and a few other countries, boast higher production but export the bulk of their forest products. By contrast, most U.S. production ends up in the huge domestic market.

For decades, per capita U.S. demand for wood products has increased steadily. In 1990, Americans consumed an average of 81 cubic feet of raw wood in the

TABLE 4-1. The Largest Forest Producing Nations			
	Roundwood Production (mill. cu. meters)	Per Capita Production (cu. meters)	Portion for Fuelwood and Charcoal (%)
United States	491	1.90	19
People's Rep. of China	291	0.24	67
India	282	0.31	91
Brazil	269	1.72	71
Russian Federation	244	1.66	23
Canada	172	5.97	4
Indonesia	185	0.97	79
Nigeria	115	1.09	93
Sweden	53.7	6.27	8
Malaysia	49.9	2.79	18
World	3,462	0.65	52

Source: 1991–1993 data from *World Resources 1996–97*, Tables 8.1, 9.3.

FIGURE **4-2.Estimated Consumption of U.S. Forest Products**

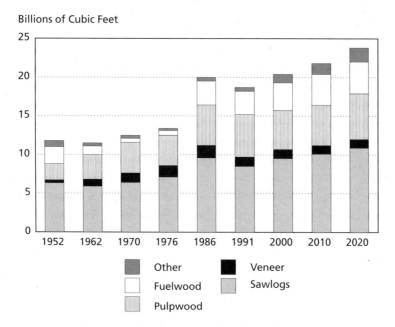

Source: Adapted from Haynes et al., 1995.

form of lumber, paper, and other forest products—33 percent more than in 1970 (U.S. Bureau of the Census, 1995). It now takes approximately 1.8 acres of timberland managed on a sustained timber-yield basis to meet these demands. Given the 490 million acres of legally available timberland productive enough to justify harvesting, the United States can now just barely meet domestic demand for wood products on a sustained-yield basis.

Overall U.S. timber consumption is best understood in terms of major uses—lumber for housing and furniture, plywood and other panels for construction, and pulp for paper and paperboard. Each is measured in different units, complicating the quick assessment of overall trends. (*See Figure 4-2.*) The rise in wood products consumption from about 12 billion cubic feet (b.c.f.) in 1950 to more than 20 b.c.f. today is steep (Haynes et al., 1995).

Lumber and Engineered Wood Products

Sawlogs for construction constitute the largest and most diverse segment of U.S. roundwood consumption. New housing construction, along with residential upkeep and remodeling, accounts for just under half of all sawlogs (Haynes et al., 1995). The rest is used for furniture, shipping pallets, and other items. Domestic lumber consumption topped 56.7 billion boardfeet in 1990 and is expected to reach 69 billion board feet by the year 2020. More than four-fifths of these lumber products draw on softwood (coniferous trees, especially Douglas-fir and southern and western pine).

This upward trend in sawnwood masks profound changes in how forest products are manufactured and used. For instance, around 1900, many homes were built of large dimension beams and solid wood flooring. Walls featured wood lath and plaster construction and solid wood trim. Today, a comparable house would be made of pine 2x4 studs in the walls, perhaps with some steel framing, and plywood flooring. Large sawlogs are increasingly scarce and expensive, while viable technological substitutes (such as steel studs, plaster board, or "engineered wood products" produced from wood chips of smaller, younger trees) are now available.

Engineered wood products constitute the fastest growing segment of the U.S. forest products industry, partly because the real prices of high-grade timber products are rising. In 1962, structural panels—including plywood, particle board, and laminates—accounted for 9.5 billion square feet ($\frac{3}{8}$ inch basis), rising to 25 billion square feet by 1990 (Haynes et al., 1995). Until the 1980s, plywood made from large-diameter softwoods was the predominant paneling material, but as older trees were used up and "oriented strand board" (OSB) and other new engineered wood products introduced, the market shifted. By the year 2010, the U.S. Forest Service predicts the use of OSB and other wafer board products will overtake plywood to command more than half of the structural panel market (Haynes et al., 1995).

Pulp and Paper Products

Demand for paper and paperboard products greatly influences the U.S. forest sector. Pulpwood—trees harvested for chips and pulp to make paper and paperboard—constitutes the second largest use of industrial roundwood in the United States. As a nation, we now consume over 90 million tons of paper and paperboard, a doubling of per capita consumption since 1960 or some 700 pounds per year for every man, woman, and child (EDF, 1995). The U.S. Forest

Service projects that paper consumption will rise by a further 44 percent to 130 million tons per year by 2020 (Haynes et al., 1995). Paper is only partly dependent on pulpwood; the rest of the fiber comes from sawdust and other wood residues, and, increasingly, from recycled paper.

Of all the forest products, paper commands the most environmental attention. Pulp and paper mills have long been recognized as major sources of air and water pollution. Paper products—tremendously versatile and inexpensive—are often equated with a consumerist "throw-away" economy. Recently, researchers have probed deeply into these issues and revealed numerous ways in which institutional and individual consumers can get more out of less paper (WWF, 1995; EDF, 1995; IIED, 1996).

Imagining life without pulp and paper products is difficult. We use paper to communicate in print to large audiences and to colleagues and friends, to store information, and to package everything from ice cream to electronic equipment. Since mid-century, paper consumption has gone hand-in-hand with economic growth, both in the United States and in much of the rest of the world.

Waste is certainly the flip side of our unrivaled paper consumption—per capita, the world's highest. Each year Americans discard 336 billion square feet of corrugated cardboard, 24 billion newspapers, 20 billion paper grocery bags, and 10 billion mail order catalogs. Collectively, pulp and paper products make up 36 percent of the country's solid-waste output, more than any other material (U.S. EPA 1995a). Roughly three-quarters of this municipal garbage is deposited in landfills. The rest is incinerated or composted. Landfills expose local communities to groundwater contamination and air pollution, and such "dumps" lower property values. Incinerators release air pollutants and ash residues that must later be cleaned up. Recycling operations can pollute the air and water and generate solid wastes, some of them toxic (OTA, 1989).

Does the issue of waste threaten the "sustainability" of the U.S. forest sector? No, but trash management does impose very tangible costs, especially to the communities, where the public shoulders responsibility for recycling and disposal infrastructure to process used paper.[*] In any case, concerns about solid waste have brought change within the forest products industry. Over the last decade, the

---

[*] Since consumers cannot easily control the junk mail, advertising inserts, and product packaging that seem to be by-products of life in modern America, the producers of all this paper that ends up in the landfill should shoulder some of the disposal costs.

FIGURE **4-3**. **Fiber Sources for U.S. Pulp and Paper**

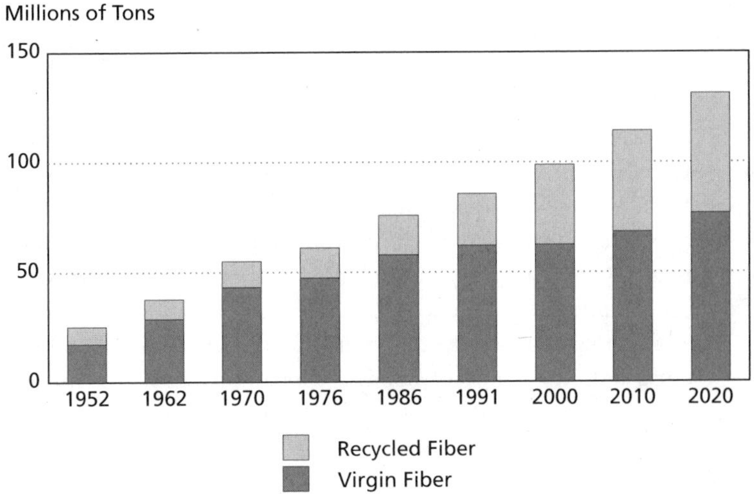

Millions of Tons

Recycled Fiber
Virgin Fiber

Source: Adapted from Hayes et al., 1995.

recycled content of new paper products has risen steadily with disposal costs. The industry already routinely recycled paper scrap internally, but the rising awareness and costs of solid waste have pushed up demand for products containing post-consumer fiber and have boosted the mix of "post-consumer" recycled content. (*See Figure 4-3.*)

The upswing in the supply of paper products with greater "post-consumer" recycled content has steered investment toward recycled-paper mills and helped change consumer preferences. The paper industry has expanded mills to accommodate recycled fiber, and some that are near sources of urban waste paper have a competitive edge over those near forests. Still, recycling tends to weaken and shorten cellulose fibers and poses other practical problems, and continuing progress in paper recycling will not totally offset the rising overall demands on U.S. forests. (*See Box 4-2.*)

Industrial pollution is the other major environmental concern facing the pulp and paper industry. Conventional air and water pollution have declined dramatically thanks to heavy investments in pollution control. Still, the industry ranks third in releases of toxic chemicals behind the chemical and primary metals sectors. In 1993, some 569 U.S. forest industry facilities reported a total of 216 million pounds of toxic chemical emissions, 89 percent of them airborne (U.S. EPA, 1995b).

Over the last decade, the pulp and paper industry has been subject to intense scrutiny as a source of dioxins—a class of extraordinarily toxic chlorinated organic pollutants—and other persistent toxic pollutants. Evidence from Europe and North America implicated chlorine-bleaching technology as the most likely culprit within the mill, and the industry has responded by changing materials, equipment, and processes. Significant uncertainty remains, partly because it is technically difficult and expensive to detect extremely low concentrations of these pollutants. New EPA regulations on pulp and paper mills sharpen the definition of minimal requirements for compliance, but are unlikely to satisfy either environmental groups or industry itself. While the contribution of the forest products industry is one of many, potential environmental impacts are serious, especially since the most controversial of these pollutants accumulate in aquatic ecosystems and the food chain.

Nontimber Forest Products

Forest products include a wide variety of wild products such as nuts, berries, mushrooms, medicinal and herbal plants, and items used in floral displays and arts and crafts. Many of these *nontimber forest products* have been used for as long as people have lived near North American forests. Others, such as the bark of the Pacific yew tree used in the anti-tumor agent taxol, are newfound. The true scope and value of nontimber forest products in the United States remains to be seen, partly because many (such as wild mushrooms or berries for home consumption) aren't bought and sold and partly because many are processed in cottage industries too numerous to count. Still, as consumer appetites for "natural" products grow and as timber-dependent communities find alternative livelihoods to offset job losses in timber harvesting, the profile of nontimber forest products and interest in their trade are rising rapidly.

Most of what is known about this part of the forest products industry is anecdotal or limited to snapshots of production and trade in one product or region. One exception is in the Pacific Northwest, where the wild edible mushroom industry and the collection of wild plant materials for floral greens and Christmas displays have been relatively well documented. Collectors now gather about a dozen species of mushrooms for commercial trade. In 1992, the wild edible mushroom harvests in Oregon, Washington, and Idaho totalled an

---

BOX 4-2. Paper Recycling: Necessary but Insufficient

Recovery of recycled paper is a modern environmental success story. Practical changes in the home, school, and workplace are diverting municipal solid waste from landfills and incinerators (U.S. EPA, 1995a). Compared to virgin wood fiber, recycled fiber requires lower chemical inputs and produces less air and water pollution at the mill (EDF, 1995). But even continued progress in paper recycling will not assure a sustainable forest sector.

Why not? For starters, pulp and paper constitute only a fraction of overall demand for forest products and most paper products manufactured from recycled fiber also require some virgin fiber. More importantly, decreasing demand for pulpwood frees forestland for other uses. As an example, when one East Coast paper manufacturer decided to build a new mill to produce newsprint from recycled paper, it financed the investment by selling off some of its forest holdings, which were then subdivided for vacation homes.

To complicate the picture even more, about one-quarter of the supply of virgin wood fiber to the pulp and paper sector comes from sawmills and other operations (Haynes et al., 1995), so increasing the content of post-consumer fiber in paper production may not relieve pressure on forest resources. This finding is underscored by U.S. Forest Service models that take into account domestic and international forecasts of supply and demand. In an intensive recycling scenario assuming a 60:40 ratio of recycled to virgin fiber, together with a real decline in waste generation, U.S. pulpwood demand would drop from 5.9 to 5.1 b.c.f. (Ince, 1994). Even this would be partly offset by a rise in sawlog demand of 0.2 b.c.f.

---

estimated 4 million pounds, for which harvesters were paid approximately $21 million, and value added through processing, marketing, and distribution contributed some $20 million more to the regional economy (Schlosser and Blattner, 1995). In 1989, the contribution to the regional economy from the wild greens, Christmas boughs, and decorations florists use was pegged at roughly $130 million (Schlosser et al., 1991). Harvesting yew bark for taxol production, fiddlehead ferns for salads, and wild herbs for medicinal use generates millions more for the region from its forests (Cohn, 1995).

The harvest of these products is not always a blessing for the forest environment and local economies. Mushroom collecting in parts of the Pacific Northwest has become so intense that scientists are now assessing its impact on the close relationship between trees and the fungi's subsurface mycorrhizae (which draw moisture and nutrients to tree roots). Others fear that overharvesting could make species scarce or extinct and take an unpredictable toll on other forest-dwelling species (Cohn, 1995). One example: the flying squirrels that the endangered northern spotted owl preys upon eat the same fungi that are being overharvested in some areas. In the Pacific Northwest, the nontimber forest product industry has bolstered the regional economy, but much of the lucrative processing and marketing takes place in other U.S. regions or even other countries (namely, the Netherlands for floral display materials and Japan for matusake mushrooms). Further, for most people, harvesting these products—mushrooms or maple syrup—is a seasonal occupation only.

Harvesting mushrooms and other nontimber forest products could never substitute for the multi-billion dollar timber industry. Still, they are probably more important to local forest economies than research and speculation suggest. At a minimum, the annual value of these products totals hundreds of millions of dollars nationally—and perhaps much more—with some markets expanding by 15 to 20 percent annually in recent years (Goldberg, 1996). If expenditures for recreational hunting and fishing or the market value of herbal medicines based on forest plants were included, the total would quickly climb into the billions. In any case, harvesting these specialty products can diversify local timber-dependent economies (Goldberg, 1996).

Public Benefits from U.S. Forests

Besides forest products, an enormous wealth of other public benefits flow from forests. They protect water quality, sequester atmospheric carbon, house biodiversity, provide recreational opportunities and aesthetic pleasures, and intangibly contribute to social, cultural, and economic development. As with forest products, demand for these "goods"—some of them newly recognized—is escalating steadily.

Since these benefits are generally unmarketed, it's difficult to accurately gauge demand for them. Some of the best information comes from the U.S. Forest Service, which periodically estimates national demand trends for recreation, fish and wildlife use, water quality, and other nontimber forest benefits. Otherwise, data is diffuse, anecdotal, or spotty where it exists at all.

Watershed Protection

Most of the high-quality surface water in the United States, and much of our groundwater, comes from forests. Since many are in rainy or mountainous areas, on average forests receive twice as much precipitation as other land areas in the

United States (Ellefson, 1992). They also store and release water four times more efficiently than nearby nonforested lands, and, by slowing rates of overland flow and enhancing infiltration rates, forests improve surface water quality and groundwater recharge rates. Finally, forests retain soil and stabilize loose rock, gravel, and dirt on slopes—good flood protection.

Conversely, where forests have been removed or poorly managed, enormous amounts of money are spent to build filtration plants, repair flood damage, or to recover fisheries devastated by habitat loss. In New York City; Seattle; Washington; and Portland, Oregon—all dependent on upstream forests to protect drinking-water quality—millions get clean water from the tap without investing billions of dollars in water-filtration plants. The city of New York is spending over $1.2 billion to protect its principal forested watersheds in the Catskill Mountains through land purchases and zoning and development agreements with local governments. The alternative would be to spend nearly $3 billion on water filtration plants that EPA would require if water quality deteriorated. Seattle and Portland face hundreds of millions in water-filtration plant investments if their water quality cannot be protected from sediment caused by logging and road building in city watersheds. Inappropriate logging practices, especially on steep slopes in the West, have been increasingly implicated as major factors in flood damage. Between 1935 and 1992, the number of large deep pools (critical habitat for salmon and steelhead trout populations) on 412 miles of monitored streams in Washington and Oregon's National Forests declined by 60 percent due to clearcutting and logging road construction on steep slopes (Thomas et al., 1993). Similar problems plague the intermountain West, especially in Idaho and Montana. Some experts point to a four-decade legacy of clearcut logging and road building on steep slopes in western Oregon, Washington, and northern Idaho to help explain the severe and costly floods and mudslides in February and December 1996 that cost the region several billion dollars and over a dozen lives (Jones and Grant, 1996; Sleeth and Bernton, 1996). If the cost to the public of flooding, watershed erosion, and fisheries habitat loss could be charged against the financial returns from clearcutting on steep slopes, clearcut logging would abruptly end in most of the West (O'Toole, 1988).

Timber and watershed management can complement each other if topography, silviculture, and harvesting practices are considered at the same time—as the Forest Service's original mission requires. But though nearly all state forest legislation and voluntary codes of Best Management Practices emphasize water protection and post-harvest reforestation, timber extraction has clearly unbalanced this equation in many places during the past 50 years, especially on steep slopes in the West.

Recreation

Even if the estimates and projections for forest recreational demand produced by the U.S. Forest Service were halved, it would still be growing faster than demand for any other forest products or services. From hiking and outdoor photography to camping, rafting, and wilderness backpacking, forest recreational activities are expected to skyrocket by 60 to 190 percent during the next 40 years. As one indication, the number of visitor-days registered in the National Forest systems is ten times higher today than it was in 1950 (Barber et al., 1994). A growing number of private forest land owners are leasing lands to hunters and fishermen. In Alabama, for instance, hunting rights on private forest lands typically lease for $3 to $5 an acre per year, and sometimes for as much as $20 (Johnson, 1993).

The economic value of forest-based recreation is difficult to estimate precisely, but sketchy evidence does suggest it is not far behind timber values on a national basis. In parts of the Northeast and in the Rocky Mountain West, forest-based recreation is clearly more valuable to regional economies than wood products.[*] Even in the Pacific Northwest during the timber boom in the late 1980s, forest-based tourism and recreation on National Forest lands in Oregon and Washington was estimated at two-thirds the value of the agency's timber sales (Barber et al., 1994). Nationwide, the market value of outdoor recreation in the National Forest system is estimated at over $6.6 billion per year—or a little over $20 per person for each of the nearly 300 million visitor-days annually (O'Toole, 1995).

Randall O'Toole, an Oregon-based forest economist, believes charging visitors more for recreational activities is one of the best ways to get the most out of public lands. This approach would help to end environmentally damaging below-cost timber sales on National Forests and alleviate chronic funding shortages at national parks. With actual recreation receipts of only $50 million annually, it's hard to argue with O'Toole's basic logic—receipts help to drive

---

[*] For example, O'Toole (1988) estimated that the value of recreational benefits on the Beaverhead National Forest in Montana outweighed benefits from logging by nearly an order of magnitude. In some cases, Forest Service personnel knew timber sales were directly diminishing more valuable recreational benefits on the forest but could do little about the false economy that has grown up around timber harvesting on public forests in the Rocky Mountains.

budgets and management priorities so recreational opportunities frequently take a back seat to commodity production. On the other hand, critics of the "let the market decide" approach argue that the gondolas, condos, amusement park rides, scenic drives, and off-road vehicle trails that might result from relying heavily on tourism and recreation demands may be no better for forest sustainability than timber management.

Expected gaps in the supply of specific recreational opportunities in coming decades are likely to vary by region. The Forest Service's 1993 RPA update suggests most future demands on National Forests can be met, but that users (especially backpackers, cross-country skiers, hikers, and mountain bikers) will have to travel farther and spend more to get what they want (USDA, 1994). In the eastern United States, shortages in the supply of forest-based recreational activities are likely to be acute near metropolitan areas. But lease agreements, access fees, and the like could increase recreational opportunities on private forestlands where public lands are limited.

Biodiversity

The United States possesses a varied array of forest ecosystems. The U.S. Forest Service uses 10 eastern forest types and 12 western forest types in its forest inventories, though ecologists typically distinguish many more. From these ecosystems Americans can extract a multitude of timber species, an even greater variety of wildlife species, recreational opportunities, and species and genetic diversity of economic value, especially important in the face of climate change.

The extensive loss of natural forest habitats and declining populations of many forest species (and their genes) suggest that biodiversity loss will become increasingly important in coming decades. Not all signs argue for the worst. Populations of white tail deer, wild turkey, elk, and a number of other large mammals have increased dramatically during the past century, thanks partly to better forest management practices and forest regeneration on cleared lands in the eastern United States (MacCleery, 1992). Still, most data suggest worrisome trends in the near future. One Nature Conservancy review of the conservation status of 20,500 species (most of which live in or near forest habitats) finds that roughly one-third of all U.S. species warrant concern. An estimated 2.1 percent of the species have become extinct (or have not been seen in several decades) during the past 200 years (TNC, 1996). These losses can't be laid to logging alone. Residential subdivisions, roadways, commercial development, and industrial pollution erode habitats and biodiversity.

Aesthetic Benefits from Forests

The aesthetic benefits forests can provide for rural and residential environments are increasingly valued by the public—witness the rapid growth of greenways, the protected open spaces and natural habitats within growing metropolitan and semi-rural areas. In cities and suburbs, greenways follow rivers, streams, canals, roads, or biking trails to link parks, natural areas, and historic sites with each other and with populated areas. In the countryside, greenways are planned natural corridors linking state parks, forests, or wildlife refuges together. In increasingly fragmented landscapes, greenways can increase property values and property tax revenues, attract tourists, and boost local spending on recreational activities (Diamond and Noonan, 1996).

The Structure of the U.S. Forest Products Industry

The product chain for lumber, panels, and paper links a complex set of landholders, loggers, primary processors, and secondary manufacturers of housing, furniture, packaging, and more. As the world's largest producer of wood fiber and forest products, the U.S. forest products industry is a significant player in the domestic economy. In 1992, sales of forest products surpassed $200 billion, making it the country's ninth largest industrial sector (U.S. Bureau of the Census, 1995). If furniture, envelopes, corrugated boxes, and other finished products are included, the industry employs about 1.6 million Americans.

The U.S. forest industry remains highly diverse in terms of company size, vertical integration, product lines, control over forest resources, and access to capital and markets. Its shape resembles that of an asymmetrical hour glass. At the top are nearly 270 million consumers who use a bewildering array of forest products from telephone poles to telephone books and from plywood to rayon. Supplying them are hundreds of thousands of retailers, including lumber yards and home-improvement stores, furniture outlets, office supply and stationary stores, and thousands of general retailers, such as supermarkets, discounters, and department stores, that sell everything from paper towels to chairs. These retailers are supplied by several thousand wholesalers who distribute various forest products. In addition, tens of thousands of institutional consumers, including fast-food franchises, school districts, government agencies, publishers, and hospitals, purchase most of their forest products directly from wholesalers. Some wholesalers' lines may come from secondary manufacturers, such as cabinet makers, specialty paper product makers, or suppliers of finished building materials.

At the waist of the hour glass are the primary manufacturers of forest products. Dominated by several dozen companies, this industry produces lumber products, engineered wood products, pulp, and bulk paper. The raw materials come from one of three categories of forest land owners: several million private, nonindustrial timberland owners who control 285 million acres of timberland (or 58 percent of the total timberland area); several hundred firms in the forest industry that control 71 million acres of timberland (14.5 percent); and state and federal government agencies that manage the remaining 130 million acres (26.5 percent). Left out of this oversimplified profile are suppliers of nonwood materials needed in manufacturing (say, chemicals for pulping), foreign exporters of wood products to U.S. markets, consumers of U.S. wood products overseas, and industries and consumers involved with nontimber forest resources.

In this multifaceted industry, more diverse strategies are required to fix problems than in more homogenous industries, such as electric utilities or automobiles. But, by the same token, efforts to improve industry performance can be tailored selectively to address the most promising opportunities first; there is no need to persuade the whole industry to move in concert.

How decision-makers respond to environmental issues at the same point along the production chain also varies enormously. For example, while the total market for paper products is huge, end-use markets range from newsprint, tissue, and freesheet paper (such as office paper and stationary) to packaging. Consumers who may want to recycle newspapers can't necessarily influence the overall environmental performance in an industry that also manufactures many other paper products. Further, many of the most serious environmental impacts occur long before the ultimate consumer comes into the picture, so even those opting for environmentally responsible products have no way to cast their vote for a cleaner, greener forest sector.

Timing is important too. A paper company planning to replace old fully depreciated mill technology to take advantage of markets for "chlorine-free" paper will be more likely to install new pulping and bleaching technologies than a company that updated its mills ten years ago and still has 20 years to go to pay off the debt. Similarly, the owner of a small nonindustrial forest is less likely to be able to undertake "green" forest planning and harvesting practices than a large corporate landowner.

The flexibility needed to make sustainability a reality is most constrained at the manufacturing level, especially in pulp and paper companies—among the most capital-intensive industries in the United States. A modern paper mill can cost one billion dollars. Much of this investment is embedded in the technology for transforming raw logs, chips, and recycled paper into large volumes of high-quality paper at the lowest possible cost. Over the past three decades, an increasing share of this enormous investment has been devoted to pollution-control technology. Once these extraordinarily expensive decisions are made, the mill has a financial incentive to operate as close to full capacity as possible.

Some notable efforts are being made to link better forest management, cleaner production, and markets for forest products. The Forest Stewardship Council, based in Oaxaca, Mexico, is establishing international criteria for certifying that forest products are sustainably produced (FSC, 1995). So far, independent evaluators accredited by FSC have certified more than 30 forest managers and companies around the world that together control more than 10 million acres. In February 1996 Several Swedish companies managing 22 million acres applied for independent certification under the FSC principles (Greenwire, 1996), six U.S. forest products companies managing over 1 million acres were granted FSC-certification as of December 1995. Certification applications are pending for many other companies.

Industry is also trying to assure consumers that forest products are being harvested and processed sustainably. In 1995, the American Forest Products Association launched the Sustainable Forestry Initiative (SFI) to devise principles for sustainable forest management and help companies monitor themselves (AFPA, 1995). Product labeling to convey information about harvesting and processing of the product itself is part of such market-based attempts to make claims of sustainable forest management credible.

The U.S. Forest Sector in a Global Context

Today, the U.S. forest sector is becoming less isolated from economic and forest-resource trends elsewhere in the world. While the stock of forests is essentially stable in the United States, demands on them will continue rising rapidly. Most demands will remain domestic, but three trends ensure that international issues will play a more important future role in U.S. forest policy and management. First, regional wood shortages will emerge in the United States, increasing wood imports and making greater investment by American forest product companies overseas necessary. On the other hand, as cheap sources of wood fiber are exhausted in Asia and other parts of the world, countries that depend heavily on wood imports are searching for supplies around the world, including in the United States. Finally, the sustainable management and conservation of forests has become an international political issue driven by concerns about biodiversity loss, climate change, deforestation, and forest degradation.

U.S. market demands for forest products are enormous compared to those in virtually any other country. Americans (less than 5 percent of the world's population) now use nearly one-third of the world's paper production and over one-fourth of all industrial roundwood (FAO, 1995). The U.S. annual average of 700 pounds of paper per capita compares with 500 pounds in Japan, 430 pounds in Canada, and about 350 pounds in the European Union (EDF, 1995). Americans use 10.7 times as much paper as Latin Americans, 14 times the average in Asia, and over 61 times the average consumption in Africa (WWF, 1995).

As demand grows and domestic supplies hit constraints, the United States will join the global search for fiber in a big way. The United States is now one of the world's largest importers of forest products, primarily from Canada: over 36 million cubic meters of sawnwood (boards, planks, beams, etc.) crossed our northern border in 1993. We are also net importers of structural panels and paper and paperboard, though the U.S. trades significantly in both directions. (*See Figure 4.4.*) This search is opening up the last large unexploited tracts of natural forest ecosystems in South America's Amazon Basin and Guyana Shield, Central Africa, the Russian Far East, and Canada (Bryant et al., 1997). Fueled principally by Japan, Korea, China, and Europe, international fiber markets are also stimulating the rapid growth of extremely productive plantations in parts of Brazil, Chile, New Zealand, and Australia—far-flung venues where American firms too are seeking low-cost fiber supplies.

Growing U.S. reliance on Canadian sawnwood, Chilean logs, tropical plywood, and other imported forest products may help to make forestry unsustainable in British Columbia, Chile, Russia, Indonesia, and elsewhere. In our quest to become more sustainable at home by, for example, protecting old-growth forests in the Pacific Northwest, we may be driving world markets to dig deeper into the unprotected forest resources of the Pacific Rim, Russia, the Amazon, and other regions, basically exporting unsustainability (WWF, 1995).

But, even as the United States is looking overseas for timber supplies as never before, other countries are buying more wood fiber from the United States, making it the world's largest exporter of softwood logs, wood chips, particles, and wood residues (sawdust and wood scraps). Our biggest customer is Japan—about 90 percent of the market for U.S. softwood logs (FAO, 1995). Since most of this wood comes from private lands (logs harvested on federal lands can't be exported) in the Pacific Northwest where logging has been sharply curtailed in the national forests to protect old-growth forests, log exports have been blamed for "exporting" jobs (Barber et al., 1994). Global demand for paper and wood panels is projected to grow steadily by 3 percent annually over the next decade (FAO, 1994), fueled mainly by the emerging economies in Asia and, to a lesser extent, in Latin America and Eastern Europe.

Since per capita GNP is an excellent predictor of per capita paper consumption, global roundwood consumption seems destined to rise as economies grow. Investments in tropical plantations, technological efficiency, and increased recycling will probably meet much of this demand, but natural forests in the United States are also likely to feel the pinch. Could this challenge the sustainability of U.S. forest resources? Perhaps. Steadily rising stumpage prices could lure growing numbers of small private landowners with little knowledge of sustainable forest management practices into the market, putting natural forest habitats, water quality, and soil productivity at risk. Hardest hit would be the rural South, where incomes are lower and export facilities are already equipped to handle large shipments of pulpwood and wood chips.

In sum, the U.S. resource base probably can't be expanded enough to meet domestic and international demands for timber products and many other goods and services, so tough choices for Americans as consumers, as stewards of forest land, and as citizens lie ahead.

SUSTAINABILITY AND THE FOREST SECTOR

What are the competing concepts of sustainability? Which work best for the forest sector? And what are the most important challenges to maintaining a diverse, healthy, and productive forest sector in the United States?

Concepts of Sustainability

The science of silviculture and the profession of forestry were founded in Germany and France over two centuries ago on a relatively simple concept that the harvest could be regulated to maintain a dynamic balance between timber growth and harvest volume from one year to the next (Westoby, 1987; Mather, 1990). Principles of sustained timber-yield management were first applied in the United States a century ago. But concepts of forest sustainability are evolving rapidly, and sustaining biodiversity, community economic vitality, and recreational opportunities now figure in current thinking for two reasons. First, beginning in the 1950s, evidence had begun to reveal that sustained timber-yield alone was insufficient or even incompatible with maintaining fish, wildlife

FIGURE **4-4. The U.S. Forest Sector in a Global Market**

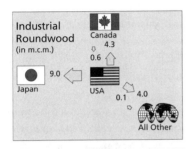

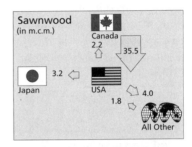

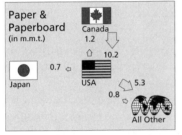

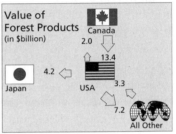

Source: Adapted from FAO Yearbook of Forest Products, 1993.

habitat, and some other nontimber forest products and services (Lyden et al., 1990). Second, the Bruntland Commission report, *Our Common Future*, prompted wider recognition of the social and economic dimensions of sustainability (WCED, 1987). Concepts of sustainability were expanded beyond such biophysical factors as timber yield, water quality, wildlife habitat, and regenerative ecosystem processes to embrace economic vitality, a fair sharing of benefits, and meeting present and future human needs.

But if sustainability concepts have grown more sophisticated, they have also become more debatable than the concept of sustained timber yield, and perhaps fuzzier. Gale and Cordray (1991) classified forest sustainability concepts into eight groups ranging from "dominant product sustainability," in which the ecosystem is manipulated to yield high commodity outputs, to "ecosystem-centered sustainability," in which natural processes are not manipulated at all. Rather than seek a single definition of sustainability, some researchers (Poore et al., 1989; Aplet et al., 1993) consider it wiser to define a broad and dynamic concept of sustainability at a large geographic scale (say, the national level) and then choose more specific goals at regional and local levels (say the watershed level).

As defined broadly here, a sustainable forest sector *maintains the productivity of forest ecosystems, provides a diverse array of products, environmental services, and social and economic opportunities for all citizens, and protects ecosystems' capacity to adapt to environmental change and disturbance.* A sustainable U.S. forest sector should also help solve such broader sustainability concerns as energy production, climate change, and industrial pollution.

Of course, many issues influence sustainability in the forest sector. Suburban development converts forest lands to nonforest uses. Clearcut logging practices invite soil erosion on steep slopes while intensive forest management eliminates older mature forest (what biologists call late successional forest stands). Pulp and paper production generate conventional pollutants that de-oxygenate our waterways and contain potentially dangerous toxins. Forest industries consume prodigious amounts of energy, and consumers of forest products generate mountains of municipal solid wastes.

To find out which of these issues are of strategic importance to a more sustainable U.S. forest sector, four sustainability criteria can be applied—*the direction of trends, the degree of reversibility, the availability of substitutes, and the nature of the risks.* (*See Box 4-3.*) Viewed this way, the most important determinants of sustainability in coming decades will be maintaining forest productivity, safeguarding biological diversity, adapting to climate change, and preventing industrial pollution. (*See Table 4-2.*)

Challenge #1: Maintaining Forest Health and Productivity

Ecosystem productivity measures forest ecosystems' capacity to sustain life and to supply people with necessities. Although the critical data needed to directly measure most aspects of forest productivity is still lacking, one long-used measure—the net annual growth of timber species—does allow us to monitor and forecast one critical and economically important aspect of sustainability—sustained timber yield harvesting.

The convergence of several trends shows why forest productivity is an important sustainability issue. First, the country's 490 million acres of timberland area is projected to continue declining slightly to about 460 million acres in 2040 (Haynes et al., 1995). Second, wood demands and harvest levels are projected to increase by more than 40 percent during this same period. Third, timber productivity has not grown in recent years while harvest levels have. Finally,

BOX 4-3. Four Criteria for Assessing Sustainability Issues

1. Direction of Trend: "Is the Situation Getting Better or Worse?"
What are the long-term quantitative changes in a forest resource, environmental condition, or other factors relevant to the productivity of forest ecosystems and the U.S. forest sector?

2. Reversibility: "Are the Consequences Permanent?"
Is the desired forest resource, ecological process, or environmental condition irrevocably altered, depleted, or lost in harvesting, manufacturing, or distributing forest products? Few processes—except species extinction—are absolutely irreversible, but in practical terms (cost, time, social and political support, etc.) many may be.

3. Substitutability: "Are There Viable Alternatives?"
Is a forest resource, product, or manufacturing process unique or does it have alternatives? Sustainability concerns will be less compelling if practical alternatives are available and economically and technologically feasible than if they aren't.

4. Risk: "What's at Stake?"
What do we know about the threats that affect the health and productivity of forest ecosystems, the forest sector, and regional economies? Significant risk and substantial uncertainty elevate sustainability concerns.

natural tree mortality rates appear to be rising, possibly signaling more widespread forest health problems.

Failure to maintain forest productivity in coming decades will force the United States to consume fewer wood products on a per capita basis, import a greater percentage of these products, establish plantations on agricultural and other lands, harvest timberlands unsustainably to meet growing demands for wood products, or possibly open public forest areas (such as watershed reserves and wilderness) now off limits to logging. Several of these roads lead away from sustainability as defined here.

For most of the past half century, increased timber productivity (annual growth minus annual natural mortality) has been one of U.S. forestry's greatest successes. While timberland area in the United States has declined slightly since 1952, productivity has increased dramatically. Overall, net growth increased approximately 2.7 percent annually between 1952 and 1991 (Powell et al., 1993). As a result, the volume of timber theoretically available for harvest grew from 11.8

TABLE 4-2. U.S. Forest Sector Sustainability at a Glance

Sustainability Challenge	Criteria 1: Trends	Criteria 2: Reversibility	Criteria 3: Substitutability	Criteria 4: Risk
Forest Productivity	Historical growth in timber productivity is slowing, and in some cases declining.	Trends can be reversed over time, provided that the causative factors such as nutrient loss and air pollution are addressed.	Slower growth in forest productivity boosts demand for wood substitutes, but many nontimber values can't be readily replaced.	Consequences of this trend include future scarcity of forest products and greater demand for nonwood products and services.
Biodiversity	Changes in forest ecosystems are contributing to endangered species.	Species loss is irreversible; redressing some causative factors like habitat loss could take decades or more.	There are no true substitutes for biodiversity, an indicator of overall ecosystem health.	Difficult to assess at-large, but in some ecosystems, loss of keystone species can lead to widespread ecological impairment.
Climate Change	Managed forests are increasingly vulnerable to changes in climate change.	Resiliency to climatic changes requires long time scales for more robust forest ecosystems.	Not all species are equally susceptible, but shifting the species mix of forests takes decades.	Diminished timber productivity, increased susceptibility to pests, fire and air pollution.
Clean Production	U.S. forest products makers face competition from less polluting manufacturers.	Conventional pollutants have short duration impacts, but accumulation of trace toxic pollutants is practically irreversible.	There are no practical means of substituting for persistent environmental pollutants.	Human and environmental health impacts are suspected, large costs of overhauling industrial technology.

billion cubic feet in 1952 to 21.6 billion cubic feet in 1991, and net annual growth is now estimated to be more than 3.5 times greater than it was in 1920 (Fedkiw, 1989).

Several factors explain this surge in timber productivity. Widespread post-harvest replanting on private lands did not begin until the 1950s, when forest practice regulations, growing demand, regional supply shortages, and other economic and policy factors stimulated greater investment in reforestation. The average age of forest stands decreased as older stands were replaced by vigorously growing younger stands. Second-growth forests began to reclaim large areas of abandoned farmland in the East, starting in the 1930s. In the 1950s and 1960s timberland owners (especially industrial owners) began to invest in intensive forest-management practices, such as thinning, tree farms, chemical control of insects and competing vegetation, and development of highly productive seedling stock. And wildfires declined from an average of 40 to 50 million acres burned annually in the 1930s to an average of between 2 and 5 million acres burned annually since the 1960s (MacCleery, 1992).

Thanks to stunning productivity growth, national harvests have remained below net annual growth for the past half century. In 1952, timber growth nationally surpassed harvests by 17 percent. By 1976, the comparable figure was 54 percent. Today, sustained-yield harvesting is the norm in all regions of the country and for all types of operations, except on forest industry lands, where softwood harvests exceed growth.

Yet, for two reasons, the surplus of growth over harvests could over coming decades become a deficit. First, this century's steady increase in timber growth is projected by the U.S. Forest Service to slow dramatically. One base-case scenario by Richard Haynes and his colleagues at the U.S. Forest Service (1995) projects only slight increases in net annual growth in future decades—averaging approximately 0.03 percent annually for all timberlands between 1991 and 2040. Alternative scenarios by the same team follow the same curve. Second, harvests are projected to increase steadily—43 percent, according to the Haynes team's projections, an annual rate of just under 1 percent. Most of this harvest increase is projected to be concentrated in the South, where the surplus of growth over harvests is already steadily disappearing. Under these projections, after around 2010, sustained yield timber management as conventionally defined will no longer apply to most U.S. timberlands.

The most recent data on timber growth suggest that even the low growth projections made by the U.S. Forest Service may be optimistic. (*See Figure 4-5.*) Between 1986 and 1991, net growth levels declined by 2 percent—the first national decline since official records began. Nearly all of the decline reflected a 4.4 percent fall-off in softwood growth. Hardwood net growth rates have slowed considerably since 1976 from an average national increase of 2.1 percent between 1952 and 1976 to an increase of only 0.16 percent annually between 1976 and 1991. On nonindustrial lands in the East, net annual growth has fallen by 4.3

FIGURE 4-5. U.S. Timber Growth and Removals, 1920–1992

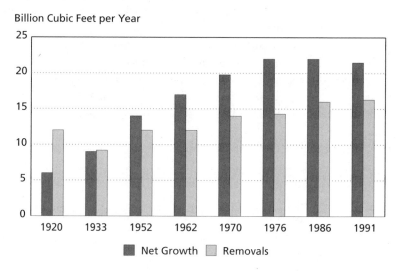

Source: Powell et al., 1993.

percent since 1976 (Powell et al., 1993)—the largest and longest productivity decline of any ownership group in the country.

The excess of harvests over growth does not mean that the United States will lose forest area or run out of wood supplies any time soon. But it does mean that the average size of harvested trees and the average age of timberlands will fall. Indeed, the average diameter of harvested trees from all timberlands decreased by over 20 percent between 1976 and 1991, although the rate of decline is projected to slow since most older stands on private timberlands have already been logged (Haynes et al., 1995). More striking is the complete loss of older forest stands on private timberlands in the Pacific Northwest and in the South. (*See Figure 4-6.*) By 2010, Haynes and colleagues project that virtually no forest stands older than 60 years will remain on forest industry lands in the Pacific Northwest; in the South, none older than 35 years on forest industry lands and very little over 45 years on nonindustrial lands. As a result, most of the available U.S. timber inventory in coming decades will be near or below minimum harvesting size—a significant threat to the survival of some rare and endangered species that live in older forest habitats. Meanwhile, the falling average diameter

FIGURE 4-6. Timberland Area by Age Class

Forest Industry Lands in Pacific Northwest

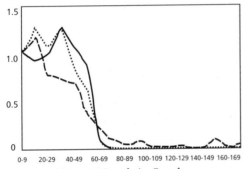

Nonindustrial Private Lands in South

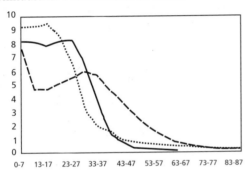

Legend

----- 1990

·········· 2010

——— 2030

Forest Industry Lands in South

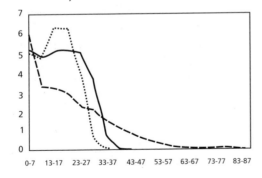

Source: Haynes et al., 1995.

of trees on private timberlands is reflected in sharply rising increases in stumpage prices for saw timber, which typically comes from bigger trees. In real terms, prices for softwood saw timber are projected to nearly triple over the next four decades while softwood pulp prices (which can be made from small diameter trees) are projected to rise by only about 30 percent.

Causes of Forest Productivity Problems

Various factors are cited to explain stagnation in net timber growth rates, though their relative contributions are difficult to determine. Declining investment in forest management, aging of second- and third-growth forest stands on private lands and on National Forests, extended drought, severe pest outbreaks and fire damage, and the conversion of highly productive timberlands to suburban or other uses all apparently play roles.

Some of these factors can be at least partially substantiated. For example, while the forest industry in the South increased per acre investments in softwood forest management by 13.2 percent between 1952 and 1992, investments on nonindustrial lands in the same region did decline by about 10 percent (Wear, 1993). Meanwhile, southern nonindustrial timberland area declined by 16 million acres, offsetting the 8 million acres added by the forest industry (Powell et al., 1993). On public lands, the average age and size of trees has increased, so net annual growth is slower (Powell et al., 1993). Severe droughts in parts of the Southeast and South central regions in the late 1980s and early 1990s could also have reduced tree growth between 1986 and 1991.

Other factors on the list are hard to verify. The country's timberland area decreased by only 0.5 percent between 1977 and 1991, and actually increased by 1 percent between 1987 and 1991, when net annual growth on a national basis declined for the first time (Powell et al., 1993). While 1986 through 1991 saw many fires, most notably in Yellowstone and the northern Rockies in 1988, most of the area burned was in National Parks and wilderness areas outside the timber base. And nationwide, the average age of forest stands on timberlands decreased, especially in the South, presumably paving the way for greater timber growth rates.

One factor in the recent downturn in timber productivity is increased timber mortality. Between 1986 and 1991, it increased by 24 percent—four times the greatest increase in any five-year period between 1952 and 1986 (Powell et al., 1993). If the high mortality rates continue, forest health, already a priority concern of the U.S. Forest Service, could become a major sustainability issue in much of the country vis-à-vis both timber growth and forest ecosystem health more generally (USDA, 1994). The impact of air pollution, forest-management practices, and climate change on forest health have been widely debated since the 1970s and 1980s, when acid precipitation was implicated in forest diebacks in the Appalachians (MacKenzie and El-Ashry, 1989). Since then, severe fires in the West, reports of high mortality in sugar maples in New England, and potentially widespread growth declines and increased mortality in selected species on the Allegheny Plateau have reignited the debate (e.g., Little, 1995; Loucks, personal communication). (*See Boxes 4-4, 4-9, and 4-10.*) While there is growing consensus that air pollution, suppression of natural fire regimes, and heavy logging of selected species have increased tree mortality in some areas, their relative importance compared to natural factors and the amount of forest area suffering significant damage are poorly understood and hotly debated. (*See Box 4-4.*) Even so, forest health problems exacerbated or caused by air pollution and climate change seem likely to increase in coming decades and reduce the availability of some forest resources and environmental services.

Responses to Forest Health and Productivity Concerns

Concerns about destructive logging practices and minimal investment in refor-estation on private lands first prompted Gifford Pinchot (and later others) to seek federal authority to regulate private forest lands (Clary, 1986). They failed, but today between 15 and 20 states do have state forest practices acts, mainly to maintain or improve timber resources and protect water quality (Boyd and Hyde, 1989). Most such regulations specify post-harvest stocking densities or require landowners to replant after harvesting. Some are also intended to maintain future inventories or restrict the location and size of harvesting operations in sensitive wildlife habitats and along rivers and streams. Some other states, especially in the South, rely on voluntary "best management practices" to encourage timberland owners to maintain productivity and protect environmental quality, and local governments are also showing increasing interest in passing forest ordinances (Martus et al., 1995; Ellefson and Cheng, 1994).

  In recent decades, economic factors have driven most of this shift within the forest industry. Increased silvicultural knowledge, federal tax credits for refores-tation, genetically improved seed stock and other technological improvements, and the prospect of long-term steady increases in demand have made investments in forest productivity attractive, even though stumpage prices fluctuate. As a result, timber productivity on industrial lands has increased faster than on private

BOX 4-4. Air Pollution and Forest Health

Air pollution has long been linked to acute tree injuries around smelters and other industrial facilities. Before pollution controls were mandated, smelters in such places as Spokane, Washington; Kellogg, Idaho; and Sudbury, Ontario; caused 100 percent tree mortality within several miles and substantial tree injury and death for up to 50 miles (MacKenzie and El-Ashry, 1989). More recently, lower levels of air pollution have been firmly linked to widespread forest declines in Europe and Asia and to more limited declines in high elevation areas in the eastern United States. More uncertain is whether air pollutants—even at the lower levels required under the 1990 Clean Air Act Amendments—exacerbate environmental stresses and insect and disease problems over wider forest areas.

In the 1970s, European and U.S. researchers linked air pollutants, especially sulfur dioxide (SO_2) and nitrogen oxides (NO_x), to the acidification of rain and snow. This in turn was linked to the acidification of freshwater lakes and streams, and researchers hypothesized that acid rain was responsible for damage to foliage, productivity declines, and increased mortality in crops and trees in Europe and North America. Research documented the decline of spruce-fir forests at high elevations in the Appalachians and low pH levels in rain, snow, and fog. Researchers also tracked atmospheric transport routes that could carry emissions to forests far removed from industrial sources and identified mechanisms by which acid deposition could damage or kill trees.

nonindustrial and public timberlands (Wear, 1993). On the other hand, in most areas owners of small nonindustrial forest lands have reduced such investments, whether from a lack of knowledge and capital or the desire to pursue nontimber management objectives. However, voluntary programs, such as the Tree Farm Program sponsored by the American Forest and Paper Association and the Federal Stewardship Incentive Program, are widely recognized for improving timber productivity on lands managed by thousands of small landowners. Several states have programs to encourage small owners to invest in forest management (for example, the Oregon Forest Stewardship Program uses lottery proceeds to provide low-interest loans to qualifying landowners to restore productivity on degraded forest lands).

Forest health problems, on the other hand, are notoriously difficult to define, diagnose, and treat. Scientists don't fully understand how forest ecosystems respond to both natural disturbances and human activities, and a huge diverse

BOX 4-4 cont.

Still, an extensive research effort during the 1980s—the National Acid Precipitation Assessment Program (NAPAP)—failed to conclude that acid precipitation caused the observed declines on the summits of Mt. Mitchell in North Carolina and Camel's Hump in Vermont. This assessment set off controversy in the scientific community. Citing evidence that up to 75 percent of the tree volume died in less than 20 years and that most natural causes of forest decline were ruled out on Mt. Mitchell and on Camel's Hump, a number of scientists involved in the project openly disagreed with the summary conclusions (Little, 1995). Recent research findings tend to support NAPAP's critics. Analyzing three decades of data from the Hubbard Brook Experimental Forest in New Hampshire, Likens and his colleagues (1996) found that SO_2 and NO_x leach calcium—a critical buffer that neutralizes acid and an important nutrient—out of forest soils, leaving them vulnerable to even relatively low levels of acidic deposition. In West Virginia, research at the Fernow Experimental Forest shows that NO_x from car exhaust is clearly linked to increased tree mortality and growth declines (Gilliam et al., 1996). While the policy responses are increasingly clear to a small but growing number of researchers—reduce SO_2 and NO_x emissions through further tightening of the Clean Air Act, get Americans to drive fewer total miles, or find a substitute for the internal combustion engine—such prospects seem dim in the near future (Kaiser, 1996).

group of public and private sector players must cooperate to mount effective defenses against insect outbreaks or air pollution from vehicles or industry. For this reason, most responses to forest health issues focus on research and monitoring. Many such programs protect the economic value of timber in limited areas but do not address the underlying causes that make trees vulnerable in the first place.

Responses to forest health issues are generally uncoordinated, reactive rather than anticipatory, and compartmentalized rather than integrated. For example, many insect and fire control programs undertaken to protect the economic value of timber in limited areas do not address potentially underlying causes of fires or pest outbreaks, such as certain forest management practices or air pollution. Moreover, responses to forest health problems are frequently controversial—from raging debates over aerial pesticide spraying in the 1970s, the role of acid deposition and forest decline in the 1980s, and in the mid-1990s, to suspension of environmental laws to expedite an emergency salvage logging program to control insect and fire threats on federal lands. Key opportunities are monitoring and treating potential forest health problems better—preventing them in the first place.

Challenge #2: Safeguarding Biodiversity

Disappearing natural habitats, growing numbers of endangered species, and the loss of genetic diversity within populations have prompted warnings from scientists that Earth's biological foundation is eroding faster than at any time in at least 65 million years (UNEP, 1995). Indeed, maintaining biodiversity has rapidly become one of the most visible and complex issues in forest management as biological loss mounts in many of the world's ecosystems, including North America's temperate forests. Although the "variety of life in the infrastructure of forests is immense, fragile, and still very poorly understood," E.O. Wilson (1993) writes that enough is known to conclude that "biodiversity is vital to healthy forests, while proper forest management is vital to the maintenance of biodiversity."

Biodiversity is an important sustainability issue for several reasons. First, it helps determine the productivity and health of individual organisms, including economically important species. For example, much of a tree's nutrient requirements are provided through complex and poorly understood interactions among microfauna in the soil (insects, fungi, and bacteria). Second, biodiversity plays an important role in the productivity and sustainability of ecosystems. While it is well demonstrated that soil fertility and productivity influence diversity, researchers have only recently demonstrated the converse is also true in grassland ecosystems and probably in other ecosystems as well (Tilman et al., 1996; Kareiva, 1996). Third, diversity is the "raw" material that fuels important advances in biotechnology and developments in such fields as pharmacology, agronomy, and industrial processes. Two cases in point from the United States: bacteria collected from hot springs in Yellowstone National Park yielded the enzyme *Taq polymerase*, an essential tool for genetic engineering that is now worth hundreds of millions of dollars to the company that patented the discovery; in the forests of the Pacific Northwest, researchers discovered *taxol*, a compound in the bark of the Pacific yew that has proven effective against breast and ovarian tumors that did not respond to other treatments.

The loss of biodiversity is also a strategic sustainability issue. Extinction and the loss of genetic diversity within a species are irreversible. There may be no substitutes for lost species or genetic resources, and some "keystone" species in each ecosystem are vital to the survival of many other species and to biological productivity. And because we don't know which species might yield the next anti-cancer compound or which species are linchpins in ecosystem functions and biological productivity, any loss of biodiversity poses risks.

Although the world's most biologically diverse ecosystems are found in the tropics, the United States is home to more species than all but Brazil, Colombia, Indonesia, Mexico, China, and a few other countries (Johnson, 1995). Well-known biologist Thomas Eisner and his colleagues (1995) estimate that just over 100,000 native species have been described in the United States; meanwhile, scientists have yet to describe thousands of additional insects and microorganisms that are likely to exist. Forest ecosystems are an important reservoir of this "black box" biodiversity and may be home to at least half of all terrestrial species found in the United States. Biodiversity measured at all levels—ecosystem, species, and genetic—shows significant signs of decline, though information is frustratingly limited.

Ecosystem Diversity

The structure and composition of U.S. forest ecosystems have changed dramatically since Europeans arrived nearly 400 years ago. Late successional and old-growth forest stands have virtually disappeared, except in the Pacific Northwest, high elevations in the intermountain West, and Alaska. By most estimates, primary (or relatively undisturbed) forests outside of Alaska have been reduced to less than 5 percent of their original extent[*]—a much smaller percentage than in most tropical countries (Postel and Ryan, 1991). Indeed, only a few forest ecosystems have significant areas remaining in a relatively natural state (Bryant, 1997; Noss et al., 1995). (*See Box 4-5.*) Just as troubling, examples of many forest ecosystem types, especially in the East and South, are not part of any national parks or other large conservation areas (Noss and Peters, 1995).

While the increase in forest area in the South, Northeast, and upper Midwest is a positive development, second-growth forests on former agricultural lands and clearcuts may take centuries to regain their original biodiversity (Duffy and Meier, 1992), and tree plantations are less biologically diverse than natural forests

---

[*]   Postel and Ryan (1991) indicate there are about 65 million hectares of primary forest in the United States, or 15 percent of all forest cover. However, most of it is in Alaska (52 million hectares), with only 13 million hectares remaining elsewhere.

BOX 4-5. "Threatened" and "Endangered" Forest Ecosystems in the U.S.

Critically Endangered Forest Ecosystems (98% decline)

Old-growth and other primary forest stands in the eastern deciduous forest region
Spruce-fir forest in the southern Appalachians (VA, NC)
Mature and old-growth white pine and red pine forests in Michigan
Longleaf pine forests and savannas in the southeastern coastal plain (MS, AL, FL)
Slash pine forests in south Florida
Loblolly pine-shortleaf hardwood forests in the west Gulf coastal plain (MS, LA, TX)
Atlantic white-cedar stands in Great Dismal Swamp (VA/NC)

Endangered Forest Ecosystems (85-98% decline)

Old-growth and other primary forest stands in all regions and states, excepting those in eastern deciduous forest region (critically endangered) and Alaska (not threatened)
Moist limestone forest in mid-Atlantic region (VA, MD)
Coastal plain Atlantic white-cedar swamp forest (VA, MD, DE, NJ, NY, CT, RI, MA)
Red spruce forests in central Appalachians (WV)
Late successional oak-hickory forest on the Cumberland Plateau (WV, TN, KY, GA)
Slash pine forest in southwest Florida

of any age (Hansen et al., 1991; Showalter and Means, 1988). Unfortunately, information about the structure and composition of forests—key to assessing biodiversity conservation—is not available for most forest areas, but the distribution of age classes or successional stages tells us something. Forests with a relatively uneven or heterogenous age structure are more likely to approximate the structure and composition of natural forest ecosystems than forests dominated by younger even-aged stands. Data from state forest inventories indicate most forests are relatively young and very few, especially on industry-owned lands, have trees of mixed ages. More than three-fourths of forest industry lands in western Washington are in even-aged stands less than 70 years old while mixed-age stands cover less than 5 percent of the area (MacClean et al., 1992).

Species Diversity

At the species level, biologists can point to some encouraging trends. Some forest-dwelling wildlife populations, for example, have been increasing. Since

BOX 4-5 continued.

Red and white pine forests in Minnesota
Coastal Redwood forests in California
Old-growth ponderosa pine forests in northern Rockies, intermountain west, eastside Cascades (MT, WY, ID, UT, NV, OR, WA, CA)
Riparian forests in desert southwest (CA, AZ, NM)
Dry tropical forest (HA)
Riparian forests in lower Rio Grande basin (TX)

Threatened Forest Ecosystems (70-84% decline)

Riparian forests nationwide (other than those listed above)
Northern hardwood forests and jack pine forests in Minnesota
Tropical hardwood hammocks on Florida keys
Live oak-pine-magnolia forest, upland longleaf pine forest, and wet mixed loblolly pine-hardwood forest in Louisiana
Mountain spruce-fir forest and pitch pine-blueberry forest in New York
Western red cedar forest in Idaho
Southern tamarack forest in Michigan and Wisconsin

Source: Adapted from Noss et al. (1995)

1930, populations of wild turkeys have risen from approximately 100,000 to nearly 4 million in 1990, elk from an estimated 60,000 to over 500,000, and whitetail deer, now considered a pest in many places, from approximately 2 million to 15 million (MacCleery, 1992). Several predators, such as coyotes and endangered bald eagles and timber wolves, are returning to parts of their former ranges thanks to stricter hunting laws, increased forest area, species adaptation to a variety of habitat conditions, and more sophisticated wildlife management (MacCleery, 1992).

Still, a growing number of species, most of them requiring special habitats, are in decline. The Nature Conservancy evaluated the conservation status of 20,500 native species in the United States and found that about one-third are at some risk of extinction, and nearly half of those have been reduced to fewer than 20 populations (TNC, 1996). The extinction of 100 species during the past 200 years has already been documented, and an additional 450 species may now be extinct (TNC, 1996). Moreover, Curtis Flather and colleagues at the U.S. Forest

Service (1994) found that over half of the 728 species listed as threatened or endangered in 1992 (compared to 965 species listed in 1996) live in or depend on forest ecosystems. In addition, four of the ten U.S. regions with the highest levels of species endangerment—the Southern Appalachians, Peninsular Florida, the eastern Gulf Coast, and coastal northern California—include extensive forest areas. Some researchers believe that the Pacific Northwest will also soon make the list (Flather et al., 1994).

In these beleaguered regions, the leading causes of species endangerment are habitat loss, followed by over-harvesting, introduced species and disease, and inadequate or poorly enforced resource management laws (Flather et al., 1994). Circumstantial evidence suggests that species endangerment in forested regions stems from multiple causes—forest clearing and logging, but also agriculture, grazing, water pollution, road building, and residential and industrial development.

Genetic Diversity

A significant and rapidly growing number of forest-related species are believed to be losing geographic range and population size, but virtually nothing is known about how these trends are affecting genetic diversity within a species. (*See Box 4-6.*) The use of genetically improved seedlings is one key to increasing timber productivity in intensively managed forests. Specifically, genetic resource reserves for timber species should be maintained *in-situ* in natural forest areas in sites representative of the species' geographic range and ecological conditions; germplasm collections representative of a species' genetic diversity should be maintained *ex-situ* for breeding; and tree plantations should contain several species as a hedge against virulent new diseases or pests.

Responses to Biodiversity Loss

In recent decades, research findings on the ecological structure, composition, and function of forest ecosystems have begun to influence forest-management practices. Promising new approaches to integrate the maintenance of key ecosystem processes, biodiversity, and other environmental quality concerns with timber management—including prescribed and controlled fires, leaving buffer strips along streams, and retaining snags and downed logs after harvest— are being tested in forests around the country. At the same time, increased public interest in nontimber values of forests—especially for fish and wildlife, recreation, and water quality—have nudged many public and private forest managers toward sustainability.

BOX 4-6. Genetic Diversity of Forest Trees

Sustainable forestry depends on measures to maintain the genetic diversity of trees so forests can respond to environmental and biological change. Unfortunately, science has done little to document human impacts on genetic diversity in forest trees, in the United States or anywhere else (Ledig, 1988 and 1986). Still, some experts speculate.

Ledig suggests that these impacts can be ranked as follows. First, the introduction of diseases, herbivores and insect pests, and nonnative tree species have had well documented and overwhelmingly negative impacts on forest biodiversity. In the United States, the American chestnut and, more recently, the American elm have been eliminated as commercial species by introduced disease, while the sugar pine, eastern white pine, western white pine, eastern hemlock, and Port Orford cedar are threatened by introduced diseases and pests. Second, environmental degradation caused by air pollution is responsible for some forest decline and will probably affect genetic structure. Atmospheric pollution and climatic warming are likely to become major threats to forest genetic diversity in the near future, particularly because forests are fragmented and migration is impeded (Ledig, 1992). Third, deforestation has reduced diversity by directly eliminating locally adapted populations, though reforestation in the United States has probably kept overall genetic diversity losses for most species low. Fourth, logging and forest management practices may affect local genetic structure, but the fact that seeds and young trees often remain following logging probably means that overall genetic diversity is minimally affected (Ledig, 1992).

Human impacts on forest composition and genetic diversity are not new. What is new is how fast changes are now occurring, perhaps exceeding the capacity of most long-lived tree species to respond (Ledig, 1992). Technologies to measure tree genetic diversity and to conserve diversity *in-situ* and *ex-situ* when it is threatened are readily available but not widely used, so assessing the genetic vulnerability of many forest trees in the United States is difficult. The uncertainty associated with genetic diversity trends, and the risks if significant genetic diversity is lost, suggest that better monitoring would be a prudent investment in maintaining future forest productivity.

Cooperative research programs involving U.S. Forest Service and university scientists have shed light on how forest ecosystems function, and "new forestry" and kindred approaches are being used to apply the findings of forest ecosystem research to forestry practice. (*See Box 4-7.*) A shift from managing forest stands for sustained timber yield to managing large and diverse forest ecosystems for a wider range of objectives has begun, just as sustained-yield forest management

---

BOX 4-7. "New Forestry"

During the 1970s and early 1980s, Jerry Franklin, a plant ecologist with the U.S. Forest Service, and colleagues at the agency's research station in Corvallis, Oregon, discovered that old-growth forest ecosystems are highly variable and biologically diverse, and that they leave a legacy of structural diversity and organic productivity in younger forests for decades after most of the old trees succumb to windstorms, fires, insects, or disease.

In the early 1980s, the Corvallis team turned their attention to forest management. A few years later, backed by research at the H.J. Andrews Experimental Forest in Oregon, they began urging foresters to reexamine conventional silvicultural wisdom in the Northwest. For decades, convention had dictated that clearcuts, intensive burning of post-harvest woody debris ("slash"), and soil scarification (exposure) were the best techniques for regenerating forests of Douglas-fir in the Pacific Northwest. But, Franklin's group argued that forest managers should better mimic the complexity of natural forest stands for the sake of biodiversity, fisheries habitats, and, ultimately, the health of the forest (e.g., Franklin and Forman, 1987). This approach to forest management eventually came to be known as New Forestry.

Three principles characterize New Forestry. First, New Forestry means leaving clusters of large green trees in harvest areas that will slowly age with the younger stand around it to provide the dead snags and rotting logs that support much of the forest's biodiversity. It also means leaving downed logs in harvested areas, since the rotting wood serves as nurseries for many tree seedlings, and allowing substantial

---

replaced "cut and run" practices earlier in this century (Aplet et al., 1993). Still, such practices won't catch on widely until they become more profitable than unsustainable practices.

Cooperative efforts by private landowners, conservation groups, and state and federal agencies to restore wild salmon in Maine through the RESTORE program signal a promising trend that could substantially improve the prospects for biodiversity and other environmental services in forests, particularly those on private lands (Keystone Center, 1996). In 1994 the Georgia-Pacific Corporation and The Nature Conservancy agreed to jointly manage 21,000 acres of Georgia-Pacific lands along North Carolina's lower Roanoke River (PCSD, 1996)—one of only two remaining large forested wetlands along the southern Atlantic Coast and home to more than 200 bird species. Timber harvesting, and all other management activities, are agreed upon by a committee composed of

BOX 4-7 continued.

amounts of wood to collect in streams to provide habitats for riparian organisms. Second, New Forestry seeks to extend rotation lengths to between 80 and 120 years, and perhaps more. Third, harvest patterns are planned on a landscape scale to preserve connections between forest stands of various ages and to maximize the contiguous area of older stands.

New Forestry has critics on both sides of the forestry-reform debate (Gregg, 1991). Those who believe that forestry works well as currently practiced see little "new" in New Forestry and claim that many of its techniques have been used in the past and found to be less effective than clearcutting and intensive soil preparation (e.g., Atkinson, 1990). At the other extreme are critics who think New Forestry doesn't go far enough and believe its loose definition allows foresters to leave a few standing trees and a snag or two in the middle of a large clearcut and call it New Forestry. And if rotation lengths aren't extended to between 200 and 400 years, some suspect that New Forestry will simply create slightly cluttered, older tree plantations. Many critics also fear that New Forestry will be invoked to justify logging in unprotected old-growth forests. Still, the U.S. Forest Service and other public land-management agencies are applying the principles of New Forestry to timber sales, timber companies are experimenting with New Forestry in limited areas, and some New Forestry concepts have found their way into forest practices regulations in states such as Oregon and Washington.

Source: Barber et al., 1994.

Georgia-Pacific and Nature Conservancy staff with advice from specialists at the U.S. Fish and Wildlife Service and North Carolina State University.

Across the United States, such partnerships are becoming more common. Researchers at the University of Michigan recently analyzed more than 100 cooperative "ecosystem management" initiatives in which landowners, government agencies, and environmental groups tried jointly to define environmental management goals and strategies to improve habitat conditions and environmental quality, many in forests managed for timber production (Yaffee et al., 1996). Compared to more traditional stand-by-stand management efforts, cooperative efforts can give individual owners better access to technical or financial assistance and information, wildlife can find more suitable habitat, and government agencies can eliminate duplication and target limited funds more efficiently.

The threat of stringent restriction in the Endangered Species Act of land use in habitats of endangered species has helped prompt new forest management approaches and new partnerships to conserve biodiversity on timberlands. In general, the federal government has tried to implement species-recovery plans on public lands and to use voluntary agreements or land swaps to develop habitat conservation plans (HCPs) where federal lands alone can't meet endangered species habitat requirements. As long as landowners honor the bargain, they can manage their lands without fear that new habitat-protection requirements will be imposed when new species are listed or regulations change. For example, voluntary agreements between forest industry landowners and the U.S. Fish and Wildlife Service are protecting sensitive habitats for red-cockaded woodpeckers and the endangered Louisiana black bear in the southeast, while HCPs are being actively used to protect habitat on private lands for northern spotted owls, marbled murrelets, and other potentially threatened species in Northern California, Oregon, and Washington.

The Endangered Species Act is likely to be revised before 2000. Concerted efforts to weaken the Act, especially as applied to private lands, could kill much of the impetus for modifying forest management practices to sustain biodiversity. While the number of voluntary partnerships on private lands is growing, it's not clear how well strictly voluntary efforts will work. An inventory and assessment of such partnerships would reveal which remaining actions might still be needed to set and follow biodiversity priorities in forest areas. In the absence of regulation, market forces would be the main incentive for biodiversity conservation and independent certification of sustainably produced forest products is one promising possibility. Still, no one has proposed a practical way to capture the full value of biodiversity in monetary terms.

Challenge #3: Adapting to Climate Change

Poring through microscopes at pollen that settled on the bottoms of lakes and ponds centuries ago, scientists have determined the species composition of local forests over thousands of years. They find that distributions of tree species and forest types in North America, have shifted rapidly and dramatically in response to warming and cooling episodes during the past 10,000 to 15,000 years. Working in forests today, scientists know that average temperature changes of as little as 1°C over several years can affect the growth and reproduction of many tree species. Should climate warming occur, the distribution, health, and economic value of U.S. forests could thus be altered. But forests can also mitigate warming by sequestering atmospheric carbon—the most abundant of greenhouse gases. In any case, forests are likely to be at the center of future climate change policies.

Indeed, the most recent assessment by the International Panel on Climate Change (IPCC) indicates growing scientific consensus that the world will warm by 0.8 to 3.5°C over the next century. For two reasons, climate change is a strategic issue for forest sustainability. One is significant risk that climate warming, as projected by the IPCC, will change species composition, increase the prevalence of insect pests and microbial pathogens, and invite more frequent and severe fires, storms, and flooding. Second, by offsetting some greenhouse gas emissions, forests and trees buffer the transitional shock in the move away from fossil fuels.

Climate Warming Impacts on U.S. Forests

Several factors influence how global warming might affect American forests. First, many tree species have a narrow temperature niche for growth and reproduction. Second, forests are extremely vulnerable to extremes in water availability, whether drought or flooding, and changes in temperature and precipitation could gradually shift suitable habitat conditions for many species and forest types. This has happened many times in the past, but the rate of warming will likely be one or two orders of magnitude faster than at any time during the past 100,000 to 200,000 years (IPCC, 1995).

These changes could upset the balance of forest ecosystems and economies. Suitable habitats for many forest species and forest types will shift faster than many species can migrate. Species that cannot keep pace will grow less, reproduce less, and eventually die out over large parts of their former range (IPCC, 1995). In general, species with wide environmental tolerances, such as early successional species, will be favored while those with specialized habitat requirements and slow migration rates will be left behind. This also means insects and disease could rapidly invade these altered forest ecosystems (IPCC, 1995).

Other negative consequences could also be in store. Water availability could decline in interior continental areas where water supply is already scarce, converting drier forests in the West to grasslands. (*See Box 4-8.*) Storms could become more frequent and intense, affecting the composition, age-distribution, and biomass of some types of forests. Finally, though temperate forests are now considered to be carbon sinks, temperature rises could speed organic decomposition and make some forests into carbon sources.

Any bright spots for U.S. forests under warming conditions are dimmer than they seemed just a few years ago. Forests could become more productive with higher concentrations of atmospheric carbon dioxide, but the 30 percent rise since 1780 didn't have that effect (IPCC, 1995). Some forests' ranges and growing

BOX 4-8. Fire

Devastating fires swept large areas following logging and agricultural clearing in the late 19th and early 20th centuries, occasionally killing hundreds of people. The 1894 Hinkley fire in northeastern Minnesota burned over 1 million acres and claimed nearly 500 lives. This and other conflagrations prompted effective fire-control efforts. Between 1920 and 1990, the average annual area burned by wildfire decreased by 90 percent (MacCleery, 1992). But fire suppression worked so well that many large forest areas are now heavily loaded with highly flammable dead wood and young firs (Clark and Sampson, 1995). Meanwhile, ecologists and forest managers now view fire as a necessary force in the natural regeneration of many forest types. Indeed, many ecologists and conservationists consider forest-management practices that have created the potential for fire a more important sustainability concern than fire itself. They advocate more use of prescriptive fires, a halt to logging what large fire-resistant pines and other species remain, limited use of thinning to remove more flammable young firs, and the use of salvage logging only in exceptional cases, where fire threatens human life or puts extensive private property at risk. Many local communities and timber companies, on the other hand, believe that fire risks justify immediate and large-scale salvage logging efforts.

In general, fires affect a relatively small part of U.S. forest area, most of it on relatively remote public lands, and burned forest ecosystems usually recover quickly. Indeed, a case can be made that prescribed burns and carefully monitored low-intensity wildfire should be used in more areas to reestablish an important ecological process and reduce the potential for severe fire damage. However, climate change may contribute to more frequent and severe fires that over time could lead to permanent forest loss in drier areas, especially in the West (IPCC, 1995).

seasons could expand if winters were warmer, but invasions of destructive temperate and subtropical insects and pathogenic fungi could offset any gains (Dobson and Carper, 1992), especially where vast single-species stands grow in polluted environs (IPCC, 1995). Because well-off countries can afford to reduce climate change's impacts through integrated fire, pest, and disease management, or reforestation (IPCC, 1995), temperate forests are likely to be least affected by climate change, and most models project little or no loss of overall temperate forest area. Still, U.S. forests could change significantly and unpredictably if global warming unfolds as projected over the next century, thus altering the mix of forest fiber available to industry and depleting forest biodiversity.

U.S. Forest Sector Strategies to Slow Climate Warming

Each year, fossil fuel combustion, the conversion of forests to farmland, and other human activities add a net 3 billion tons of carbon to the atmosphere (Trexler, 1991). As a result, the global carbon cycle is now out of balance—atmospheric carbon dioxide concentrations have risen by over 30 percent in just 250 years. In the long term, this trend can't be stopped or reversed unless energy appetites for fossil fuels are curbed and tropical deforestation slowed. In the short term, tree planting, forest protection, and sustainable forest management to sequester carbon can slow and mitigate climate change (Trexler and Haugen, 1995). Most strategies for using forests to respond to climate change aim to increase the carbon-storage capacity of terrestrial ecosystems by planting trees or by slowing carbon release (which occurs as a plant decays) through forest conservation and sustainable management. The most practical of such admittedly partial solutions are to plant more trees on suitable agricultural lands and in urban areas, increase timber productivity on managed timberlands, conserve carbon-rich old-growth forests, and use wood for durable products and more climate-friendly substitutes for disposable products (e.g., more electronics and less paper). Such strategies will win more adherents if they generate other benefits as well, whether rural income and employment, fuel for biomass energy plants, urban shade and amenities, wood fiber, or better water quality and wildlife habitat.

Evaluating eight biological options for increasing the uptake and storage of carbon in the United States, researcher Mark Trexler determined that though these options could together yield more than 2 billion tons of carbon stored annually (slightly more than U.S. CO_2 emissions from fossil fuel use in 1994) in practice only a fraction of this amount—200 to 400 million tons a year—could be captured. The most productive option is likely to be converting marginal agricultural lands to tree cover, followed by using wood and other organic materials for biomass energy production, and boosting timber productivity on managed timberlands. The price tag would be high: $37/ton on average for a total of roughly $12 billion, though other researchers peg them at half or less of this sum (Moulton and Richards, 1990). In general, improving energy efficiency, developing renewable energy alternatives, or simply increasing prices or taxes on energy consumption are cheaper ways to prevent or offset CO_2 emissions.

Still, forestry options to offset CO_2 emissions are worth considering. The marginal cost for the first 100 million tons of carbon benefits or so will be low (since the cheapest opportunities are exploited first) and may compare to the marginal cost of tree planting and other forest management options. Also, relative to most nonforestry options, forest management and tree planting can generate benefits beyond carbon sequestration, including higher farm income, better water quality, higher wood fiber yields, and better habitat for hunting and fishing. In addition, because of these "joint benefits," energy producers and consumers may not have to pay all of the costs of sequestering a ton of carbon by themselves: voluntary utility contributions, consumer premiums, or consumption taxes could all induce forest managers or land owners to plant more trees or to improve forest—and carbon—management practices that don't make economic sense now.

Responses to Climate Change in the U.S. Forest Sector

Domestic forests and forest policy have so far played only minor roles in any U.S. strategy to mitigate climate change. Instead, U.S. electric utilities have invested in tree planting and improved forest management in other countries where costs are lower, including Costa Rica, Guatemala, Ecuador, Russia, Ukraine, and elsewhere. As Trexler (1991) suggests, the practical potential of U.S. forests to mitigate CO_2 emissions is still very difficult to assess, and any large-scale initiatives would probably require major public policy changes and could spark major land use conflicts.

Fortunately, such "two-fers" exist. Planting trees in cities reduces energy use, stores carbon, beautifies the landscape, and could even yield new sources of wood fiber for furniture, pulp, and other uses. Similarly, growing more trees on marginal agricultural lands can generate new sources of wood fiber, help farmers fight erosion and water quality problems, and add carbon-storage capacity.

As for preparing U.S. forests for climate change, not much is being done. In fact, current trends—establishing large single-species tree plantations; fragmenting natural forest stands through road, residential, or tree farm development; and abiding by forest-management practices that upset natural forest fire regimes and deplete biological diversity—all make U.S. forests more vulnerable (IPCC, 1995).

Challenge #4: Preventing Industrial Pollution

The United Nations Environment Programme defines clean production as the "continuous application of an integrated preventive environmental strategy to processes and products to reduce risks to humans and the environment" (UNEP, 1994). This includes "conserving raw materials and energy, eliminating toxic raw materials, and reducing the quantity and toxicity of all emissions and wastes before they leave a process [and] reducing impacts along the entire life-cycle of the product, from raw material extraction to the ultimate disposal of the product." Converting these concepts into practical business decisions is a vital and high-stakes challenge for the U.S. forest sector.

Obviously, the forest sector is just one of many sources of persistent pollutants. Viewed in isolation, the contribution of the forest products industry or any given facility is inevitably small, and the industry as a whole has made great strides in pollution control in the last three decades. But this is a weak rationale for complacency. Irreversible environmental contamination—the price of inaction—can undermine the long-term economic competitiveness of the U.S. forest sector, as well as the sustainability of firms allied with forest products. Environmental problems—both real and perceived—already influence the relative consumer appeal of products and the way regulators, consumers, and investors see companies (Schmeidheiny and Zorraquín, 1996).

Pollution from the U.S. Forest Sector in Context

Compared to the steel mills of the coal-fired industrial revolution or the petrochemical complexes of postwar America, the forest products industry may seem a minor threat to human health and environmental quality. But the manufacture of forest products—particularly the chemical processing of wood fibers at pulp and paper mills—involves massive emissions of airborne particulates, sulfur dioxide, and a host of other air and water pollutants.

Just after the turn of the century, the United States began to take advantage of its cheap and plentiful forest resources. With the introduction of the kraft process, a new pulping technology from Sweden, wood fiber quickly displaced rags and cotton as the fiber of choice for U.S. paper makers. This technology produces an especially strong pulp and is well suited to the resinous pines of the southern United States. But this process—which today accounts for more than 80 percent of all wood pulp capacity—yields a dark brown pulp (U.S. EPA, 1995b). Unbleached kraft paper is fine for grocery bags but doesn't meet demands for office paper, packaging, and other applications, so various chlorine-based bleaching technologies were developed.

As the demand for paper products rose, pulp and paper mills grew into significant sources of local air and water pollution. Downstream from sprawling mill towns, untreated effluents sapped water bodies of oxygen, killing fish and overwhelming nature's assimilative capacities. In the air, soot and particulates rained down on neighboring communities. Yet, by the 1950s, as awareness of the health and environmental costs of industrial pollution grew, pulp and paper mills began employing biological treatment facilities to reduce pollutant loadings to water-ways.

Strong federal legislation of the early 1970s solidified the U.S. commitment to controlling industrial air and water pollution. The Clean Air Act of 1970 established national air quality standards and limited the release of particulates, sulfur dioxide, and total reduced sulfur. The Clean Water Act of 1972 triggered technology-based standard-setting to control the discharge of conventional water pollutants (U.S. EPA, 1995b). The forest products industry responded by upping investments in pollution-control equipment, making noteworthy headway against gross pollution and reducing the release of suspended solids in water and airborne particulates.

These important gains notwithstanding, the forest products industry still places third behind chemicals and primary metals in the total quantity of toxic chemicals released, as measured by the Toxics Release Inventory (U.S. EPA, 1996). With its enormous energy requirements, it also contributes to regional and global air pollution. Pulp and paper alone accounts for roughly 10 percent of total industrial energy use in the United States. Sulfur dioxide (SO_2) from burning fossil fuels in industrial boilers and furnaces declined in the pulp and paper sector from 875 million tons in 1980 to 600 million tons in 1990, thanks largely to fuel switching and gains in energy efficiency (Pinkerton, 1993). But, over the same period, oxides of nitrogen (NO_x) rose slightly from 274 to 306 million tons as overall paper production rose by approximately 30 percent.

Preventing Persistent Pollution: The Unfinished Agenda

In general, these trends support the view that conventional air and water pollution from the forest product sector is generally on the decline, both in the aggregate and per unit of product. Yet, this picture is incomplete. Success in reducing pollution stems primarily from better pollution control, which works but costs industry large sums and obliges government to monitor firms and assure compliance. In 1991, the pulp and paper sector devoted 14 percent of its capital investment to pollution abatement on top of more than $1.1 billion in annual operating costs (OTA, 1994). Even this grand sum understates the true economic impact on the industry since few firms properly account for their environmental outlays (Ditz et al., 1995). Further, attacking pollution at just one stage of the manufacturing chain ignores the associated environmental burden of consumer demand for products and services. Indeed, the amount of toxic chemicals released in connection with a paper cup far exceeds the amount attributable to its manufacture alone (Lave et al., 1995). The environmental impacts of coatings, dyes, bleaching agents, and other chemical inputs should be included in an environmental assessment of such "forest" products. In this more holistic view, cleaning up the mill—while a significant technological improvement—could still fall far short of assuring clean products.

Finally, the attention of many environmental scientists, regulators, and activists has moved from conventional air and water pollutants to trace contaminants, particularly chlorinated organic compounds. While the sources of gross and trace pollution are similar, the problems are qualitatively different and great uncertainties surround the causes and specific effects of these pollutants. Some contaminants are released at levels that almost defy detection, but—new ecological and epidemiological research findings suggest—could still have pernicious effects on human and ecosystem health (Colborn et al., 1995; Repetto and Baliga, 1996). (*See Box 4-9.*) If they do, prevention makes even more sense than pollution control or clean-up.

Responses to Persistent Pollutants

Reducing environmental releases of persistent pollutants, such as chlorinated organic chemicals, is becoming the *de facto* policy of the pulp and paper sector. Regulatory actions outside the United States and steady pressure by environmental groups and some green-minded consumers are pushing business toward cleaner production, but the waters are uncharted for much of the forest products industry. As dioxin emissions slip below the limits of detection by conventional monitoring at most modern mills, the industry has strongly resisted any wholesale prohibition on the use of chlorine compounds, which could cost as much as $75 million for a large integrated mill (Beckenstein et al., 1996). Meanwhile, advocates for chlorine-free processing underscore the prudence of adopting totally chlorine-free technology. While the controversy boils, many firms in the forest industry are already curbing their releases of chlorinated dioxins and other persistent pollutants.

Even a radical transformation of pulp and paper manufacturing can't eliminate all exposures to industrial toxics. But in the long run, cleaner paper production will reduce industry's environmental burden, and failing to "green" the manufacturing process will put forest products at an increasing disadvantage relative to a host of fiber and nonfiber alternatives.

The "precautionary principle"—choosing to err on the side of prevention over remediation—should guide private investment decisions just as it guides

BOX 4-9. Uncertainties about Chlorinated Pollutants

In spite of dramatic improvement in the control of conventional air and water pollution, the pulp and paper sector continues to draw scrutiny as a source of toxic and persistent chemicals. Recent research about noncancer health effects of broad classes of chlorinated organic compounds, together with ever more sophisticated analytic techniques for detecting trace pollutants, is forcing the industry and regulators to reconsider environmental problems once thought solved.

Dioxins and furans, a group of more than two hundred related chemicals, are perhaps the most notorious of these pollutants. Dioxins and other chlorinated organic pollutants are produced in tiny quantities as a by-product of many combustion and chemical processes. The regulation of dioxins has incited fierce scientific, legal, and popular controversy in issues ranging from the clean-up of Superfund sites, to the siting of waste incinerators, to the use of Agent Orange in Vietnam. Even so, surprisingly little is yet understood about the origins, movement, and consequences of these pollutants. EPA's 1987 National Dioxin Study found unexpectedly high concentrations of dioxins in fish downstream from the 57 pulp and paper mills sampled. This and other follow-up studies pointed to the bleaching process, where the cellulose-rich pulp is treated with elemental chlorine, chlorine dioxide, or other bleaching agents. In response, industry has moved to reduce the use of elemental chlorine in favor of chlorine dioxide and to modify pulping and delignification processes, often at significant cost (Kinstrey, 1993).

EPA's recent five-volume reassessment of dioxin and related chemicals casts new light on the sources of these chemicals, human exposure, and health effects (U.S. EPA 1994). On the basis of this evidence, the pulp and paper industry is not the dominant source of these long-lived chemicals. Total environmental releases of dioxin from the sector are estimated at far less than a pound per year, a quantity that defies practical monitoring. But, the true sources and amounts of persistent organic pollutants and the practical impact on human and ecosystem health remain uncertain. Most research has concentrated on links with cancer, but a variety of other impacts, including developmental, reproductive, and immunologic impacts, have recently been uncovered (U.S. EPA, 1994; Davis and Bradlow, 1995). Clearly, if dioxin-like compounds are responsible for such effects, they work their menace at extraordinarily low exposures.

public environmental policy-makers. Increasingly, paper firms will be able to differentiate their products from competitors' through cleaner technology. Already, over half of Finnish pulp capacity has been weaned from chlorine use to tap into emerging market demands for chlorine-free products.

During the last years of the Bush Administration, EPA tried to coordinate separate regulatory efforts on air and water pollution with potentially large economic impacts on the pulp and paper industry. The American Forest and Paper Association has criticized the resulting draft "Cluster Rule" as insufficiently integrated, overly stringent, and too costly, while some environmental activists who want all chlorine compounds eliminated view the draft rule as a weak compromise. Yet, on both sides of the issue many participants applaud EPA's willingness to craft rules that transcend the old legislative boundaries of air, water, and land.

The prospect of tougher regulatory requirements under traditional air and water pollution laws have induced changes in pulping technology and spurred investments in cleaner technology. By 1994, roughly one-third of the more than 90 kraft mills in the United States had switched to elemental-chlorine free bleaching (Beckenstein et al., 1996). This step, along with extended pulping and oxygen delignification, can bring dioxin emissions down below detectable limits. Arguably, technological modifications such as these will eventually open the door for recycling wastewater and help realize the long-term goal of a "closed loop" mill.

Searching for New Ways to Regulate

The bramble of federal and state regulations on industrial pollution is an outgrowth of a quarter century of piecemeal solutions. The Clean Air and Clean Water Acts, the Resource Conservation and Recovery Act, and the broad-reaching "Superfund" legislation all take aim at just a part of the overall problem of managing potentially hazardous materials. And complying with the thousands of regulations drafted by federal and state governments under these laws is cumbersome and costly.

An enduring criticism of U.S. regulatory policy on industrial pollution is that limits on releases to air, water, and land ignore local realities. Plant managers often complain that they spend too much reducing already low emissions while bigger contributors outside industry go unchecked. So-called "command and control" approaches to regulation prescribe specific industry responses, even where firms might readily achieve much higher environmental performance using other means (Porter and van der Linde, 1995; Amoco/EPA 1992).

The Great Lakes Water Quality Initiative exemplifies a regional approach to controlling persistent pollutants from multiple industrial and other sectors, but U.S. and Canadian forest products companies have balked at the prospective costs.

For an estimated annual cost of $60 to $376 million, emissions (weighted in proportion to toxicity) would fall from the current level of 35 million pounds by 7.6 million pounds, a reduction of more than 20 percent (Renner, 1995). Is this collaborative effort by U.S. EPA, the Council of Great Lakes Governors, local industries, and other stakeholders a workable strategy for reversing the accumulation of persistent pollutants? Naturally, since this crosses political jurisdictions, a range of industry sectors, and other sources of Great Lakes pollution, it has attracted criticism. One is the cost-effectiveness of the recommendations. Another is that, in the end, regulators will still rely on specific discharge limits and standards required under the existing regulatory structure. Nonetheless, the Great Lakes Water Quality Initiative builds on the available scientific understanding of bioaccumulation, with an eye toward averting long-term or irreversible impairment of the watershed and the fish and wildlife that depend on it.

To our north, the Government of Ontario has established 2002 as the date for the "virtual elimination" of chlorine-containing pollutants. The recent efforts of the International Joint Commission, also motivated by concerns over the accumulation of persistent pollutants in the Great Lakes ecosystem, suggest that the shift toward totally chlorine-free technology in Scandinavia could spread to North America (Durnil, 1995). More generally, some U.S. firms are beginning to accept the challenge of a minimum impact mill in their R&D and investment decisions (Erickson, 1995).

Today, many voices among regulated industries, environmental officials, and public interest advocates are calling for a serious rethinking of the basic structure and function of the U.S. approach to pollution control (PCSD, 1996). Recent EPA initiatives have tested novel approaches to environmental management. Project XL (for Excellence and Leadership) may eventually lead to better, more cost-effective environmental performance at the community or facility level, and EPA's "Common Sense Initiative" is a consensus search for "cleaner, cheaper, smarter" approaches to environmental objectives. So far, both experiments have proven frustrating to firms, regulators, and public participants, but progress within either initiative may afford insights in the design of new regulatory regimes more attuned to achieving environmental results while respecting practical business needs and legitimate public interests.

BUILDING A MORE SUSTAINABLE FOREST SECTOR

How can the U.S. forest sector handle the sustainability challenges outlined here and still stay competitive? It helps enormously to have a vision of a forest sector that meets our future needs for traditional commodities, delivers valuable environmental goods and services, and contributes broadly to a more sustainable and livable United States. Realizing that vision will require the determined effort of many over a transitional period of years. Along the way, we will need some basic indicators to signal progress toward this goal.

Envisioning a More Sustainable Forest Sector

Fundamentally, what are we seeking to sustain? In the multidimensional U.S. forest sector, this simple question defies easy answers. But any vision for a sustainable U.S. forest must:

◆ Generate a wide range of *basic products and services* that we use every day—from lumber, paper, and engineered wood products to wildlife and fisheries, watershed protection, and energy.

◆ Enhance the *quality of life* of all Americans by diversifying economies and creating better paying jobs in rural forest areas, supplying more educational and recreational opportunities, and making communities, neighborhoods, and workplaces near forests more attractive and livable.

◆ Provide the United States with opportunities for *adapting to global changes* in the environment and in the marketplace by managing forests as carbon stores, biodiversity reservoirs, suppliers of wood and other internationally scarce forest products, and proving grounds for clean technologies.

Sustainability on the Land

If the forest landscape is to supply timber and other forest products, protect biodiversity, provide recreation, and maintain such vital ecological functions as nutrient cycling, carbon storage, and watershed protection, forest management and use must be planned at the right geographic scale and all related activities surrounding the forest—from agricultural practices to suburban development—must be considered.

Science and experience both indicate that small patches of forest are not large enough to simultaneously generate wood products, sustain ecological functions, and protect biodiversity. But "ecoregions"—a distinctive geographic area defined by species composition, ecological processes, and climate—are. The U.S. Forest

Service, The Nature Conservancy, and states from California to Minnesota have developed such ecoregional classifications to help guide natural resources planning. (For example, The Nature Conservancy has identified 64 ecoregions in the United States, approximately half of which are extensively forested.)

With wide public participation, state and federal agencies will use a variety of planning and policy tools to create a "mixed" landscape of protected areas, plantations, and other intensively managed forests, and areas managed for diverse economic and environmental benefits. All planning and management activities will be informed by the best available scientific information and social, environmental, and economic expertise.

Property and estate tax revisions, land swaps, acquisitions, conservation easements, low-interest loans, applied research, technical assistance, and other policies and incentives will be used to implement the goals and objectives identified in participatory planning. For example, land swaps and purchases of conservation easements could help conserve forest areas with important biodiversity, watershed, and recreation opportunities. Tax revisions for large landowners and low-interest loans and technical assistance for small landowners would increase timber production on a smaller land base well suited for intensive forestry. On remaining lands, environmental protection objectives will become more important. Long-rotation, uneven-aged forest management for high-quality saw timber, which is compatible with environmental protection goals, will increase. This will become increasingly feasible as consumers demand wood products from sustainably managed sources and as landowners and managers find ways to profit from preserving environmental benefits, whether through carbon-sequestration funds, recreation fees, watershed investment financed by water users, or other means.

The nation as a whole will meet a greater proportion of its wood fiber demands—especially for pulp, paper, and engineered wood products—from a smaller land base. While the area of public forest lands will remain relatively stable, strategic acquisitions and land swaps between private landowners and state and federal agencies will make the U.S. network of protected forest ecosystems more representative of all forest types.

Such a vision will be realized partly by addressing threats and opportunities outside the forest. The loss of highly productive or environmentally sensitive forest lands to residential, commercial, and infrastructure developments will be discouraged through higher property taxes for conversion, the use of tradable development rights and land banks, and incentives for concentrating new development in built-up areas. Policy reforms, tax incentives, and private capital will encourage farmers and investors to establish short-rotation, intensively managed tree plantations on marginal agricultural lands. In metropolitan areas, more trees will be planted to provide shading and wind breaks (thus reducing energy consumption) and to absorb air pollutants. Urban forestry will also play a role in reclaiming abandoned industrial and commercial sites, thus augmenting both bioenergy and paper mill supplies.

Climate change will become a bigger factor in forest policy and management. Actions that increase carbon storage in the forest sector will be encouraged through incentives from government and private utilities as the United States stabilizes and ultimately reduces net greenhouse gas emissions. Climate change risks to the health and productivity of forest ecosystems will be minimized by increasing the genetic diversity of tree species in plantations and gradually locating plantations within a matrix of mature and regenerating natural forest ecosystems to lower risks from disease or catastrophic fire.

Sustainability at the Mill

Long-term sustainability will require continuing progress to wring more of what society wants from the resource base. To be sure, the simple economics of scarcity will fuel price increases and technological innovation regardless of public policy changes or business decisions. But using cleaner and more efficient processing technologies, doing more with less wood fiber, incorporating nonwood fibers into traditional applications, weighing the merits of alternatives to forest products, and reshaping demand will all make the U.S. forest sector more sustainable.

"Eco-efficiency" will become a watchword in business and public policy-making. By one definition, "eco-efficiency is reached by the delivery of competitively priced goods and services that satisfy human needs and bring quality of life, while progressively reducing ecological impacts and resource intensity throughout the life cycle, to a level at least in line with the Earth's estimated carrying capacity" (BCSD, 1994). In a narrower technical sense, eco-efficiency can be thought of as a ratio of outputs to inputs: as boardfeet of lumber per cubic foot of roundwood, or as pounds of paper per ton of pulp. But notions of both inputs (wood fiber as well as water, chemicals, energy, labor, capital, etc.) and outputs (forest products as well as jobs, profits, community benefits, ecosystem health, etc.) will expand dramatically.

New technologies—whether product and process innovations or alternative materials—will reduce pressures on forest resources and minimize pollution risks. "Closed loop" manufacturing technologies will become the industry standard.

A more diverse array of wood and nonwood fibers will be used in manufacturing. And finished products will use less raw material but perform at least as well.

Continuing pressure from the public and government regulators over risks posed by the use of toxic chemicals in paper and engineered-wood products will steer new capital investments toward chlorine-free and zero-emission manufacturing. As one instance, the use of oxygen de-lignification and other processes that virtually eliminate the use of chlorine compounds and allow wastewater to be recycled will be widespread. The use of such technologies will dramatically reduce companies' regulatory compliance costs and liability risks and attract financing from capital markets, leaving laggard firms starved for outside capital. As the capital stock turns over, the forest sector will release less and less toxic pollution.

As U.S. forests come under growing pressures, agricultural fibers—including kenaf, industrial hemp, rice straw, and possibly others—will once again assume a more important role in the manufacture of products we now associate almost entirely with wood. New processing technologies will allow industry to use a more flexible mix of both wood and agricultural fibers to produce paper and composite board and meet increased demand. Less capital-intensive processing technologies and fewer chemical inputs will be needed to process agricultural fibers, which don't contain the high lignin component of wood fiber. The paper industry will be more flexible in the face of changing markets—drawing relatively more from agricultural stocks when stumpage prices are high or more from forests when agricultural stocks are limited—and thus less vulnerable to cyclical swings in profitability.

New markets for agricultural fiber will improve rural economies hurt by a fall-off in demand for tobacco, cotton, and other crops. Investments by farmers, landowners, packaging manufacturers, and the forest products industry itself in new or different technologies—including agricultural practices, materials processing, and product design and marketing—will pry market share away from tree fiber alone. Investors in nonwood capacity will bear some financial risk initially, but as the supply-side pressures build, the appeal of alternative fiber supplies will grow.

Rapid advances in packaging and electronic information technologies will lead businesses and consumers to try substitutes for traditional paper products. Cheap and durable composites—for example, from agricultural by-products and such abundant materials as limestone—will be used in new packaging materials.

The revolution in information technologies for communicating and archiving—for years, an upward push on paper consumption—will make some uses of paper unnecessary (Young, 1993). The Internet, electronic mail, online telephone directories, and new, low-cost, high-speed information technologies will be commonplace in American households. A new technical and organizational infrastructure for direct electronic data interchange (EDI) will end some traditional forms of paper-based communication and record-keeping within firms. These substitutions will allow Americans to meet their needs economically without converting most unprotected natural forest areas into tree plantations.

Sustainability in the Marketplace of Products, Services, and Ideas

Ultimately, any vision of a sustainable forest sector depends on choices made by consumers and citizens, and several trends now gathering force will flower. Younger people—better informed and more concerned about environmental quality issues than their parents—will mature and become influential consumers in the marketplace, increasingly demanding and using credible information on environmental performance to differentiate products and companies. Along with education, personal safety, and pocketbook issues, environmental quality will be seen as a cornerstone of quality of life and of political life. Citizens will demand from industry more responsibility and accountability; from government, more help implementing goals defined through citizen-controlled decision-making and less bureaucracy.

Consumers committed to sustainability will demand a fuller accounting in the marketplace for forest benefits traditionally viewed as free goods. As a result, landowners and resource managers will discover business opportunities for nontimber forest benefits. For example, urban water users will recognize the wisdom of investing a portion of their water fees in forest management to protect watershed quality. At the same time, independent verification of environmental performance on forest lands and at the mill will become an industry standard, influencing consumer choices and investor decisions. The forest products industry will hire and train professionals with a broad grasp of sustainable forest management and eco-efficiency and extend technical assistance to the loggers and smaller landowners who supply their mills.

Weary of costly and divisive conflicts over natural resource management, citizens will demand more cooperation among government, industry, and environmental groups. Regional forest planning will feature prominent participation by all stakeholders, who will expect government to support collaboratively defined goals by providing technical assistance and funding to regional and local institutions and by tailoring tax and regulatory policies to regional conditions. Landowners, industry, and environmental groups will find ways to cooperate that yield mutually beneficial outcomes. Where conflicts persist, the threat of direct ballot initiatives and referenda at state and local levels will force compromise or settle conflicts by majority rule. Such processes will educate the public about the trade-offs between conflicting goals, so the electorate will be more informed and deeply engaged in forest management issues than ever before.

In this vision, the forest sector still provides logging and mill jobs—and some communities remain dependent on traditional forest sector economies. But there will be fewer jobs in these occupations than there are today while emerging markets for specialty wood products, recreation, nontimber forest products, and activities that generate environmental services provide more. Private landowners, public forest managers, and forest communities will integrate nontimber resources more fully into their land use decisions and use forests for their most highly valued or distinct economic, social, and environmental benefits. Refugees from overcrowded metropolitan areas will gravitate to areas with diverse forest landscapes and their entrepreneurial skills (aided by powerful and cheap information technologies) will breathe new life into many rural forest economies. Overwhelmed by their popularity, these areas will face new problems—how to control growth, how to protect longtime residents from being displaced, and how to preserve aesthetic benefits, biodiversity, and water quality. Finally, to ensure that the United States and its enormous appetite for forest products does not export its sustainability problems to other parts of the world, citizens will support efforts to reduce demand at home and help promote more sustainable forest management abroad.

Making the Transition to Sustainability

This vision of a more sustainable forest sector will amount to little more than a fanciful daydream if it is not connected to the here and now. The shift from current practices will not be simple or free, and every principal player along the forest products chain—and others as well—will feel these changes (Smith, 1997). Private landowners, for example, could experience financial losses if they accept longer rotation times, more selective harvesting, and greater investments in habitat conservation. Manufacturers of forest products could face higher costs for fiber, steady pressures to upgrade their processing technologies, and more demanding communities and customers. As harvesting patterns and underlying cost structures change, some workers might lose their jobs. Investors might see these measures as higher costs for lower returns, and consumers could find higher prices on the products that they demand and turn to substitutes from other sectors.

On the other hand, consider the costs associated with *not* making sustainability a reality. If sustainable forest management practices had been the norm in the Pacific Northwest over the past two decades, for example, American taxpayers, forest workers, and others wouldn't be paying the costs of these excesses today. Boom-and-bust economic cycles, impaired salmon fisheries, and large taxpayer expenditures to restore degraded ecosystems, retrain displaced mill workers, and finance crash economic diversification programs aren't cheap. Yet, such outcomes can be expected so long as we mine the forest's capital, rather than live off its dividends.

Accurately predicting the cost of putting the U.S. forest sector onto a sustainable path won't be possible until public policy and private sector decision-makers agree on what sustainable forestry practice is. Right now, the debate hangs on overly general concepts, and specific problems—one endangered species in a particular forest tract, one pulp mill in a particular community—get more attention than the big picture. How recommended actions would affect the parties called upon to take them is hard to foresee too. What will it cost to transform a pulp and paper business into a model of eco-efficiency? How will future markets treat products derived from more sustainable forest management?

For all the discussions about sustainability, scant information is publicly available on its likely costs and benefits. Firms are generally in the best position to estimate the potential costs of going the extra mile. But cost figures—of necessity, confidential business information—rarely see the light of day, except in efforts to lobby Congress or show how burdensome proposed regulations would be. A great many firms—flying blind—have not even attempted the calculation.

The business and policy dimensions of a hypothetical "sustainable corporation" were sketched out for the Canadian forest sector in a provocative book by Daniel Rubenstein of the Canadian Auditor-General's Office (1994). The story, a case study disguised as a corporate drama, follows the interplay of a motivated business executive and his top financial manager as they sort through the practical challenges of sustainability. A fictional integrated forest products company, forced by a range of factors to reconsider its dependence on the natural capital of the forest and other resources consumed in the manufacture of lumber, pulp, and paper products, begins sifting through mountains of literature on environmental accounting, sustainable development, and the role of the firm. Senior managers think critically about the company's shareholders, employees, customers, and various publics and then map out and assign costs to actions based on sustainable forest-management principles and a "zero impact" pulp mill. The specifics—such as longer timber rotations, accommodating nontimber demands on the resource base, investing in new process technologies, alternative marketing strategies, and so on—require the expertise and judgment of the firm's managers, foresters, economists, engineers, accountants, and others.

Rubenstein spins out the potential financial implications for both the giant leap toward sustainability and a more measured stride in the same direction. Reduced to the bare logic of return on investment, each option has a negative impact on profitability. For example, the fictional firm, which holds to traditional sustained timber-yield forestry and operates an average pulp mill, is expected to earn a rate of return of 22.7 percent. Shifting either part or all of the way toward broad forest sustainability reduces returns to 20.9 percent and 18.9 percent, respectively. In addition, moving from the middle of the pack on pulp operations to a position of industry leadership reduces return on investment by 4.3 percent (or 3.3 percent, if combined with improvements in forest management).

Such impacts on the bottom line would deter most Boards of Directors or investors from jumping for sustainability, but Rubenstein builds a case for the positive opportunities these actions can create. For one, investments in increasing sustainability can strengthen a company's bid for future timber rights on public lands. Similarly, voluntarily cleaning up the manufacturing process helps the firm deal with regulators, local communities, and environmental activists. Still, to many in the private sector, such benefits may appear intangible or elusive, and a company has little motivation to worry about costs they don't have to pay.

How *could* environmental regulations, timber concessions, product certification, tax policy, or other instruments be modified to make sustainability pay? Independent certification is one market-based mechanism that allows consumers of forest products to help pay the freight for more sustainable forest practices. Since 1993, the Forest Stewardship Council (FSC) has promoted good forest management worldwide by linking "green" consumers with producers seeking to improve their forest-management practices, obtain better market access, and get higher revenues. Harnessing market forces to reach specific environmental goals, the FSC evaluates, accredits, and monitors timber-certification organizations that inspect forest operations, and FSC's label vouches for timber produced according to strict guidelines (*See Box 4-10.*)

By 1996, the FSC had certified 21 forests covering nine million acres worldwide. Several Swedish forestry companies—representing 38 percent of the country's 58 million acres of forest—announced their intention to adopt the FSC criteria. In the United States, the number of independent forest-certification groups is growing and several small and medium-sized forest companies, including Collins Pine in California and Seven Islands in Maine, have been certified. In debate over the plight of tropical forests, certification may prove too little too late. But in the U.S. forest sector, the chain connecting forests, commodities, manufacturers, end users, and waste is still largely contained within national borders, so the benefits of certification show up close to where the costs are paid.

The landowners, firms, regions, and countries at a comparative disadvantage in the shift to sustainable forest use comprise a bloc unlikely to support actions like those called for here. But consider the converse. Who might win big? Ordinarily, consumers facing higher prices due to increased environmental regulations are viewed as suffering economic losses. But theirs is not necessarily money down the drain. Investing in cleaner water from forested watersheds, reducing stresses on forest ecosystems, and preventing industrial pollution through design changes, all lower the ultimate economic tab and increase the prospects of attracting new industries and jobs.

As firms position themselves to capture the benefits of sustainability, they should consider the advantages that often accrue to "first movers" in an industry. Recognizing the potential market for totally chlorine-free bleaching technology, Union Camp in the United States and Södra in Sweden have developed manufacturing processes both to use in their own operations and to license to others (Bonifant, 1994). Aracruz Celulose, the Brazilian pulp exporter, has championed hardwood eucalyptus plantations on degraded pasture and coffee lands, capitalizing on the enormous natural advantages of the tropical climate without accelerating the loss of primary forest cover (Aracruz, 1996). Collins Pine's decision to push the envelope on certification illustrates how corporate strategy can reflect sustainability challenges. Vision Paper (the leading commercial producer of kenaf-based paper) and Fox River Paper now offer "tree-free" paper products—entering and bolstering a new market niche.

Jockeying for competitive advantage can reinforce sustainability in the forest sector if prospective cost savings or revenue gains are large enough to compensate the company for abandoning the status quo and if industry leaders and their

BOX 4-10. Forest Stewardship Council Principles for Sustainable Forest Management

Compliance with Laws and FSC Principles. Forest management shall respect all applicable laws of the country in which they occur, and international treaties and agreements to which the country is a signatory, and comply with all FSC Principles and Criteria.

Tenure and Use Rights and Responsibilities. Long-term tenure and use rights to the land and forest resources shall be clearly defined, documented, and legally established.

Indigenous Peoples' Rights. The legal and customary rights of indigenous peoples to own, use, and manage their lands, territories, and resources shall be recognized and respected.

Community Relations and Workers' Rights. Forest management operations shall maintain or enhance the long-term social and economic well-being of forest workers and local communities.

Benefits from the Forest. Forest management operations shall encourage the efficient use of the forest's multiple products and services to ensure economic viability and a wide range of environmental and social benefits.

Environmental Impact. Forest management shall conserve biodiversity and its associated values, water resources, soils, and unique and fragile ecosystems and landscapes, and, by so doing, maintain the ecological functions and integrity of the forest.

Management Plan. A management plan—appropriate to the scale and intensity of the operations—shall be written, implemented, and kept up-to-date. The long-term objectives of management, and the means of achieving them, shall be clearly stated.

Monitoring and Assessment. Monitoring shall be conducted—appropriate to the scale and intensity of forest management—to assess the condition of the forest, yields of forest products, chain of custody, management activities, and their social and environmental impacts.

Maintenance of Natural Forests. Primary forests, well-developed secondary forests, and sites of major environmental, social, or cultural significance shall be conserved. Such areas shall not be replaced by tree plantations or other land uses following harvest.

Plantations. Plantations shall complement, not replace, natural forests. Plantations should reduce pressures on natural forests.

Source: FSC, 1996.

customers drive the industry to redefine what is feasible and ultimately what is acceptable. Can this logic be applied on the public sector as well? If so, who benefits? Which regions or states? How can the public profit from protecting watersheds and biodiversity and averting rapid climate change? Obviously, it's easy to think of questions and hard to imagine answers. But it is certain that this vision of a more sustainable forest sector will be easier to realize if the demand-pull of competition can be utilized—far more catalytic and fruitful than relying on rules and regulations against the worst practices.

Indicators of Progress Toward the Vision

How do we know whether we are moving toward such a vision? Indicators of the condition within the U.S. forest sector—vital to measuring progress—can help. Already, the United States and other non-European countries with temperate and boreal forests have agreed to implement national criteria and devise indicators for sustainable forest management through the so-called Montreal Process (European countries are making similar headway through the Helsinki Process.) Seven key criteria, spanning such issues as biodiversity and forest health and productivity, are each tracked by anywhere from 3 to 20 illustrative indicators. The United States and other countries that signed the Santiago Declaration in 1995 (see Canadian Forest Service, 1995) are now testing these indicators to determine the steps still needed to implement them fully and how often they should be reported. If this work is taken seriously, countries can deepen their understanding of how sustainably their temperate and boreal forests are being managed. The key will be in overcoming some foreseeable shortcomings: lack of connection to a vision with clearly defined goals and a failure to cover the manufacturing portion of the sector. Meanwhile, as this process plays out, the indicators provided in Table 4-3 could be used to assess conditions and inform policy and management decisions.

Can all of these desires be satisfied? Maybe not. No substitute for action, indicators are like gauges on a dashboard, and it takes more than just one or two to define overall trends. The confidence or importance attached to any specific indicator depends on the quality of the data and the degree to which it is grounded in credible data, research, or experience. Even so, a relatively short set of indicators is absolutely vital to monitoring and adjusting forest policies and practices. As the United States starts applying the criteria agreed to in the Montreal Process, these indicators could serve as milestones along the path toward a more sustainable forest future.

TABLE 4-3. Indicators of a Sustainable U.S. Forest Sector

On the Land

area of natural forest ecosystems	increasing
productivity of timber species	increasing
ratio of timber harvest to net annual growth	declining
proportion of ecoregions in late successional classes	increasing
proportion of ecoregions in plantations	declining
sedimentation loadings in streams and rivers	declining
carbon storage in trees and forests	increasing
species and genetic diversity in plantations	increasing
fragmentation of natural forest ecosystems	declining
trees in urban and agricultural areas	increasing

At the Mill

production, use, and release of persistent toxics	declining
fossil fuel use throughout the forest products lifecycle	declining
efficiency in use of virgin tree fiber	increasing
recycling of paper and wood products	increasing
use of nonwood fiber in paper products	increasing
disclosure of environmental performance	increasing

In the Marketplace

markets for nontimber forest products and services	increasing
opportunities for forest recreation	increasing
jobs and wages in forest communities	increasing
per capita consumption of wood fiber	declining
public-private partnerships to meet sustainability goals	declining

STEPS ON A CRITICAL PATH TOWARD SUSTAINABILITY

Federal and other policies have improved forest practices, reduced air and water pollution from processing mills, and helped protect fish and wildlife habitats. But many are also blunt, costly instruments that fall short of goals, sometimes creating new problems, and offering the proverbial pound of cure instead of the ounce of prevention. Changes in how forests are managed and products are manufactured, in where markets for these goods and services are, in how knowledge is created and used, and even in how we think about forests will be at least as important as policy changes in making the U.S. forest sector more sustainable. Innovation, experimentation, and leadership by government, communities, resource managers, manufacturers, consumers, and others are vital to sustainability. All of this is predicated on the idea that options should be preserved before the resource deteriorates irremediably and, basically, that one good turn will catalyze another.

The transition to a sustainable U.S. forest sector will take decades, and success is not guaranteed. But constructive practical action is possible today, and the ten steps outlined here represent a sound start toward sustaining the diverse, valuable, and irreplaceable resources found in America's forest sector.

1. Develop and Implement Regional or State Sustainable Forest Sector Plans

Future-oriented planning mechanisms are needed to help public and private decision-makers, individuals, and institutions make informed choices today. Many states plan for the long-term future of utilities, transportation, tourism, and general economic development, but few do the same for the forest sector. Instead, forest sector planning is an *ad hoc,* piecemeal process usually centered on timber production and carried out mostly by individual public and private landowners. Too often, it is a reactive process that leads to blanket quick-fixes that compound the very problems they seek to address, sparking conflict rather than consensus.* Smart planning for a sustainable forest sector should keep options alive, minimize potential problems and conflicts, and use public and private sector resources more efficiently to address priority issues. This is the opposite of dictating to individual landowners, manufacturers, or government agencies once a crisis looms.

Certain basic features characterize a common sense approach to sustainable forest sector planning, even though each state's approach will ultimately be unique. First, planning should be broad enough to address all major forest products and services. Second, the planning process should welcome—but not force—the participation of all who use, manage, or benefit from the state's forests. The public should be engaged in hearings, dialogue, roundtables, and electronic

---

* For example, ballot proposals to ban virtually all clearcutting in Maine and California (both of which were voted down) might have complicated efforts to manage forests for biodiversity conservation objectives. And, the suspension of the 1995 "Timber Salvage Rider" enacted by Congress to expedite federal timber salvage sales (principally by suspending environmental laws and citizen appeals) likely exacerbated forest health problems in some areas.

debate (GIS, CD-ROM, World Wide Web) to pinpoint priority problems, identify preferred solutions and opportunities, and build a goal or vision statement. Third, the planning should be based on the best available ecological, social, and economic information on the state's forest sector. Fourth, the planning should culminate in specific binding commitments by government agencies, large corporations, nongovernmental organizations, and other participants to act. Finally, after the planning process, indicators should be used to track progress and provide the necessary input for evaluation and adaptation. Ideally, state and regional planning efforts would use an ecoregional approach to planning. *(See pages 249–50.)*

State and federal governments should lead the broad and diverse participation needed to get wide support for a sustainable forest sector, as well as set implementation policies. Still, given declining public budgets and wide distrust of government institutions, governments will be more effective if they act as conveners, facilitators, partners, and sources of financial, technical, and information assistance. Regional and state forest assessments by the Northern Forest Lands Council, Minnesota, Wisconsin, New Hampshire, and California bear this out.

2.   Establish a National Network of Demonstration Sustainable Forests

For two reasons, relatively little data is available to compare "sustainable" forest management practices with more conventional practices under realistic conditions. First, the technical aspects of more sustainable forest practices in research forests—most managed by the USFS or universities—have been overemphasized compared to day-to-day operational requirements, the costs and benefits of implementing them, and revenue flows over time. Also, though "new forestry" and other approaches to integrating nontimber benefits into management are being tried on industry-owned lands, cost and revenue projections are proprietary. A national network of perhaps a dozen sustainable forest demonstration sites managed by partnerships of federal, state, industrial, and nongovernmental organizations could generate realistic and public information on the comparative costs and benefits of conventional and more sustainable forest management regimes.

Each forest hub in such a network would reveal the capital, training, planning, monitoring, and opportunity costs of implementing various sustainable forest-management practices. Each of the model forests would be large enough to represent a region's typical timberlands and to allow managers to simulate in real time and space the capital, land, and labor constraints faced by industrial and nonindustrial landowners. Private land owners, public resource managers, conservationists, and policy-makers would all be encouraged to keep up with the activities and findings of the demonstration forests and all information on environmental impacts, operating costs, revenue flows, and management plans would be publicly available.

For landowners and forest managers, demonstration forests would provide practical information and guidance on best practices and ways to cut or share costs, find partners, or get more information. Conservation groups could observe new techniques first-hand, find practical ideas and guidance on conservation goals and objectives, and reach landowners and forest managers in a nonconfrontational setting. Researchers and policy-makers could identify leverage points in policy, needed public institutional reforms, effective incentives, and the costs of achieving and funding opportunities for public sustainable forest policy goals.

Already, an international network of "model forests" is being promoted by the Canadian Government (Canadian Forest Service, 1996). Responding to regional problems, the program will develop large working models of sustainable forest management through partnerships of federal and provincial governments, timber industries, environmental groups, indigenous peoples, and local communities. The goal of any U.S. network should not be just to test new ideas, but to document, simulate, and live with real-world constraints so the findings are useful and credible to resource managers. The costs would be reasonable, especially if shared by the federal (USFS, BLM, USFWS) and state agencies, and if forest products companies and conservation groups make in-kind contributions. Canada's national 10-forest network, for example, operates with a budget of approximately U.S. $7.5 million (and another $4.5 million of in-kind contributions)—about 5 percent of the 1995 USFS research budget.

3. Slow Fragmentation and Enhance Stewardship of Private Forest Lands Through Tax Reforms

Taxes—especially estate taxes and property taxes—make it hard to maintain large, contiguous, and diverse tracts of forest habitat on the private land, where most of the country's forests and most of the endangered species are found.[*]

---

[*] The Nature Conservancy estimates that about 50 percent of all species listed under the Endangered Species Act are found exclusively or principally on private lands.

Federal estate taxes (which now range between 37 and 55 percent of assessed market value) force many heirs of private forest land to sell or subdivide their property to settle the estate, thus fragmenting larger, more environmentally valuable forest habitats into smaller and less environmentally valuable parcels. County property taxes are typically based on assessments of the land's market value, which usually rises as residential and recreational developments encroach. Moreover, most property taxes treat all forest lands the same, whether they are managed exclusively for timber production or less intensively for greater public benefit.[*]

While not a new idea, reform of federal estate tax law can keep more forests intact at far lower cost than land purchases or conservation easements (Keystone Center, 1996). In its many permutations, the basic ingredients of estate tax reform are clear. Estate tax rates would be lowered if the lands are retained as forest habitat by their new owners, further reducing or eliminating estate taxes for heirs who enhance forest conservation. If a Department of Interior or state natural resource management agency assessment reveals the property is sold or the conservation agreement or certification breached, heirs should be subject to full federal estate tax liabilities. According to the Keystone Dialogue on Incentives for Biodiversity Conservation on Private Lands, such reforms would cost the Treasury somewhere between $4 million to $17 million annually in lost tax revenues, a very modest investment in the conservation of private forest land.

Property tax reforms could reward landowners who do more than the law requires to maintain or enhance public benefits from private forest lands. Property tax rates could be stabilized for landowners who maintain their forest lands as open space or productive timberlands and reduced if they get their forest practices independently certified as sustainable. This approach would work best in urbanizing areas or popular recreation sites, where pressures to develop forest lands are highest. Alternatively, a public benefits rating system reduces property taxes for landowners if their land is managed to sustain or enhance public benefits. King County, Washington, which includes Seattle, has created the first such program in the United States. Every piece of property gets rated, and the cumulative score determines the tax reduction—typically, 50 to 90 percent for that portion of the property providing public benefits (Keystone, 1996). For example, privately owned lands that include aquifer-protection areas, shorelines, active recreation

---

[*]   In addition, property taxes must be paid annually, though forest land—unlike agricultural lands or other commercial lands—may not generate income for decades.

areas, significant habitat for salmon, trail linkages, active farmland, or designated historic sites are designated as "High Priority Uses" which reduces their rating for tax purposes. Native plant sites, significant geological features, and buffer zones to public lands also receive ratings that reduce property tax liabilities. This system can answer local needs and generate more benefits than using the same public funds to buy land.

Tax-reduction schemes like these won't work well where tax rates are relatively low and stable or where local governments are short on funding. But since some of the public benefits are likely to accrue to state residents and even all citizens, sharing program costs with state and federal agencies may sometimes be justified.

4. Restore and Enhance Timber Productivity on Degraded Lands Through Innovative Financing Mechanisms

On many private timberlands, productivity is not as high as it could be. In western Oregon, more than 150,000 acres of privately owned timberland have been degraded by fire, reckless logging practices, or failure to reforest after harvesting (Barber et al., 1994). Although it's difficult to estimate how much private timberland nationally is degraded, surely several tens of millions of acres are. Many landowners lack the capital needed to restore such lands or can't wait many years for such investments to yield significant revenues.

Low-interest loans, cost-share agreements, profit-sharing, and other financing from both public and private sources could encourage such investments. By issuing bonds, state and local governments could fund low-interest loans for qualified landowners. In Oregon, state lottery revenues capitalize a revolving low-interest loan fund to help small landowners restore degraded timberlands or improve forest management. Under the Forest Legacy provision of the Farm Bill, the federal government shares the investment costs with landowners with less than 1,000 acres who improve forest management practices according to a preapproved plan.

From the private sector, profit-sharing arrangements might attract capital from such long-term investors as insurance companies and pension funds, which would be promised tax benefits for losses incurred in early years before revenues come in and profits at harvest time. Manufacturers of wood products might share such investment costs in exchange for long-term or future timber supply contracts.

5. **Protect and Restore Critically Endangered Forest Ecosystems Through Targeted Incentive Programs, Land Acquisitions, and Land Swaps**

As most forest ecosystems in the United States have been radically transformed during the past two centuries, the extent and quality of many types of natural forest habitat have declined—witness the growing number of rare, threatened, or endangered species. Yet, such losses are not inevitable; nor does conservation have to be costly or disruptive. A wide range of technical and scientific, legal, and economic tools and strategies can be used to integrate conservation objectives into forest management.

A vital step is to determine which forest types most need conservation help. Surprisingly, no such national inventory has ever been taken. State-level forest inventories, state "natural heritage" programs, and cooperative federal/state "gap analysis" projects, however, provide some of the necessary data. Although the National Biological Service was intended to carry out such analyses, political sensitivities over landowner concerns and limited budgets have blocked most action on this front. More effective for the immediate future would be a federal/state/private partnership that would identify endangered forest ecosystems and evaluate conservation options, especially if an independent agency such as the National Research Council were to establish criteria and indicators first.

The choice of conservation options for qualifying forest ecosystems would depend on who owns the land. Public lands would be assessed in terms of their ability to improve the conservation status of the forest type, and needed conservation actions would focus on these lands first—both because such actions would be less expensive and because government agencies generally have a public trust responsibility to manage public lands sustainably. For critically endangered forest ecosystems on private lands, a joint state or regional task force of state and federal agencies, private landowners, conservation groups, and others could propose voluntary incentives, whether tax breaks, cost-share programs, land swap possibilities, management partnerships, or trust funds for purchasing conservation easements or land parcels.

Conservation need not always mean expanding protected areas or establishing new ones. Restoring key ecological processes (such as low-intensity ground fires) can be encouraged on private lands through technical assistance, state or federal assumption of liability, and education and training programs. Or technical assistance from state forest agencies could help landowners find ways to encourage hardwood species to regenerate naturally once a low-productivity pine plantation is harvested in an endangered forest type. Such actions would qualify the landowner for various tax or stewardship incentive programs and can help meet conservation objectives by adding to protected areas.

6. Encourage Forestry Efforts within the United States to Sequester Carbon, Increase Fiber Supplies, and Enhance Rural Development

The best opportunities for addressing sustainability challenges address more than one problem. As an example, one response to intensifying land use conflicts within forest areas is to draw more fiber from outside of natural forest areas and discourage infrastructure and housing development within them. Similarly, several U.S. utility companies have already financed forestry activities in at least six countries to offset emissions of CO_2 from new fossil-fuel power plants. Such strategies may not be economically justified by forest benefits alone, but might if agricultural development, carbon sequestration, and renewable energy benefits were part of the investment calculus.

A domestic carbon sequestration program might work like this: a utility building a power plant would estimate how much CO_2 the plant would emit and invite bids from forest-management companies, farmers, or even public land-management agencies to provide the desired "carbon offset." To qualify, a bidder would demonstrate that offset funds would finance carbon storage that would not otherwise take place. (For example, farmers could propose tree-planting programs on surplus agricultural lands not covered by 1995 Farm Bill incentive programs.) Private land owners lacking capital could propose to restore degraded forest lands. Public agencies might even allow utilities to bid on timber sales and claim credit for the net carbon stored over the life of a power plant by not harvesting the wood.

Several obstacles complicate innovative programs like these. For one, utility companies have no assurance that they will receive credit for their carbon-offset investments if new limits on carbon dioxide get implemented later. A short-term solution is for EPA to issue guidelines on sound forest-carbon offset activities and for the federal government to guarantee that investments conforming to the guidelines would be eligible for incentives or regulatory breaks should it ever establish a program to reduce utilities' CO_2 emissions. Of course, utility or other financing of carbon offsets is unlikely unless the United States and other governments take decisive steps to curb CO_2 emissions under the international climate change treaty.

Another obstacle is the relatively high price per ton of forest carbon sequestered in the United States compared to investments in tropical countries, where land and labor are cheaper and trees often grow faster. One way out might seek "matching grants" from public forestry, agriculture, or rural development programs with compatible mandates or objectives. Or utilities could create partnerships with wood product manufacturers or other investors to develop new fiber supplies on marginal agricultural lands or abandoned industrial sites. To fully gauge such opportunities, utilities and forest products companies should jointly finance an independent assessment to determine the conditions needed to make shared investments work.

Finally, carbon-offset activities on a large scale would not necessarily make the forest sector more sustainable. Converting low-productivity natural forest stands to higher productivity planted forests, for example, could yield carbon benefits and increase fiber supplies but erode biodiversity, water quality, and other environmental services. In fact, increased production could even drive down prices, thus forcing more private forest lands onto the auction block and inviting subdivision and conversion to other uses. The U.S. Forest Service and the National Research Council should jointly assess such risks and benefits. Meanwhile, valuing the contribution forests make to carbon storage should encourage better forest-management and make the U.S. forest sector more sustainable over the long term.

7. Make the Environmental Performance of Forest Companies and Their Products More Open to Public Scrutiny

The success of the Toxics Release Inventory (TRI), which contains information from 24,000 U.S. industrial plants, demonstrates that public disclosure can help motivate and reward firms that reduce industrial pollution. Unfortunately, comparable information on other pertinent environmental dimensions of the forest products industry is not widely available. The industry has wide-ranging efforts under way to reduce manufacturing's impacts. Changes in pulping and bleaching technology have drawn the most praise from environmentalists, but other process modifications reduce the release of persistent pollutants and lower water consumption and energy use per unit of production. Yet, local communities, consumers, investors, and researchers receive precious little credible information, so firms that go "beyond compliance" cannot take credit for their success.

What can be done about this blind spot? The Paper Task Force encourages purchasers of paper products to query suppliers about their operations' environmental footprint. Within the industry itself, many firms have tracked the development of international standards, including ISO 14000, EMAS in Europe, and other fledgling schemes to get companies to establish environmental policies, set goals and monitor progress, and strive for continuous improvement. Firms can also convey this information to stakeholders and stockholders. Meanwhile, federal and state environmental authorities are increasingly willing to consider regulatory initiatives that give firms greater flexibility in exchange for greater public accountability. Better environmental performance measurement and reporting makes such experiments more credible and sure-footed.

Firms and trade associations in the forest products industry that are committed to environmental excellence can redefine the yardsticks by which they would be held accountable, preferably in collaboration with community groups and other stakeholders. These metrics should then be applied across the industry to allow business decisions-makers and their many stakeholders to benchmark progress in environmental performance.

8. Integrate Sustainability in Corporate Goals, Planning, and Operations

The day-to-day realities of harvest schedules, production targets, and customer demands can make grand thinking about sustainability seem like a luxury. But to stay competitive here and abroad, forest sector firms need to take trends in timber productivity, biodiversity, climate change, and clean production into account in their business decisions. Smart firms will do even more. By anticipating change, they can search out opportunities for competitive advantage and ways to preserve their options. Of course, facing up to these challenges entails financial risk. But those who ignore these fundamental economic driving forces only make themselves more vulnerable and undercut the sector's overall competitiveness.

Facing both real opportunities and risks, what should a forest company do? To find some answers, Weyerhaeuser is putting its expertise in forest management, manufacturing, product design, and marketing to the test. Multidisciplinary teams within the company are assessing trends and options across the forest product chain and looking for ways to dovetail its responses to changes in timberland, industrial technology, and customer requirements. The exercise is also helping strengthen internal linkages across business units, functional areas, and professional disciplines that have grown up in this vertically integrated company over the past half century.

Smaller firms would be hard-pressed to muster the internal resources of a Weyerhaeuser to think through the implications of such broad emerging trends. But most smaller firms have fewer options to assess in the first place and many can capitalize better than huge firms can on the opportunities created by the tilt toward sustainability. For example, by leading the pack in timber certification, California-based Collins Pine caught the attention of Home Depot which now distributes Collins' products.

The challenges of forest sustainability will affect any business that buys from, sells to, or otherwise depends on the U.S. forest sector. Alternative uses of forest resources that contribute to the challenges laid out here will generate value and create new possibilities for firms and investors with the vision and commitment to make it happen.

9. Cultivate a More Robust Concept of Sustainability in U.S. Forestry Education

If Americans want decision-makers in the U.S. forest sector to think creatively about sustainability challenges, then tomorrow's professionals must be educated differently. While there are centers of excellence in forestry research at several U.S. universities, most foresters entering the workforce today are scarcely exposed to the fundamentals of sustainability.

One strategy is to integrate sustainability issues into the mainstream curriculum, in courses ranging from silviculture to paper engineering to resource economics. The Management Institute for Environment and Business (MEB) pioneered efforts to bring environmental challenges into business schools curricula[*] by creating case studies and other teaching materials for use in finance, accounting, and marketing. Following MEB's lead, professional forestry programs might ask whether their graduates will have the breadth of knowledge needed to understand and assess the criteria for sustainable management outlined by the Forest Stewardship Council. Such foresight should pay off since rising demand for certification of sustainably managed forests should mean jobs for professionals qualified to assess practices on-the-ground.

---

[*] The Management Institute for Environment and Business merged with the World Resources Institute in October 1996.

A few schools have already begun to shift in this direction. With MacArthur Foundation support, the College of Forestry at Oregon State University is bringing researchers, students, and cooperative extension staff together to reshape the university's role in the sector through a "New Forestry" initiative. But the onus of change shouldn't fall on universities alone. By recruiting graduates with these skills, public and private employers will reinforce the value of this education and better adapt to social and economic change.

10. Bolster U.S. International Leadership to Improve the Sustainability of Forest Management Worldwide

When it comes to sustaining forest ecosystems and maintaining healthy forest economies, no country—including the United States—is an island. In the short term, domestic action taken to protect nontimber forest resources, such as protecting the last old-growth forest ecosystems in the Pacific Northwest, may increase logging pressures in other countries where sustainability concerns may be even more pronounced. Accordingly, the United States should help supplier countries develop the capacity to manage their own forest resources sustainably since U.S. demand helps fuel deforestation.

In the longer term, expanding international trade in forest products and capital could place tremendous and unsustainable demands on U.S. forests too. In this case, it is in the U.S. interest to provide international leadership to make forest management more sustainable worldwide. Given the enormous environmental, social, and economic benefits the world's forests provide not only locally and nationally, all countries have a stake in sustainable forest management everywhere.

To provide international leadership on sustainable forest management, examples will go farther than rhetoric or financial assistance. In many respects, the United States is already clearly a world leader in tree planting, using wood fiber efficiently, recycling, and reducing air and water pollution, to name but a few. Still, notable gaps remain in operating procedures and policies that Americans would not want to see other countries emulate. These include little or no legal protection for many forest types, particularly in the East and South; continued logging in the country's last remnants of old-growth forest ecosystems; clearcut logging on steep mountain slopes, particularly in the West; and below-cost timber sales on public lands where environmental costs outweigh economic benefits.

Since April 1995, the United States has had an obligation under the Montreal Process to implement comprehensive criteria and indicators of sustainable forest management. These measures should form the basis for a reformed forest data and inventory system that would replace the largely *ad-hoc* system used to inform long-term planning under the RPA program. Implementing these criteria and indicators aggressively will make the United States a leader in sustainable forest practices in coming years.

Finally, the United States has much to offer the world in terms of technical know-how, legal and policy development, and private sector management and marketing skills. Working with host countries, U.S.-based companies can provide leadership by applying the same environmental standards in overseas operations that they follow at home. Firms can also build markets for sustainably managed tropical forest products and the capacity of local companies and forest managers to supply such markets. The U.S. government can work with governments of developing countries to build their capacity to design, monitor, and enforce sound concession policies; respect established and traditional tenurial regimes; implement forest inventory and monitoring programs; and exchange useful ideas, practices, and data and information with colleagues in other developing countries. The United States should also ensure that future international policies do not diminish provisions of the Convention on Biological Diversity and the Framework Convention on Climate Change, which have already been negotiated to maintain the irreplaceable biodiversity and climate change benefits that forests provide.

CONCLUSIONS

The U.S. forest sector—including the country's extensive forest and timberland areas, their watershed, wildlife habitat, and other nontimber resources, and the jobs and communities they support—has changed enormously during the 20th century. Since the 1920s, after a long era of agricultural land clearing and unsustainable logging, forests have rebounded in the eastern United States, timber productivity has increased nationwide, a handful of once dwindling wildlife species has recovered, and wood products manufacturing has become more efficient and less polluting.

Despite these successes, the ability of U.S. forest lands to satisfy steadily increasing appetites for fiber plus other demands—to provide places to live, to protect wildlife, and to help offset our consumption of fossil fuels—is limited. There are signs that we are encroaching on these limits: after decades of steady improvement, forest productivity growth is beginning to tail off; biodiversity is being eroded by the loss and fragmentation of natural forest ecosystems; climate change threatens forest health and productivity; and forest product manufacturers are under competitive pressures to prevent industrial pollution. Each of these challenges is exacerbated by escalating international demands for the many valuable goods and services from forests. Any vision of a more sustainable U.S. forest sector in coming decades must address these challenges.

In WRI's vision of a sustainable forest sector, Americans will derive more value from the forest sector—more forest products and more environmental services. For example, pulp and wood chip production will be concentrated on a smaller but more productive land base, leaving most timberlands to grow longer rotation saw timber and provide nontimber forest benefits. A new generation of manufacturing technologies will virtually eliminate pollution at paper mills and wood-processing plants. Growing markets for specialty wood products, recreation, nontimber forest products, and for such environmental services as clean water and carbon storage will open up new opportunities to replace traditional logging and mill jobs lost to technological change and other factors. Besides supporting more diverse rural economies, the forest sector will supply more educational and recreational opportunities and make communities, neighborhoods, and work-places more attractive and livable. Finally, consideration of climate change impacts will increasingly help shape forest management and policy decisions—from where tree plantations are established to how we can best conserve and use genetic diversity.

A more integrated approach to using and managing forest resources through participatory planning informed by the best science and experience is central to a more sustainable forest sector in the United States. An ambitious program—using land swaps, targeted tax benefits, conservation easements, and purchases—is needed to conserve underprotected forest ecosystems, particularly in the eastern and southern United States. On private lands, tax and policy incentives should be revised to concentrate more intensive fiber production on a smaller land base and encourage long-term management of other forest areas for wildlife, recreation, long rotation saw timber, and other specialty forest products. Greater public disclosure of corporate environmental performance in land management, manufacturing, and products is needed to make sustainability a competitive advantage in the marketplace. Finally, citizens—the beneficiaries of a more sustainable forest sector—must get more involved in planning and policy decisions, guiding public

and private decision-makers and making them more responsive to community
and national needs.

REFERENCES

AFPA, 1995. *Sustainable Forestry.* Brochure on the Sustainable Forestry Initiative. American
Forest and Paper Association, Washington, D.C.

Amoco/EPA, 1992. *Pollution Prevention Project Yorktown, Virginia: Project Summary.*
Amoco/U.S. Environmental Protection Agency, Washington, D.C.

Aplet, G.H., N. Johnson, J.T. Olson, and V.A. Sample, 1993. *Defining Sustainable Forestry.*
Island Press, Washington, D.C.

Aracruz Celulose, S.A., 1996. *The Eucalyptus and Sustainable Pulp Production: Facts and
Figures.* Rio de Janeiro, Brazil.

Atkinson, W. 1990. "Another view of New Forestry." *Forest Watch* 11(2): 12–15.

Barber, C.V., N.C. Johnson, and E. Hafild, 1994. *Breaking the Logjam: Obstacles to Forest Policy
Reform in Indonesia and the United States.* World Resources Institute, Washington, D.C.

Beckenstein, A., B. Webb, F. Long, and B. Marcus, 1996. "Chlorine and the Paper Industry."
In: Beckenstein, et al. (eds.). *Stakeholder Negotiations: Exercises in Sustainable
Development.* Irwin Press, Chicago.

Bonifant, Ben, 1994. *Competitive Implications of Environmental Regulation of Chlorinated
Organic Releases in the Pulp and Paper Industry,* Management Institute for
Environment and Business, Washington, D.C.

Boyd, R.G., and W.F. Hyde, 1989. *Forestry Sector Intervention: The Impacts of Public Regulation
on Social Welfare.* Iowa State University Press, Ames, IA.

Bryant, D., D. Nielsen, and L. Tangley, 1997. *The Last Frontier Forests: Ecosystems & Economies
on the Edge,* World Resources Institute, Washington, D.C.

Business Council for Sustainable Development, 1994. *Getting Eco-Efficient: Report of the
Business Council for Sustainable Development,* First Antwerp Eco-Efficiency Workshop.

Canadian Forest Service, 1996. *Criteria and Indicators for the Sustainable Management of
Temperate and Boreal Forests.* Canadian Forest Service, Hull, Quebec.

——— , *International Model Forest Network.* Canadian Forest Service, Hull, Quebec.

Clark, L.R., and R.N. Sampson, 1995. *Forest Ecosystem Health: A Science and Policy Reader.*
American Forests, Washington, D.C.

Clary, D.A., 1986. *Timber and the Forest Service.* University of Kansas Press, Lawrence, KS.

Cohn, L., 1995. "The Growing Market for Special Forest Products: An Opportunity and
Challenge." *Forest Perspectives* 4(4):4–9.

Colborn, Theo, John Peterson Myers, and Dianne Dumanoski, 1995. *Our Stolen Future: Are
We Threatening our Fertility, Intelligence and Survival?* Dutton Signet.

Davis, Devra, and H. Leon Bradlow, 1995. "Can Environmental Estrogens Cause Breast Cancer?," *Scientific American*, p. 166, October.

Davis, M.B., 1996. "Eastern Old-growth Forests: Extent and Location." In: M.B. Davis, (ed.), *Eastern Old-Growth Forests: Prospects for Rediscovery and Regrowth,* pp. 18-34. Island Press, Washington, D.C.

Diamond, H.L., and P.F. Noonan, 1996. *Land Use in America.* Island Press, Washington, D.C.

Ditz, D., J. Ranganathan and R.D. Banks (eds.), 1995. *Green Ledgers: Case Studies in Corporate Environmental Accounting.* World Resources Institute, Washington, D.C.

Dobson, A., and R. Carper, 1992. "Global Warming and Potential Changes in Host-Parasite and Disease-Vector Relationships." In: R.L. Peters and T.E. Lovejoy (eds.), *Global Warming and Biodiversity.* Yale University Press, New Haven, CT.

Duffy, D.C., and A.J. Meier, 1992. "Do Appalachian Herbaceous Species Ever Recover from Clearcutting?" *Conservation Biology* 6: 196-201.

Durnil, Gordon K., 1995. *The Making of a Conservative Environmentalist,* Indiana University Press, Bloomington, IN.

EDF, 1995. *Paper Task Force Recommendations for Purchasing and Using Environmentally Preferable Paper,* The Paper Task Force: Duke University, Environmental Defense Fund, Johnson & Johnson, McDonald's, The Prudential Insurance Company of America, Time Inc., New York.

Eisner, T., J. Lubchenco, E.O. Wilson, D.S. Wilcove, and M.J. Bean. 1995. "Building a scientifically sound policy for protecting endangered species." *Science* 268: 1231-1232.

Ellefson, P.V., 1992. *Forest Resource Policy: Process, Participants, and Programs.* McGraw-Hill, Inc., New York.

Ellefson, P.V., and A.S. Cheng. 1994. "State Forest Practices Regulation of Private Forestry Comes of Age." *Journal of Forestry* 92(5): 34-39.

Erickson, R., 1995. "Closing up the Bleach Plant: Striving for a Minimum-Impact Mill." Unpublished paper. The Weyerhaeuser Company, Tacoma, WA.

FAO, 1995. *FAO Yearbook: Forest Products, 1982–1993.* United Nations Food and Agriculture Organization, Rome.

——, 1994. *Paper and Pulp Towards 2010—An Executive Summary.* United Nations Food and Agriculture Organization, Rome.

Fedkiw, J., 1989. *The Evolving Use and Management of the Nation's Forests, Grasslands, Croplands, and Related Resources.* U.S. Forest Service General Technical Report RM-175, U.S. Department of Agriculture, Fort Collins, CO.

Flather, C.H., L.A. Joyce, and C.A. Bloomgarden, 1994. *Species Endangerment Patterns in the United States.* General Technical Report RM-241, U.S. Department of Agriculture Forest Service, Fort Collins, CO.

FSC, 1995. *Principles of Forest Management,* Forest Stewardship Council, Oaxaca, Mexico.

Gale, R.P. and S.M. Cordray, 1991. "Eight Answers." *Journal of Forestry* 89(5): 31-36.

Gilliam, F.S., M.B. Adams, and B.M. Yurish, 1996. "Ecosystem Nutrient Responses to Chronic Nitrogen Inputs at Fernow Experimental Forest, West Virginia." *Canadian Journal of Forest Research* 26(2):196-204.

Goldberg, C., 1996. "From Necessity, New Products from the Forest." *New York Times,* March 24, 1996.

Greenwire, 1996. "Ecolabeling: Swedish Firms Seek Forest Certification." *Greenwire* 5 (202): February 29.

Gregg, N.T., 1991. "Will 'New Forestry' Save Old Forests?" *American Forests* 97: 49-53.

Hansen, A.J., T.A. Spies, F.J. Swanson, and J.L. Ohmann. 1991. "Conserving Biodiversity in Managed Forests." *BioScience* 41: 382-392.

Haynes, R.W., D.M. Adams, and J.R. Mills, 1995. *The 1993 RPA Timber Assessment Update,* U.S.D.A. Forest Service General Technical Report RM-259, Fort Collins, CO: USDA, March.

IIED, 1996. *Towards a Sustainable Paper Cycle: Final Report.* International Institute for Environment and Development, London, and World Business Council for Sustainable Development, Geneva.

Ince, P.J., 1994. *Recycling and Long-Range Timber Outlook.* U.S. Department of Agriculture Forest Service, Washington, D.C.

IPCC, 1995. *IPPC Working Group II Second Assessment Report.* International Panel on Climate Change.

Johnson, N.C., 1995. *Biodiversity in the Balance: Geographic Approaches to Setting Conservation Priorities.* Biodiversity Support Program, Washington, D.C.

Johnson, R., 1993. "Supplemental Sources of Income for Southern Timberland Owners." *Journal of Forestry* 93(3): 22-25.

Jones, J.A., and G.E. Grant, 1996, "Peak Flow Responses to Clear-Cutting and Roads in Small and Large Basins, Western Cascades, Oregon." *Water Resources Research* 32: 959-966.

Kaiser, J., 1996. "Acid Rain's Dirty Business: Stealing Minerals from the Soil." *Science* 272:198.

Kareiva, P. 1996. "Diversity and Stability on the Prairie." *Nature* 379: 673.

Keystone Center, 1996. *The Keystone National Policy Dialogue on Ecosystem Management.* Final Report, October 1996. The Keystone Center, Keystone, CO.

Kinstrey, R.B., 1993. "An Overview of Strategies for Reducing the Environmental Impact of Bleach Plant Effluents," *TAPPI Journal* 76(3).

Lave, L., E. Cobas-Flores, C. Hendrickson, and F.C. McMichael, 1995. "Using Input-Output Analysis to Estimate Economy-Wide Discharges," *Environmental Science and Technology,* Vol. 29, pp. 420A–426A, September, 1995.

Ledig, F.T., 1992. "Human Impacts on Genetic Diversity in Forest Ecosystems." *Oikos,* 63:87–91.

———, 1988. "The Conservation of Diversity in Forest Trees." *Bioscience* 38:471-479.

———, 1986. "Conservation Strategies for Forest Gene Resources." *Forest Ecology and Management* 14:77-90.

Likens, G.E., C.T. Driscoll, and D.C. Buso, 1996. "Long-Term Effects of Acid Rain: Response and Recovery of a Forest Ecosystem." *Science* 272:244-246.

Little, C.E., 1995. *The Dying of the Trees: The Pandemic in America's Forests.* Viking Press, New York.

Lyden, F.J., B.W. Twight, and E.T. Tuchmann, 1990. "Citizen Participation in Long-Range Planning: The RPA Experience." *Natural Resources Journal* 30:123-138.

MacClean, C.D., P.M. Bassett, and G. Yeary, 1992. *Timber Resources Statistics for Western Washington.* USFS Resource Bulletin PNW-RB-191. U.S. Department of Agriculture Forest Service, Portland, OR.

MacCleery, D.W., 1992. *American Forests: A History of Resilience and Recovery.* U.S. Department of Agriculture Forest Service and Forest History Society, Durham, NC.

MacKenzie, J.J., and M.T. El-Ashry, 1989. *Air Pollution's Toll on Forests and Crops.* Yale University Press, New Haven, CT.

Martus, C.E., H.L. Haney Jr., and W.C. Siegel, 1995. "Local Forest Regulatory Ordinances: Trends in the Eastern United States." *Journal of Forestry* 93(6): 27-31.

Mather, A.S., 1990. *Global Forest Resources.* Timber Press, Portland, OR.

Miner, R., and J. Unwin, 1991. "Progress in Reducing Water Use and Wastewater Loads in the U.S. Paper Industry." *Tappi Journal.* August 1991: 127-131.

Moulton, R.J., and K.R. Richards, 1990. *Costs of Sequestering Carbon through Tree Planting and Forest Management in the United States.* General Technical Report WO-58. U.S. Department of Agriculture Forest Service, Washington, D.C.

Noss, R.F., and A.Y. Cooperrider, 1994. *Saving Nature's Legacy: Protecting and Restoring Biodiversity.* Island Press, Washington, D.C.

Noss, R.F. and R.L. Peters, 1995. *Endangered Ecosystems: A Status Report on America's Vanishing Habitat and Wildlife.* Defenders of Wildlife, Washington, D.C.

Noss, R.F., E.T. LaRoe III, and J.M. Scott, 1995. *Endangered Ecosystems of the United States: A Preliminary Assessment of Loss and Degradation.* Biological Report 28, U.S. Department of Interior National Biological Service, Washington, D.C.

OTA, 1994. *Industry, Technology, and the Environment: Competitive Challenges and Business Opportunities*, U.S. Congress, Office of Technology Assessment, Washington, D.C.

——, 1989. *Facing America's Trash: What Next for Municipal Solid Waste?*, U.S. Congress, Office of Technology Assessment, Washington, D.C.

O'Toole, R., 1995. "The Reinvented Forest Service Budget: Still the Wrong Incentives." *Different Drummer,* 2 (2): 12–19.

——, 1988. *Reforming the Forest Service.* Island Press, Washington, D.C.

PCSD, 1996. *Sustainable America: A New Consensus for Prosperity, Opportunity, and a Healthy Environment.* The President's Council for Sustainable Development, Washington, D.C.

Pinkerton, J.E., 1993. "Emissions of SO_2 and NO_x from Pulp and Paper Mills." *Journal of the Air and Waste Management Association* 43: 1404-1406.

Poore, D., P. Burgess, J. Palmer, S. Rietbergen, and T. Synnott. 1989. *No Timber Without Trees: Sustainability in the Tropical Forest.* Earthscan Publications, London.

Porter, Michael, and Claas van der Linde, 1995. "Green and Competitive: Ending the Stalemate," *Harvard Business Review*, pp. 120-134, September-October.

Postel, S., and J.C. Ryan. 1991. "Reforming Forestry." In: L.Brown (ed.), *State of the World 1991,* pp. 74-92. W.W. Norton and Company, New York.

Powell, D.S., J.L. Faulkner, D.R. Darr, Z. Zhu, and D.W. MacCleery, 1993. *Forest Resources of the United States, 1992.* U.S. Forest Service General Technical Report RM-234, U.S. Department of Agriculture, Fort Collins, CO.

Renner, R., 1995. "EPA Great Lakes Guidance Hits a Squall." *Environmental Science & Technology* 29(9):416A-419A.

Repetto, Robert, and Sanjay Baliga, 1996. *Pesticides and the Immune System: The Public Health Risks.* World Resources Institute, Washington, D.C.

Rubenstein, Daniel B., 1994. *Environmental Accounting for the Sustainable Corporation: Strategies and Techniques.* Quorum Books, Westport, CT.

Sample, A.V., and D. LeMaster, 1992. *Assessing the Employment Impacts of the Proposed Measures to Protect the Spotted Owl.* American Forestry Association, Washington, D.C.

Schlosser, W.E., and K.A. Blatner, 1995. "The Wild Edible Mushroom Industry of Washington, Oregon, and Idaho: A 1992 Survey." *Journal of Forestry* 93:31-37.

Schlosser, W.E., K.A. Blatner, and R.C. Chapman, 1991. "Economic and Marketing Implications of Special Forest Products Harvest in the Coastal Pacific Northwest." *Western Journal of Applied Forestry* 6(3):67-72.

Schmidheiny, Stephan, and Federico J.L. Zorraquín, 1996. *Financing Change: The Financial Community, Eco-efficiency, and Sustainable Development*, MIT Press, Cambridge, MA.

Showalter, T.D., and J.E. Means, 1988. "Pest Response to Simplification of Forest Landscapes." *Northwest Environmental Journal* 4: 342-343.

Sleeth, P.D., and H. Bernton, 1996. "Swollen Streams Tied to Logging." *The Oregonian*, February 28, 1996.

Smith, Maureen, 1997. *The U.S. Paper Industry and Sustainable Production: An Argument for Restructuring*, MIT Press, Cambridge, MA.

Thomas, J.W., M.G. Raphael, et al., 1993. *Viability Assessments and Management Considerations for Species Associated with Late Successional and Old-Growth Forest of the Pacific Northwest.* Report of USDA Forest Service Scientific Analysis Team to U.S. District Court Judge William L. Dwyer, U.S. Department of Agriculture Forest Service, Washington, D.C.

Tilman, D., D. Wedin, and J. Knops, 1996. "Productivity and Sustianability Influenced by Biodiversity in Grassland Ecosystems." *Nature* 379: 718-720.

TNC, 1996. *Priorities for Conservation: 1996 Annual Report Card for U.S. Plant and Animal Species.* The Nature Conservancy, Arlington, VA.

Trexler, M.C., 1991. *Minding the Carbon Store: Weighing U.S. Strategies to Slow Global Warming.* World Resources Institute, Washington, D.C.

Trexler, M.C. and C. Haugen, 1995. *Keeping It Green: Tropical Forestry Opportunities for Mitigating Climate Change.* World Resources Institute, Washington, D.C.

Twardus, D., M. Miller-Weeks, and A. Gillespie, 1995. *Forest Health Assessment for the Northeastern Area, 1993.* U.S. Department of Agriculture Forest Service, Radnor, PA.

United Nations Environment Programme, 1995. *Global Biodiversity Assessment.* United Nations Environment Programme. Cambridge University Press, Cambridge, U.K.

——, 1994. "What is Cleaner Production and the Cleaner Production Programme?", *Industry and Environment*, October-December.

U.S. Bureau of the Census, 1995. *Statistical Abstract of the United States 1995.* U.S. Government Printing Office, Washington, D.C.

USDA, 1994. *RPA Assessment of the Forest and Rangeland Situation in the United States — 1993 Update.* U.S. Department of Agriculture Forest Service, Washington, D.C.

U.S. EPA, 1996. *1994 Toxic Release Inventory: Public Data Release.* Office of Pollution Prevention and Toxics, U.S. Environmental Protection Agency, Washington, D.C.

——, 1995a. *Characteristics of Municipal Solid Waste in the United States: 1994 Update.* U.S. Environmental Protection Agency, Washington, D.C.

——, 1995b. *Profile of the Pulp and Paper Industry.* Office of Compliance, Sector Notebook Project. U.S. Environmental Protection Agency, Washington, D.C.

——, 1994. *Estimating Exposure to Dioxin-Like Compounds. Volume I: Executive Summary.* U.S. Environmental Protection Agency, Washington, D.C.

WCED, 1987. *Our Common Future*, Report of the World Commission on Environment and Development, Oxford University Press, Oxford, UK.

Wear, D.N., 1993. *Private Forest Investment and Softwood Production in the U.S. South.* General Technical Report RM-237. U.S. Department of Agriculture Forest Service, Fort Collins, CO.

Westoby, J., 1987. *The Purpose of Forests.* Basil Blackwell, Oxford, UK.

Wilson, E.O., 1993. *The Diversity of Life.* Harvard University Press, Cambridge, MA.

WRI, 1996. *World Resources 1996-97: A Guide to the Global Environment.* A joint publication published in collaboration with the United Nations Environment Programme, United Nations Development Programme, and the World Bank. Oxford University Press, New York.

WWF, 1995. *Pulp Fact: The Environmental and Social Impacts of the Pulp and Paper Industry,* World Wide Fund for Nature, Gland, Switzerland, December.

Yaffee, S.L., 1994. *The Wisdom of the Spotted Owl: Policy Lessons for a New Century.* Island Press, Washington, D.C.

Yaffee, S.L., A.F. Phillips, I.C. Frentz, P.W. Hardy, S.M. Malecki, and B.E. Thorpe, 1996. *Ecosystem Management in the United States: An Assessment of Current Experience.* Island Press, Washington, D.C.

Young, John E. 1993. *Global Network: Computers in a Sustainable Society.* Worldwatch Paper 115, Worldwatch Institute, Washington, D.C. September.

5.
POWER TO CHOOSE
Sustainability in the Evolving Electricity Industry

Keith Kozloff

A VISION OF THE FUTURE

A major industrial sector of the U.S. economy, the U.S. electricity industry provides a vital service—a high-quality, versatile, affordable, and reliable energy source to fuel American economic production and consumption. Electricity production has improved its environmental performance over the last twenty years, substantially cutting releases of several menacing pollutants. But, increased reliance on fossil fuels will make these reassuring trends more difficult and more costly to maintain, and emissions of combustion-related compounds that aren't currently controlled are growing.

Another vision of the electricity industry's future points clearly in a more sustainable direction. Technological transformations sharply reduce environmental insults without interrupting or downgrading service. Over the next 50 years, virtually all of our current electricity-producing and electricity-consuming equipment will be replaced, providing the opportunity to phase in more efficient and less polluting equipment as the industry's operations and structure change. Informed by the following vision, this transformation could be both dramatic and cost-effective.

> The way that electricity will be produced and used is very different than it was in the 1990s. Electricity grids still exist, but less power is generated from large central stations. Many households and businesses generate their own power using micro-turbines, micro-cogeneration, or fuel cells powered by natural gas, gasified biomass, or hydrogen. Photovoltaic cells are incorporated into exterior building materials, including roofing shingles,

A Vision of an Environmentally Sustainable United States

and accompanied by on site electricity storage. Most homes and businesses with on site generation remain connected to the grid and sell surplus power back to it.

All homes and communities are designed to take full advantage of solar space conditioning and lighting through architecture, site orientation, and landscaping. "Smart" buildings incorporate integrated energy-management systems using real-time-of-use electric meters, occupancy sensors, and equipment controls to minimize electricity requirements. Electric appliances use less electricity than any models available today.

To serve large population centers, renewably generated power is transmitted from the hinterlands using high-efficiency transmission lines. Large-scale wind farms and biomass feedstock production supplement income for rural landowners. In sunny regions, solar-thermal electric installations also feed the grid. Biomass, hydro, and geothermal generation provide power on demand to complement solar and wind resources when the sun does not shine and the wind does not blow.

This broad portfolio of sustainable energy technologies will be cost-competitive because public and private investment in commercializing emerging technologies combines with a favorable policy environment. Together, various renewable energy technologies meet half of all power requirements. Fossil fuel generators made more efficient by extracting useful energy from waste heat for additional power production meet the rest. Waste heat warms urban centers.

Greater reliance on renewable energy technologies combines with greater efficiency in both power production and end use, so emissions of carbon dioxide and conventional pollutants from power production have declined from 1990 levels. Because other countries followed the example of the U.S. electricity industry in reducing atmospheric releases, global carbon emissions hold steady.

Technological transformation is not the only factor responsible for these changes. Various stakeholders (local, state, and federal policy-makers; the private sector; consumers) have seized opportunities afforded by changes in the electricity industry's structure and operation to guide power production in a sustainable direction.

Just as efficiency, equity, and environmental goals have been balanced in regulating the utility industry, states also balance these social goals as the industry becomes more competitive. Not all jurisdictions incorporate sustainability in the same way, and many details remain blurry. Nonetheless, state and federal governments ensure that:

- Electricity service providers have the incentive and ability to enter the market and offer the full range of sustainable energy technologies on both the supply and demand side.
- The environmental costs and risks of alternative ways of meeting electricity needs are fairly evaluated by the industry in making decisions about resource acquisition and use.
- The industry provides price signals to consumers based on life-cycle social costs and benefits.
- The industry has incentives to invest in technology development and commercialization that create long-term benefits.

Meanwhile, greater reliance on small-scale technologies is paralleled by more decentralized decision-making. Through local governments, consumers have more choice over which electricity services are provided. Communities solicit proposals from potential providers of electricity-distribution services, much as they issue cable TV franchises. Besides rate requirements, local governments require competing providers to develop packages of services that respond to constituent preferences. As a result, some minimum percentage of power is renewably generated, and energy-efficiency programs are implemented, while electric rates remain stable.

The shifts described in this vision are no more dramatic than those that have already occurred in response to driving forces inside and outside the electricity industry. Farsighted government and community leadership and cooperation can complement market forces to make sustainability a fact. Still, moving toward sustainability will require maintaining a long-term commitment in the face of political swings while new policies and ways of doing business take hold.

The rest of this chapter presents the case for policy measures to move the electricity sector along this path to sustainability. Some can be initiated immediately, but the analysis of current trends in electricity production show why market forces alone can't induce investments in sustainability, especially as the industry's structure evolves.

TRENDS IN THE ELECTRICITY INDUSTRY

Since its early days, technological, institutional, and market forces have governed trends in the electricity industry. Our mix of electricity-producing and consuming technologies has never been static, nor even predictable. Nor will the industry's future be a simple surprise-free extrapolation of the present. Still, salient underlying trends help reveal whether the industry is moving toward or away from sustainability.

Supply

The physical structure of each utility is dominated by large central power stations from which high-voltage transmission lines bring power first to distribution networks and ultimately to customers' homes and businesses. A utility system typically contains baseload capacity (generators that run all the time), intermediate service capacity (turned on when electricity demand rises during the day), and peaking capacity (used only for short periods, such as when air conditioning loads become very high). The power output of some generators must be increased or decreased on short notice so the system as a whole can meet demand that fluctuates by season, day, and minute throughout the service area.

Electricity production is the largest single use of primary energy in the United States. Of the 89 quads of total primary energy[*] consumed in the United States in 1994, some 36 percent went to electricity production. Electricity is the largest consumer of coal (85 percent of the volume delivered in 1994). While the electricity industry competes with the natural gas industry in meeting some energy needs, electricity production itself is a significant user of natural gas (15 percent in 1994, excluding government utilities) (EIA, 1996a).

Over the past two decades, different types of electric generation have lost or gained market share due to technological changes, fuel-price shocks, policy constraints, and other factors. As of 1994, some 52 percent of all power came from coal, 20 percent from nuclear energy, 14 percent from natural gas, 11 percent from hydro sources and other renewables, and 3 percent from oil (EIA, 1996a). Projections (through 2015) still show natural gas gaining market share in power production at the expense of nuclear power while those of coal, petroleum, and renewable resources change little. *(See Figure 5-1.)*

The combustion of fossil fuels for power generation will increase in the coming decades, and coal-fired generation can be expected to gain market share as the 21st century progresses. If current trends hold, the nation's aging nuclear capacity will be replaced by gas generation. Conventional U.S. natural gas reserves

---

[*]   A quad is a quadrillion British thermal units; one Btu is the quantity of heat needed to raise the temperature of 1 pound of water 1 degree F.

FIGURE **5-1. Baseline Projections of Electricity Generation by Fuel Type**

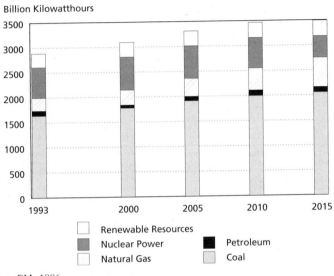

Billion Kilowatthours

Legend
Renewable Resources
Nuclear Power — Petroleum
Natural Gas — Coal

Source: EIA, 1996a.

are projected to be depleted sometime in the middle of the next century, though *total* supplies (which includes those harder to tap) are much larger in this country and Canada. U.S. coal stocks are so extensive that it will take at least 200 years to exhaust them even if gasified coal is substituted for natural gas in power generation and heating.

Individual generators are growing smaller but more efficient. Economies of scale dominated the industry's thinking until recently; now *advanced combustion* technologies, such as turbines developed from jet airplane turbines, allow modular generating units to achieve high fuel efficiencies.[*] Another advanced combustion process (combined cycle) uses the heat produced during fuel combustion to boost overall power generation. More limited efficiency improvements are possible for coal generation using "fluidized bed combustion."

---

[*]   Gas combustion turbine efficiency for electricity production will reach 60 percent before long, making gas economical as a fuel for baseload generation.

Demand

Electricity demand will continue to grow indefinitely, but offsetting influences will come into play. Power demand is greatly affected by population and economic growth, efficiency improvements, and changes in how various sectors use energy. The single most important factor is growth in the number of customers (Hirst, 1994).

On one hand, the versatility of electricity as an energy carrier spurs the economy to become increasingly electrified. Electro-technologies are substituted for other technologies that provide similar goods and services (electronic mail and faxes vs. conventional correspondence, or home offices vs. commuting). Electricity can also replace other sources of energy in major "end uses" (such as residential space heating). New electricity-consuming machinery and appliances (say, VCRs and CD-ROM players) are being purchased all the time. As prices fall, market penetration (and electricity demand) increases. Aside from commercial water heating, the fastest growing commercial and residential uses of electricity are personal computers, other office equipment, and other equipment (microwave ovens, second televisions, dishwashers, and miscellaneous commercial equipment) that have not yet saturated their markets. If three-quarters of today's cars in the United States were electric, national electricity consumption would be about 10 percent higher (James MacKenzie, personal communication, December 1996).

Broad demographic and economic changes are also pushing demand upward. Trends toward more people, fewer people per household, and larger dwellings are fueling growth in total electricity demand. Structural changes in the economy, such as the increasing relative importance of the commercial sector in the overall economy, also boost electricity use.

On the other hand, electrical equipment is becoming more efficient. Technological improvements are reducing the amount of primary energy required to get work done. More efficient household appliances are but one example.

Some of the same trends in information-processing and communications technologies that are stimulating electrification are improving efficiency at the same time. For example, as the cost of microprocessors (which have tiny electricity requirements) falls, it gets easier to reduce electricity consumption in appliances, industrial processes, lights, motors, and in heating, ventilating, and air-conditioning (HVAC) systems, etc.

Taken together, these trends spell continued growth in electricity consumption but at lower rates of growth than before the 1970s oil shocks. Even as

FIGURE 5-2. **Annual Growth in Electricity Consumption by End-Use (percent) for 1994–2015**

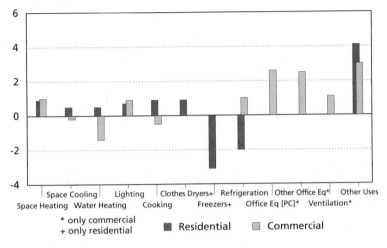

Source: EIA, 1996a.

electricity gets substituted for other fuels in some uses, its share of all primary energy consumption is projected to stay about the same through 2015. Efficiency improvements will reduce total electricity consumption of refrigerators, freezers, and lighting (though lighting will remain the largest single commercial use of electricity). *(See Figure 5-2.)* The average annual growth from 1994 to 2015 in purchased electricity (excluding cogeneration) is projected to be 1.1 percent in the industrial sector, 1.7 percent in the residential sector, and 1.2 percent in the commercial sector (EIA, 1996a).

Cost

The cost of electricity to various customer classes will be relatively stable since both fuel supplies and generating capacity are abundant. Electricity costs are expected to remain affordable for all but the poorest Americans, who are already finding it expensive. Compared to most other industrialized countries (including our major trade competitors), average U.S. electricity prices are low. *(See Figure 5-3.)* In real terms, electricity rates have dropped back to early 1970s levels, after rising in the early 1980s. *(See Figure 5-4.)* Over the next few decades, prevailing

FIGURE 5-3. Worldwide Disparity in Electricity Prices
1992 Industrial Prices—U.S. cents/kWh

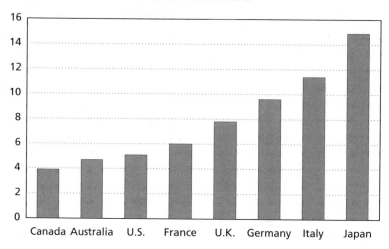

Source: Ontario Hydro, 1994.

FIGURE 5-4. Annual Increase in Average Household Income
and Residential Electricity Rates Relative to Base Year (1973=1)

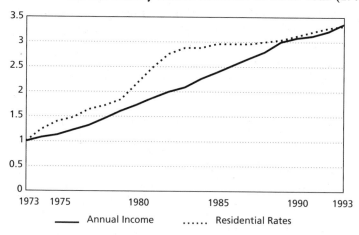

Sources: U.S. Bureau of Census and EIA, 1995a.

BOX 5-1. Affordability Trends

Electricity will remain affordable for most U.S. consumers under current trends. From 1960 to 1970, real electricity prices decreased by 30 percent while from 1973 through 1982 electricity prices rose much faster than overall inflation as fuel, capital, and operation and maintenance expenses increased. Relative to 1973, percentage increases in average household income have kept pace with percentage increases in electricity rates. (*See Figure 5-4.*) From 1994 to 2015, real electricity prices are projected to decline slightly while those of other energy sources grow slowly (EIA, 1996a).

For lower income groups, especially those living in high-cost utility service areas, the picture is bleaker. As a share of total expenditures for households in different income levels, low-income people spend proportionately more for electricity than people in higher income brackets. Moreover, as a percentage of household expenditures, electricity costs consistently increased from 1972 through 1993, doubling for some household groups. (*See Figure 5-5.*)

Some portion of the increased household expenditures allocated to electricity is due to increased consumption. Average residential consumption increased over this period by 22 percent, while commercial consumption (which includes apartment buildings) increased by 31 percent (EEI, 1994). Electricity cost data by income group do not show how much hikes in total electricity bills result from increased costs for basic electricity services (lighting, refrigeration, cooling, washing, etc.) versus growth in consumption as the use of new types of electricity-using equipment spreads.

electricity supply and demand trends aren't expected to drive up electricity rates much. *(See Box 5-1.)*

Not all of the costs of electricity consumption show up in monthly electric bills. Indirect costs from unsustainable electricity trends can, for example, show up in medical bills—conversely, in one recent period when air emission control investments added to electricity rates, decreased costs from illnesses related to fossil fuel emissions offset the loss in the industry's productivity (Repetto, 1990).

Environmental Performance

The full electricity fuel cycle—from fuel extraction to end-use consumption—has many economic, ecological, and human health impacts. For example, coal-burning entails human health risks and land degradation from mining, coal dust emissions from rail transport, various air emissions during combustion,

FIGURE **5-5. Percentage of Annual Expenditures per Household Spent on Electricity per Income Category**

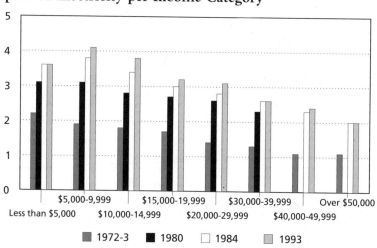

Source: EIA, unpublished computer data, 1995.

gangly power lines, and electromagnetic fields from electrical appliances. Nuclear generation produces various radioactive wastes: tailings from uranium mining, processing by-products, spent fuel, gaseous emissions, and radioactive reactor parts.

Some positive trends can be spotted in the war against conventional pollutants. But it is not certain that they can be maintained if a greater share of electricity comes from fossil fuels, and several other environmental trends are negative. Relying more on new plants based on coal and other fossil fuels, rising demand, and greater use of today's relatively dirtier plants will drive emission levels up.

One problem is the increasing atmospheric concentration of carbon dioxide, the principal gas implicated in global warming. Already the largest category of greenhouse gas emissions from the United States (the largest emitter of global greenhouse gases), annual carbon releases from power generation are projected to increase from 490.6 million metric tons in 1993 to 655.7 million metric tons in 2015. *(See Figure 5-6.)* In addition, methane (another greenhouse gas) is released in the extraction, processing, and burning of fossil fuels.

Even if global carbon dioxide emissions were stabilized at today's level, atmospheric concentrations would continue to increase, reaching nearly twice

FIGURE 5-6. U.S. Electric Industry Carbon Emissions

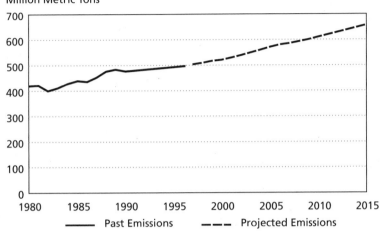

Million Metric Tons

Sources: EIA, 1996a, and U.S. EPA, 1995c.

the preindustrial concentration by the year 2100 (IPCC, 1996). To keep climate change at bay, major reductions in emissions are clearly needed. Specifically, holding the global concentration of carbon dioxide to twice the preindustrial level would require gradually reducing carbon emissions by about 75 percent below today's levels.

Net greenhouse gas emissions from the U.S. electricity industry are projected to increase despite voluntary actions taken under the Clinton Administration's Climate Change Action Plan (CCAP), which calls for emission reductions by the year 2000. Not surprisingly, this plan emphasizes supply- and demand-side efficiency improvements and so-called carbon offsets[*] in the United States and elsewhere that can be implemented relatively quickly. U.S. utility-sponsored tree planting and other carbon-offset projects in other countries represent steps in the right direction, but are unlikely to reduce the long-term growth in carbon emissions enough to stabilize atmospheric carbon concentrations at safe levels.

---

[*]   An action such as tree planting that can remove or otherwise avoid emitting a given amount of carbon dioxide into the atmosphere.

FIGURE **5-7. Electric Industry Emissions of SO$_2$ and NO$_x$ as a Percentage of Total National Emissions**

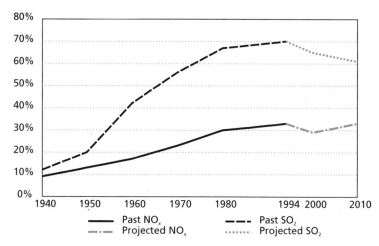

Source: EPA, 1995c.

Current trends related to other air emissions have been improving, but for how long it is hard to say. On the positive side, the electricity industry no longer contributes a growing share of the total U.S. SO$_2$, NO$_x$, and particulate emissions. *(See Figure 5-7.)* Indeed, total emissions of SO$_2$ and particulates are projected to decline thanks to "emission caps" mandated by the 1990 Clean Air Act Amendments, the emission limits of EPA's New Source Performance Standards, and higher power-generation efficiencies as old stock is replaced by new. Near-term levels of NO$_x$ are below those believed to acutely damage vegetation anytime soon or to toxify soils. *(See Figure 5-8.)*

On the negative side, at current levels of allowable emissions, residual emissions of acid rain "precursors" continue to threaten ecosystem functioning, and NO$_x$ emissions are still growing. Scientific analyses indicate that both sulfur and nitrogen deposition play an important role in acidifying surface waters. Episodic acidification occurs when pulses of acidic water enter lakes and streams after storms or spring snowmelt—especially dangerous if the surface water chemistry changes while fish are spawning or reproducing. Moreover, in terrestrial ecosystems, adding large amounts of nitrogen could change the mix of species, especially in mature forests and wetlands (U.S. EPA, 1995b).

FIGURE 5-8. U.S. Electric Industry NO$_x$ and SO$_2$ Emissions, 1940–2010

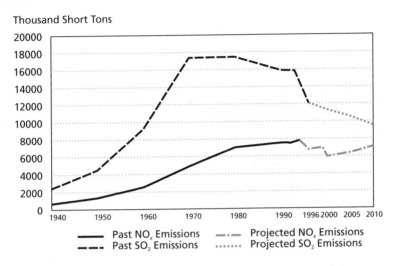

Thousand Short Tons

——— Past NO$_x$ Emissions   —·— Projected NO$_x$ Emissions
– – – Past SO$_2$ Emissions   ········ Projected SO$_2$ Emissions

Source: EPA, 1995c.

Small particles and nitrogen oxides also threaten public health. Nitrogen dioxide can irritate the lungs and lower resistance to respiratory infections. Exactly what short-term exposure does remains unclear, but continued or frequent exposure above normal concentrations may cause respiratory disease in children, and so-called "fine particulate matter" has also been implicated in potentially fatal respiratory disorders. After controlling for other risk factors, a comprehensive review of recent U.S. and European studies found a strong correlation between mortality and fine particulate air pollutants, including sulfates (Dockery and Pope, 1994; Dockery et al., 1993). If the nation followed current California standards (which are stricter than federal standards) for reducing concentrations of particulate matter, health benefits would total some $6 billion to $18 billion per year (Chestnut, 1995). Even more stringent controls on airborne particulate matter may be warranted considering the ill effects of the smallest particles on human lungs. While power plants contribute only a minor share of overall particulate emissions, they are the biggest contributors of fine particulates in the eastern states (Kidney, 1996).

FIGURE 5-9. **Estimated Mercury Emissions from Economic Activities (tons/year)**

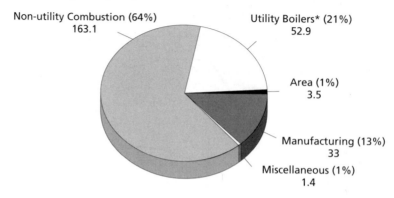

Non-utility Combustion (64%)
163.1

Utility Boilers* (21%)
52.9

Area (1%)
3.5

Manufacturing (13%)
33

Miscellaneous (1%)
1.4

*includes independent power producers

Source: EPA, 1994a.

Trace emissions of chemicals from power plants include heavy metals, organic compounds, and radionuclides. Power plants are the primary source of the organic methylmercury that people inhale and that accumulates in fish and other seafood. *(See Figure 5-9.)* This neurotoxin can be deadly at extremely high doses and wreak neurological damage at lower doses. As with other airborne toxics, exposure to mercury from power plants is at very low levels over a long period, and few studies shed light on the long-term health effects of this type of exposure.

The health risks from hazardous air pollutants depend on several factors, not all of which are well understood—power plant stack emissions, the dispersion of exhaust plumes in the atmosphere, the pathways of health effects, population patterns in the affected region, and demographic and lifestyle characteristics. The Electric Power Research Institute concludes that both cancer and noncancer health risks from these emissions are tiny (Lamarre, 1995), but EPA has identified 67 hazardous air emissions whose risks warrant further study (*Environment Week,* 1995).

Organizational Structure

Current changes in the industry's organizational structure could affect sustainability. Today, most electricity is provided by vertically integrated utilities (those combining generation, transmission, and distribution). Retail sales by these monopolies are regulated by state Public Utility Commissions (PUCs) while wholesale generation and transmission services are regulated by the Federal Energy Regulatory Commission. Individual utilities have organized themselves into regional councils that coordinate power flows to keep supply reliable. Publicly owned utilities also play a significant role in the power industry, especially in small towns and rural areas. Private independent power producers are the fastest growing source of new capacity. Between 1986 and 1995, their contribution averaged 43 percent of all new capacity and ranged as high as 61 percent (RCG/Hagler Bailly, 1996).

Responding to the 1992 federal Energy Policy Act (EPACT), the electric power industry is becoming more competitive. This drive toward competitiveness has gotten a boost from the experiences of other U.S. economic sectors and from other countries. The U.S. telephone and natural gas industries have been restructured over the past ten years to make them more competitive. In Norway, the United Kingdom, New Zealand, and other OECD countries, restructuring of the power sectors is also under way or completed.

One driving force is industrial customers' belief that competition would lower electric rates. It's certainly true that electricity rates vary widely across the country and even among utilities in the same state. *(See Figure 5-10.)* Also true is that electric power from new capacity sometimes costs less than that from older units. Much new capacity is gas-fired, which is cheap because gas prices are low and because low-cost, high-efficiency generating technology is available. Still, whether competition can really lower rates for some customers without raising rates for others remains an open question.

How retail electric services are restructured will be determined by each state. Momentum is building in more than 30 state public utility commissions and 15 state legislatures for profound changes in how electricity services are provided (*Solar Letter*, 1995b). However, because the rules under which the industry operates in one state will affect electricity suppliers in others, the federal government will undoubtedly impose some uniform guidelines to avoid the energy equivalent of a Tower of Babel.

In April 1995, the Federal Energy Regulatory Commission (FERC) issued rules requiring utilities to decouple power generation costs from transmission costs and to charge consumers a separate price for each. Utilities controlling interstate transmission lines must now allow competitors access to their transmission lines and charge them no more than the price they pay for using their

FIGURE **5-10. Average Utility Revenue by State (cents per KWh)**

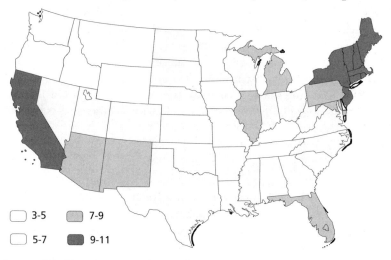

☐ 3-5	☐ 7-9		
☐ 5-7	■ 9-11		

Source: EIA, *Electric Power Annual,* December 1994.

own lines. FERC also ruled that utilities could sell power well beyond their traditional service areas, allowing low-cost producers to compete with higher cost producers on the latter's own turf (Sindelar, 1996).

While many variations are possible, two basic industry structures may emerge:

Wholesale Competition for Power Generation. In this model, robust competition reigns in wholesale electricity markets. Options range from mandatory competitive procurement of new generating capacity to divestiture of utility generating units from the regulated part of the company (transmission and distribution). Utilities could still dispatch power from their own control centers or from a larger utility-owned power pool. Over time, nonutility generation becomes the main source of the wholesale power that utilities retail to their customers. *(See Figure 5-11.)*

Retail Competition. Under this model, individual customers select their own power suppliers. Vertically integrated electric utilities that operate monopolistically within defined geographic areas (with profits regulated by state PUCs) cease to exist. Electric distribution companies emerge as regulated monopolies distributing electricity that customers are free to buy from alternative suppliers either inside or outside the distribution company's service territory. Some or all

FIGURE 5-11. Wholesale Competition and Retail Regulation

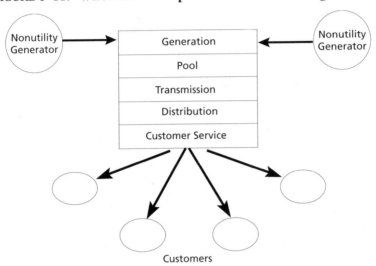

Source: Galen et al., 1995.

customers can make their own power-supply arrangements with a competitor of their franchised utility. *(See Figure 5-12.)* Retail customers pay a service fee to maintain the distribution system. Separate transmission companies transmit power and perhaps also operate wholesale bulk "power pools." Alternatively, separate companies operate such pools. Second by second, control centers would schedule, dispatch, and control power plants and the transmission system, making the flows of money and electrons independent of each other. Prices for transmission and distribution services would continue to be regulated by the Federal Energy Regulatory Commission and public utility commissions, respectively (Galen et al., 1995; Daycock, 1996).

A Transition Phase

Anticipating future changes, electric utilities have already begun to change their business practices. In this "virtual restructuring" phase, many utilities act as though they already face wholesale or retail competition even though they don't yet. In this transition of a decade or so, inherent uncertainties make utilities reluctant to make major capital commitments that could lead to higher rates,

FIGURE 5-12. Structure of Electric Utility Industry Under Full
Competition Scenario

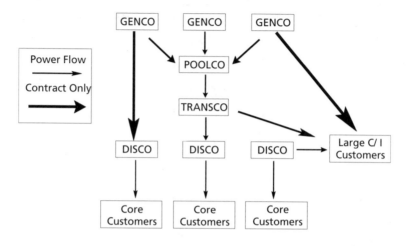

Source: Bauer, 1994.

even though the investments could lower rates in the long run. This reluctance
will slow the turnover of generation capacity. In the near term, new construction
will be aimed at eliminating transmission and distribution bottlenecks.

In such a transition phase, who pays for expensive power plants built by utilities
during the 1980s and early 1990s that deliver power at above-market costs? The
need to resolve this issue will delay restructuring. A "stranded investment" might be
a generating unit still being paid off, for instance, or a contractual obligation to buy
power or fuel supplies at rates higher than what the market later dictates. National
estimates for stranded costs range as high as $135 billion (Wamsted, 1995). If stranded
nuclear assets were sold off, buyers would demand federal insurance protection to
guarantee disposal of wastes and decommissioning costs, and the utility might be
required to bid on its own plants. Alternatively, the federal government will have to
assume ownership of stranded nuclear assets, as the United Kingdom has.

Restructuring's Ripple Effects

Industry restructuring will change the choice of generation technologies used
to provide electricity services. Power suppliers facing retail competition will try

to create generation portfolios that minimize near-term costs. Competitive electricity generation markets are likely to favor older plants that are at least mostly paid for already since they can produce cheaper power than new plants can. To hold down expenditures, utilities will probably extend the life of inexpensive old plants while shutting down expensive generators, such as many U.S. nuclear plants.

As older, dirtier coal plants get used more and expensive nuclear facilities are mothballed, the environment will feel the brunt. Since the vintage and location of plants affect their allowable emissions, competition could give an advantage to plants with higher NO_x, particulate, and CO_2 emissions per kWh. Even if the capacity-utilization rate increased by only 3 percentage points, NO_x emissions would rise by 24.6 percent of the Clean Air Act targets, thus raising abatement costs. If 6,000 MW of nuclear capacity is prematurely retired because of their higher costs, subsequent rises in NO_x emissions would equal 5 percent of CAAA targets. Such capacity shifts will have similar effects on CO_2 emissions (Lee and Darani, 1995).

At the same time, electricity prices may become more volatile, especially to the extent that spot markets develop and customers can switch suppliers. As in the telephone industry, competition may mean itemized charges, which make customers' bills more complicated. Whether additional information makes consumers more cost conscious or more frustrated depends on whether they think itemization helps them control expenses better.

As for costs, increased competition can be expected to lead to greater efficiencies and downward pressures on electricity rates in the long run. Retail competition is likely to reduce the price disparities among and within states as regions with cheap excess power sell it elsewhere.

Clearly, there will be winners and losers under any new regime. Stranded costs must be paid by some combination of ratepayers, shareholders, and, in the case of nuclear capacity, possibly taxpayers. During the transition, current customers will pay more. Moreover, between 1982 and 1992, average industrial rates fell by 25 percent in 1992 dollars, and further cuts are unlikely (NRDC study). In general, customers who now pay less than their fair share of utility costs may see their rates increase during the transition.

Utility restructuring could prompt new efforts to promote increased electricity use. The Cleveland Electric Illuminating Company's new pilot program will award customers points for electricity use that can be redeemed for electric products (Wamsted, 1996). Meanwhile, the U.S. power industry is also becoming increasingly international in the wake of the Energy Policy Act (EPACT). Nonregulated generation subsidiaries created by about 50 U.S. utilities are now looking at overseas markets, especially in developing countries where power demand is surging. These firms owe much of their ability to construct and manage power projects to the experience gained by their regulated parent companies, which are accustomed to working with independent power producers but also favor fossil fuel technologies. By the same token, U.S. utilities are beginning to buy up systems in other countries as power systems are privatized.

MOVING THE ELECTRICITY INDUSTRY TOWARD SUSTAINABILITY

What are the implications for sustainability of the electricity industry trends just described? According to the Brundtland Commission and others, sustainability requires that today's economic production and consumption activities do not restrict the well-being of current or future generations. Adhering to this simple principle of intergenerational equity would alter the decision-making rules used in the electricity industry by requiring power producers to make sure that future generations will have access to affordable and reliable electricity services and no long-term harm will be done to human, economic, and ecological health.

Progress has been made toward this principle, and emission trends for some pollutants are down. But some baseline trends in electricity supply and demand threaten sustainability. Of the many industry trends, those that threaten sustainability are long-lasting and expensive or impractical to reverse. Such trends also affect the quality or quantity of critical ecological or economic resources for which no ready substitutes exist. Measured by these yardsticks, growing carbon emissions pose the greatest threat to sustainability. The share of total U.S. carbon emissions from electric power production is projected to increase from 36 percent to 38 percent through 2015, and emissions will grow in absolute terms too (EIA, 1996a). In addition, despite improvements in pollution controls on power plants, emissions of nitrogen oxides are expected to increase as fossil fuels replace other power sources in electricity generation. Choosing coal—our most abundant fossil fuel source—for power generation poses the biggest threats to both climate and to human and ecosystem health because burning it releases relatively larger amounts of carbon dioxide, sulfur dioxide, particulates, and air toxics.

Needed Technological Shifts

Fossil fuel consumption can be reduced in several ways. First, electricity can be produced more efficiently. Second, the electricity required for each unit of light, heat, etc., can be reduced. Finally, electricity generation can shift away from dependence on fossil fuels.

Producing Electricity More Efficiently

Energy efficiency can be improved in power generation where, on average, only about one-third of the primary energy consumed reaches electricity-using equipment in homes and businesses. The thermal efficiency of large central station power plants peaked 20 years ago. In fossil fuel plants, most of the heat produced in burning fuel is released into the environment rather than used to produce power or heat structures. In fact, the 21 quads of heat vented from U.S. fossil and nuclear plants is more than all the homes and commercial buildings in the country use each year.

One way to boost energy efficiency is to capture waste heat to increase electrical output, provide space heating, or meet other needs. *District energy systems* (DESs) deliver heat or cooling from a central unit through underground pipes to buildings in central cities. (Waste heat from fuel combustion can't be used unless power plants are located close to heating needs.) They both recover waste heat and cut the amount of energy needed for space conditioning in buildings. At one time, most power plants were in central cities, where it was easy to capture waste heat. (In Scandinavia, about 40 percent of the buildings are served by district-heating systems.) In recent years, district cooling projects have proliferated in several major downtown areas (Seeley, 1995), and the total U.S. market for district steam, hot water, and chilled water could rise by an estimated 50 percent between now and 2010 (Teotia et al., 1994). District energy systems are most easily added when streets are repaired. One constraint to the use of DESs is the difficulty of organizing customers. Without a strong consumer base, power developers have little incentive to build district energy systems.

Another option is the fuel cell, which produces electricity efficiently and quietly through a chemical process with very low emissions. (When hydrogen is the fuel, the only emissions are water and some waste heat.) Because fuel cells come in small units, they can produce power on site. They can be stacked to serve apartments, neighborhoods, and other medium-sized loads or sized to meet a house's full electric needs. Still expensive, fuel cells are inching their way into commercial applications.

FIGURE 5-13. The Distributed Utility Paradigm

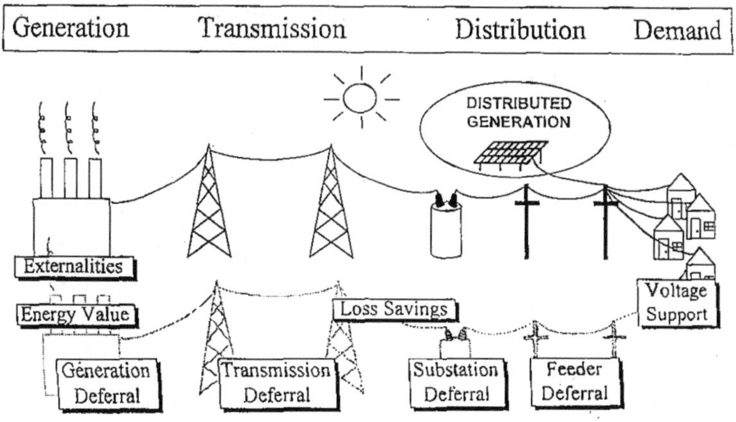

Source: Hoff and Wenger, 1995.

An even newer approach—called distributed generation—is locating small power generators and energy-saving measures close to where power is needed. Such installations are particularly valuable where service costs run high. *(See Figure 5-13 and Box 5-2.)* No technological breakthroughs are needed to displace some central station capacity, and investments in transmission and distribution system components can be scaled back under this approach. At the extreme, individual households and businesses could generate their own power.

The potential reduction in carbon emissions from improving the efficiency of power generation appears large, even though comprehensive national estimates are unavailable. For example, a cogenerated district-heating system could provide about a 50 percent greater carbon savings over conventional fossil generation (Krause et al., 1994a). If just 22 percent of the new generating capacity between 1995 and 2010 were built as cogenerating district-energy systems, annual energy savings would total about half a quad, cutting annual carbon emissions by about 5 percent (Spurr, 1994).

Reducing Waste in Electricity Consumption

Changes in both electricity-using equipment and in how such equipment is used can lead to major power savings. Look just at lighting, for example. Turning off

BOX 5-2. Distributed Generation

Distributed power generation not only displaces some central station generation; it also confers benefits within the entire utility system. Because renewable energy technologies, storage, and energy-saving technologies can be deployed at or near the location where electricity is actually needed, they can obviate investment in expanding transmission or distribution (T&D) systems, improve the use of existing T&D assets, and (because of lowered energy losses) reduce the amount of central station generation needed to maintain reliability standards. The cost of upgrading transmission and distribution system capacity is particularly high where local peak demands are sharp. Depending on location, the grid-support benefits of distributed applications may even exceed their capacity and fuel benefits. In one case, a 500-Kw photovoltaic installation yielded $70,000/year of such traditional benefits but also reduced electrical line and equipment losses, extended the life of distribution equipment, and decreased scheduled equipment-maintenance requirements (Siemens Solar Industries, 1995).

Transmission and distribution investments have traditionally been developed to serve large central generation stations and to reliably deliver enough power to meet consumer needs. These systems are sized to meet peak loads and industry standards of electric service reliability. Although generation capacity requirements long drove transmission and distribution investments, since 1988 generation has constituted less than half of total utility construction expenditures (EEI, 1994)—a trend expected to continue as long as utilities have surplus capacity.

The "distributed utility" concept can also be applied in homes or businesses on the customer's side of the meter. Rooftop grid-connected photovoltaic systems, some versions of which mimic roofing material, could eventually be installed in millions of homes. Solar domestic water heaters, passive solar design, ground-source heat pumps, and other renewable technologies can also meet demand cost-effectively (Byrne et al., 1992; Maycock, 1992; Stein, 1992). For remote or otherwise hard-to-serve customers, employing stand-alone renewables often costs less than extending distribution lines.

lights in unused rooms, replacing incandescent bulbs with fluorescent fixtures, and switching from area to task lighting would all save power. More efficient technologies for delivering light, cooling, or motion can be substituted as long as the service quality remains acceptable. (Some consumers prefer incandescent over fluorescent light, for instance.) While the average efficiency of electrical "end uses" has already improved, technical innovations and equipment turnover could further decrease power demand from buildings, equipment, and appliances. Even though automatic building and equipment controls have in some cases replaced the need for people to work at conserving energy, individual electricity consumption behavior is still a driving force in this technological shift. Consumers can choose more or less energy-efficient equipment, remember or forget to switch on energy-saving devices, and be rigid or flexible about the quality of energy services.

Additional electricity savings can come from designing, siting, and constructing buildings better. Energy efficiency features can be incorporated into new buildings and equipment so that they do not need to be retrofitted. Proper building orientation, subdivision design, and landscaping can reduce heating and cooling energy needs. Solar water heating, passive solar heating and cooling, and daylighting can shave electricity requirements, sometimes at little or no extra cost.

How such end-use efficiency improvements affect air emissions depends on how supply and demand change. For example, quantifying SO_2 emission reductions from managing air conditioning demand means comparing cuts in total system demand throughout the day and year with the operation of individual power plants serving the system (Fenichel, 1993). The time and place of kilowatt hour savings are less critical for carbon emissions (which are long-lived in the atmosphere) than for sulfur or nitrogen oxides, as long as fossil generation is being displaced. Carbon savings from efficiency improvements, however, depend on lifetime energy savings and the types of generation that are offset. Efficiency improvements that reduce peak demand more than energy consumption (for instance, automatic air conditioner shut-offs) have little direct climate benefit.

Estimated reductions in electricity consumption from end-use efficiency improvements range widely. Potential savings from end-use efficiency improvements based on so-called "bottom up" engineering models tend to be much higher than estimates based on "top down" macroeconomic models. According to one electricity industry estimate, the electricity-saving technologies in residential, industrial, and commercial use could save 24 to 44 percent of the base-case electricity consumption in the year 2000: 800,000 to 1,436,000 GWh (Gellings, 1991).

Nonfossil Power Generation

Solar energy resources consist of flowing water, wind, solar radiation, and organic materials that can be burned. *Photovoltaic* cells convert solar radiation directly into electricity. *Solar thermal electric* technologies produce electricity by concentrating sunlight onto a working fluid or engine. (Parabolic troughs, for instance, concentrate sunlight as much as 80-fold.) *Wind* machines convert the motion of wind into rotational energy that drives electric generators. Combustion of solid or gasified *biomass* (mainly wood, wood wastes and by-products, agricultural debris, and energy crops) provides both heating and electricity. (Municipal solid waste contains biomass too, but also petroleum-based products.) *Hydropower*, a mainstay of the U.S. industrial revolution, remains the most important renewable resource used for generating electricity. *Ocean thermal energy conversion* taps the temperature difference between the warm surface layer of the ocean and deep cold layers. *Direct solar energy* is also used for heating and cooling buildings and for water heating. Finally, sunlight directed to building interiors reduces the need for artificial lighting.

In addition to solar energy, *geothermal* heat trapped up to 3,000 feet below Earth's surface can be used to generate electricity. At present, hydrothermal energy (steam, hot water, or hot brine located within 900 feet of Earth's surface) is the primary form used commercially, though hot dry rock represents a large potential resource. *Ground-source heat pumps* that exploit the temperature difference between the earth and surface are used to heat and cool increasing numbers of residential and commercial buildings.

Nuclear fission generates electricity through controlled nuclear reactions that heat water to drive steam turbines. As of 1991, nuclear power constituted 68 percent of the nonfossil power generation in the United States, compared to 30 percent for hydroelectricity and 2 percent for other renewables (U.S. Council for Energy Awareness). Other nuclear technologies that are not yet commercial (or demonstrated) are breeder and fusion reactors.

Each of these climate-friendly energy-supply technologies has relative advantages and disadvantages vis-à-vis sustainability, cost, and potential for avoiding carbon emissions. Three particularly important categories of carbonless technologies are intermittent solar technologies that depend on natural energy flows (i.e., wind and solar radiation), biomass, and nuclear fission.

Intermittent Renewables. Some renewable energy flows (i.e., solar radiation, wind, flowing water) have natural rhythms that vary seasonally, daily, or hourly. Enough energy must thus be collected when these resources are available to meet electricity needs when they aren't. The closer that the power output of renewable technologies matches periods when electricity demand is high, the more value they have.

Intermittent resources can be combined to stabilize power output over an area. Also, resources can be matched to a utility's peak load. Depending on the fit between the output of an intermittent power source and the utility system's load, intermittent capacity might constitute as much as 35 percent of a system's total capacity as long as the rest is easy to dispatch.

Developers of intermittent capacity could buy back-up power on contract or develop complementary generation sources. Most efficient is pairing a highly dispatchable source with an intermittent one—for example, wind and biomass resources, though some biomass feedstocks are harvested seasonally and must be stored in the field or on site. Exploiting such opportunities requires analyzing renewable energy flows, patterns of electricity demand, generation requirements, energy-storage technologies, and transmission capacity—and having the technical and financial wherewithal needed to develop various power sources.

Among the renewable options that could complement intermittent resources, biomass power generation is likely to expand most. As intermittent renewable resources come to constitute a larger share of a system's capacity, reducing electrical demand may be a cost-effective way to maintain reliability while integrating intermittent resource flows into a generating system. While demand management alone is unlikely to make a system work using only intermittent resources, it can play a more important role than it has to date. Patterns of electricity demand are not immutable. Many utility customers have agreed to have air conditioners or other equipment automatically shut down during certain periods of peak demand while others have demonstrated their willingness to trade off lower levels of reliability for lower rates (Lamarre, 1995).

Relying on intermittent renewables to meet half or more of a system's generation requirements means storing their output until needed. Energy can be stored using batteries, compressed air reservoirs, flywheels, and small-scale pumped hydro. According to one study, baseload power could be provided from wind combined with an energy-storage system based on compressed air for about 25 percent more than the delivered cost of wind-generated electricity without storage today (Cavallo, 1995). If batteries or other more expensive storage technologies had to be deployed today, however, the total cost of energy from renewables would be significantly higher. Fortunately, the costs of several storage technologies will probably drop sharply over the next decade.

Investments in new transmission capacity may be needed to move renewable power to population centers. Beyond the cost of generating electricity at the plant gate is the cost of transmitting power to load centers. The better the fit

FIGURE 5-14. **Low Carbon Electric Generation Scenario**

Billion kWh

Legend:
- Imports
- Intermittent renewables
- Biomass
- Geothermal
- Nuclear
- Hydro
- Fossil fuels

X-axis: 1985, 2025, 2050

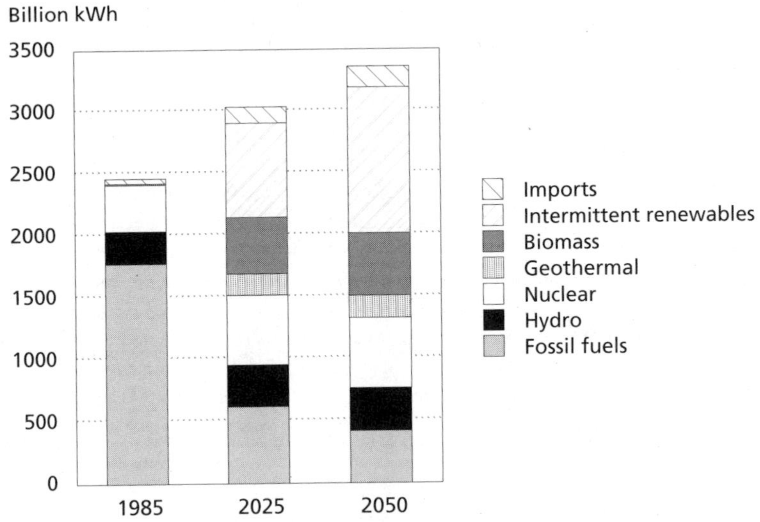

Source: Johansson et al., 1993.

between regional patterns of electric power demand and renewable energy flows, the less power must be brought in at some added cost from outside the region. While more evenly dispersed around the country than fossil fuels, renewable energy flows are often either too site-specific or too diffuse to meet all the power demands of big cities. For example, in the densely populated Northeast, renewably generated power might have to be imported from the Midwest to supplement local renewable resources. To break up interregional transmission bottlenecks, transmission capacity would need to be expanded at key points.

A low-carbon electric generation scenario developed for the Intergovernmental Panel on Climate Change suggests that, excluding hydro, renewables could meet about half of U.S. electric generation needs by 2025 in a more energy-efficient economy. (Johansson et al., 1993). *(See Figure 5-14.)* A recent study of renewables' electricity potential in the European Community came to a similar conclusion: the share of total power from renewables is 23 to 42 percent if keeping costs low is the sole concern, but increases to over 70 percent if carbon-minimization benefits are included in the calculations (Krause et al., 1995).

BOX 5-3. Sustainability and Biomass Power Production

The environmental benefits of expanded biomass energy use in the United States depend on which biomass feedstocks are produced, how they are grown, and which land parcels are converted to grow energy crops.

With respect to net carbon emissions, the question is how much carbon does each feedstock sequester, how quickly, and for how long? When agricultural or forest product residues decompose, they release carbon gradually. When residues displace fossil fuels for power generation, most of the carbon emission savings are those that would have otherwise occurred anyway as the residues decomposed. (For example, dead trees in forests release 2 to 3 percent of their stored carbon each year.) As for crops grown for energy, the longer the rotation, the greater the climate benefits. The more biomass left standing above and below ground, the better; and the less external fossil fuel input used (such as fertilizer), the greater the net carbon benefits.

Production practices also affect environmental benefits. Benefits are greatest when land now in annual row crops is converted to perennial biomass feedstock production. Landowners squeezed between the prices that utilities are willing to pay for feedstocks and the opportunity costs of producing biomass may not be able to afford to try environmentally superior production practices that require extra costs. Moreover, federal subsidies to farmers for growing "program" food crops (such as corn) raise the opportunity cost of growing biomass energy crops, and an increasingly competitive electric power industry may be unwilling to pay for feedstocks or power from biomass generators. As a result, markets may remain dominated by residue feedstocks, which generally cost less than dedicated feedstocks grown especially for energy.

Biomass. Biomass power generation can deliver power on demand, and some type of biomass resources are widely available across the country. In addition, technological innovations in gasification and aero-derivative turbines promise to bring power costs down, and capital costs are less than those of other renewable energy options.

Despite these advantages, questions surround the sustainability of large-scale dependence on biomass for power production. *(See Box 5-3.)* To displace significant amounts of fossil fuels, biomass generation will need feedstocks grown especially as energy crops, not just residues from forestry and farming. Here, land requirements become an issue. The United States has some 36 million acres of land retired for the long term under the Conservation Reserve Program that could be used for biomass feedstocks using sustainable production practices. But whether such surplus acreage will remain available for energy-related use depends on domestic and world food demand and productivity trends. Right now, trends in food demand suggest that 7 percent to 60 percent of total U.S. cropland could be available for biomass energy crops in 2030 (National Biofuels Roundtable, 1994).

Nuclear Fission. Nuclear generation currently displaces substantial carbon emissions, and greater dependence on this power source could displace much more. However, nuclear generation fails the test of sustainability for other reasons. The long-lived radioactive isotopes produced from normal operation must be kept isolated from humans and ecosystems for many generations, and accidents in which radioisotopes with long half-lives are released into the environment could threaten current and future generations.

Even if health and ecosystem threats can be contained, using nuclear power is foisting many costs on generations to come. The full costs of the so-called "back end" of the nuclear fuel cycle (waste disposal and plant decommissioning) are still speculative, but some very large bills are likely to be paid by future generations that did not benefit from the electricity produced. *(See Box 5-4.)* In a more competitive climate, neither utilities nor independent developers would willingly shoulder the risks associated with nuclear power plants, so new facilities are unlikely to be built.

Timing of Technology Transformations

All three of the general approaches to reducing carbon emissions will probably be exploited because relying solely on any single one poses technical or cost problems. As electricity-using equipment provides more and more services to the growing U.S. economy, greater efficiency improvements in generation and end uses would continually be required to keep carbon emissions tolerable. While carbon-less generation resources could meet all power needs, most such options currently cost more than providing equivalent electricity services using fossil fuels. An abrupt switch to noncarbon generation would make electricity unaffordable, but various sustainable energy technologies can be phased in as capital stock turns over to keep any increases in electricity prices to a minimum. The relative contribution of each technology to reducing carbon emissions is hard to predict since innovations and cost reductions over coming decades can't be known.

The U.S. electricity industry has invested heavily in generation, transmission, and distribution systems that simply can't be replaced overnight. The value of

BOX 5-4. Nuclear Fuel Cycle Costs

Average nuclear generation costs currently exceed costs from several fossil and nonrenewable technologies. Capital costs, perennially underestimated, have increased steadily since the first plant and peaked at $2,346/KW real dollars for plants constructed starting in 1974-75. Operation and maintenance costs for 44 plants commissioned before 1989 doubled between 1980 and 1990. While low operating costs were initially touted as nuclear power's advantage over fossil fueled plants, by 1992 their respective operation and maintenance costs had about equalized (Gielecki and Hewlett, 1994). Assuming operation at maximum feasible capacity factors, the cost of power from currently commercial pressurized water reactors in France and Germany (with more standardized designs than those built in the United States) ranges from 5 cents/kWh to 9 cents/kWh (1989 dollars).

These estimates do not reflect the full fuel-cycle costs of nuclear power since commercial experience in nuclear waste disposal and plant decommissioning is scant. Spent-fuel disposal for existing U.S. reactors is estimated to cost $26 to $35 billion (1988 dollars), and permanent storage facilities will not be available until 2010 at the earliest. For aging nuclear units with high operation and maintenance costs and components in need of replacement, owners must either pay a premium over other sources of power to keep them operating or face uncertain costs to decommission them. Cost estimates for dismantling reactors have increased from $59/KW in 1976 to $450/KW in 1994, more than a sevenfold real increase (Tellus Institute, 1994).

In the absence of permanent waste-storage facilities, some utilities are reconfiguring on site storage of spent fuel to allow continued operation. But this move represents only a small downpayment on the ultimate costs of permanent disposal.

fixed capital stock (excluding government utilities) was $810 billion in 1992—about 5 percent of the nation's fixed private capital and only slightly less than the fixed capital for all nondurable goods manufacturing. Fossil fuel extraction, some dedicated to the electricity industry, constitutes another $352 billion worth of capital stock, while manufacturing plants for electric and electronic equipment represent $162 billion in fixed capital (U.S. Bureau of Economic Analysis, 1993).

Changing technology is less disruptive and costly when a capital investment reaches the end of its useful life and must be replaced, so the rates of turnover of capital stock will largely determine how quickly the electricity industry can become sustainable. As of 1995, about 75,000 megawatts of capacity were at least 40 years old. Total capacity reaching 40 years of age will continue to increase

BOX 5-4 continued.

By early in the next century, about 30 plants will have exhausted their spent fuel pool-storage capacity (*Energy Daily,* 5-9-94). If cost pressures force premature shutdown, decommissioning and waste-storage trust funds (currently endowed by taxing the output of the plants) will be strapped.

High operating and maintenance costs, along with looming safety problems, will probably force many U.S. reactors to close before their 40-year licenses expire. No commercial nuclear reactor has yet operated for 40 years. The oldest, Big Rock Point in Michigan, is 32 years old and is already economically uncompetitive with other sources of electricity. As of January 1994, more than half of the reactors operating in the United States failed to produce electricity more cheaply than other power sources available to their owners.

To lower the costs of future nuclear plants, new reactors would incorporate standardized designs, passive safety features, and other advanced technologies and designs. How much power from the next generation of reactors will cost is highly speculative since no working demonstrations of advanced designs are currently planned in the United States (although one is under construction in Japan). Standardizing and simplifying reactor design could reduce capital and operating costs significantly. According to one projection, advanced designs might produce power at the plant gate of about 5 cents/kWh in 1989 dollars, not counting government subsidies (Krause et al, 1994b). Nonetheless, this projection does not reflect either government subsidies or market externalities.

On balance, then, it is difficult to imagine circumstances in which nuclear power could become a competitive private enterprise without government subsidies.

through 2015, and the biggest share of these old plants are coal-fired. *(See Figure 5-1.)* If new coal-based generation is installed, the results will be dramatically different than if alternative energy sources are harnessed.

INDUSTRY INCENTIVES FOR SUSTAINABLE ENERGY INVESTMENTS: A CASE STUDY

Will market forces alone trigger the technological transformations needed to dramatically reduce fossil fuel use? Probably not. Climate risks simply are not adequately incorporated into the electricity industry's investment decisions. So how can firms in particular and the electric industry in general make prudent investment decisions and take other actions with future carbon constraints in mind? And what incentives would prompt socially optimal

FIGURE 5-15. **NSP Baseload Capacity Reaching 40 Years of Age (MW)**

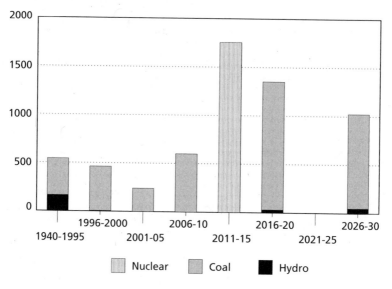

Source: EIA, unpublished computer data, 1995.

decisions? An analysis of one electric utility sheds light on these important questions.

Profile of an Electricity Supplier

Northern States Power Company (NSP) is a vertically integrated utility that serves parts of Minnesota, Wisconsin, and North Dakota.[*] NSP's 1995 15-year resource plan shows that customer peak load will probably increase by 2 to 3 percent per year through 2010, so the company will need to acquire 3,800 to 6,000 MW of additional resources by 2010. To meet this need, NSP plans to reduce demand by 1,054 MW by 2010 through energy-efficiency programs, acquire generation resources through competitive bidding to meet long-term supply needs, and buy power as needed over the short term. NSP's generation

---

[*]   A technical appendix on the NSP case study is available from World Resources Institute.

BOX 5-5. The Climate Problem: Abundant Uncertainties

The uncertainty surrounding the maximum level of greenhouse gas emissions that can be allowed without destabilizing Earth's climate will be resolved only after some decades. Neither the magnitude of climate change from human activities nor the severity of associated environmental and economic effects is known. The best type, magnitude, and timing of policies for controlling carbon emissions is also uncertain. Better information will gradually reduce uncertainty, though when understanding will become firm enough to trigger strong action is itself unknown.

Increases or decreases in greenhouse gas emissions may not be felt for decades, though these emissions have a gradual but long-lasting effect on atmospheric concentrations. Because the oceans can absorb heat, atmospheric concentrations do not cause immediate warming. Once politicians decide that action is needed, they must still design technical and social fixes, establish mechanisms for international coordination, and translate new treaties into national laws. The pace of implementation depends on how quickly the private sector is required to replace current capital stock with more climate-friendly technologies.

Costs are yet another uncertainty. Too much can be spent in the near term to mitigate carbon emissions or too little. On the one hand, the accumulation of carbon in the atmosphere may irreversibly alter the species composition of an ecosystem. On the other, equipment for removing carbon dioxide from stack gases constitutes an irreversible investment that cannot be recouped if it turns out not to be needed.

---

portfolio resembles that of the national mix in terms of the types and ages of technology in use. *(See Figure 5-15.)* Its electricity rates fall in the middle range of utilities within the Upper Midwest and have declined in real terms over the past decade. A subsidiary is developing independent power projects for the competitive wholesale market, and an expected merger between NSP and a Wisconsin utility will help the company compete for retail customers.

Responding to Climate Risks

Since investments to meet electricity needs must be made without knowing how much carbon emissions will be restricted years or even decades from now, evaluating alternative investment strategies is anything but straightforward. *(See Box 5-5.)* In NSP's resource plan, the utility poses three strategies for mitigating the risk of future climate change regulations. The first, dubbed "no regrets," means that NSP will pursue only those measures (such as low-cost ways of helping customers improve energy efficiency) that are cost-effective when evaluated on the basis of benefits other than climate. The second strategy is to require NSP's wholesale power suppliers to accept the risk of future climate regulation, which NSP already does when it acquires new production capacity. The third strategy is to increase the share of generation resources in NSP's system that do not emit greenhouse gases.

NSP prefers the first two options, arguing that shifting its generating mix in anticipation of future climate policy is risky. Uncertainty over the nature, timing, and magnitude of any policy that might be imposed makes it possible that any shift in NSP's generation portfolio could entail irreversible costs that later prove unnecessary. (Interestingly, NSP has adopted risk-management techniques to address the possibility that its nuclear capacity will be shut down prematurely, but has not applied them to the risk of future climate-related regulations.)

As a participant in the national Climate Challenge program, NSP is voluntarily committed to trying to reduce greenhouse gas emissions by acquiring 100 MW of wind capacity, adding on site storage of spent nuclear fuel to avoid premature shut-down, and managing demand (through recycling appliances, offering customer rebates for highly efficient motors and lights, or sponsoring retrofits to make buildings more energy efficient). However, the NSP/U.S. Department of Energy agreement does not specify either greenhouse gas reduction targets or time tables.

Although the percentage of NSP's generating portfolio based on renewable resources is projected to decrease, NSP is acquiring some renewable capacity. To comply with a 1994 Minnesota law that allows utilities to temporarily store additional used nuclear fuel on site if they also use more biomass and wind power, NSP must acquire 400 MW of wind generation and 125 MW of biomass generation by 2002. The cost of the first 100 MW of wind capacity is projected to be 3 cents/kWh over 30 years. Another 50 MW of biomass power may be supplied by farm-grown feedstock.

To analyze the trade-offs among the various strategies that NSP could pick to address climate risks, two ways to meet future electric service demand were modeled. In each, the mix of resources was chosen to minimize the costs. The period of analysis is 1995-2050—long enough to guarantee full capital depreciation of NSP's current generating portfolio. In one strategy ("wait and see"), NSP takes no account of the risk of future carbon restrictions, but alters its investment path if such restrictions come about. In the meantime, most near-term new capacity is fossil-based. Under the other strategy ("act then learn"), NSP anticipates possible carbon-emission restrictions and seeks the lowest cost mix

FIGURE **5-16. Climate Strategy Alternatives with Future Learning**

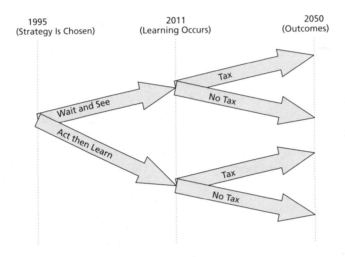

of resources to do the job, modifying its resource acquisitions accordingly. If carbon restrictions are not imposed, NSP's approach to resource acquisition reverts to the first strategy. *(See Figure 5-16.)*

Many other assumptions are made in this exercise. Under all scenarios, NSP's nuclear capacity is retired and its hydropower contracts end early in the next century. Uncertainty regarding carbon policy is resolved in 2011; either a carbon tax is phased in starting then (with full implementation in 2020) or it is deemed unnecessary. No major changes in the performance or cost of available technologies occurs over the study period. NSP accepts technology costs as uncontrollable factors, and the utility ranks technologies according to their life-cycle costs, so intermittent renewable capacity (wind) is limited to about 20 percent of NSP's generation portfolio, and substantial quantities of biomass power generation are needed to stabilize NSP's carbon emissions.

Carbon restrictions, modeled as carbon taxes on fuels, potentially affect resource acquisitions, the composition of the generating portfolio, and electricity demand. A tax of $100/ton of carbon (real dollars), set high to heighten the distinction between the two strategies, is applied to the carbon emissions of various generation options and may affect which options cost least. Relatively easy to model in a utility decision framework, carbon taxes have been one of the most widely debated policy tools for restricting carbon emissions. In fact, in early 1996 a bill was introduced into the Minnesota legislature that would phase in a $50/ton tax on carbon emissions over five years.

The other approach is a cap on carbon emissions, which allows firms to trade emission rights. A key difference between the two strategies is the relative certainty of costs and benefits in advance of implementation. Under a tax, costs are more certain than the resulting reduction in carbon emissions. Under a cap, emission reductions are certain by definition, but compliance costs vary by firm. Taxes are simpler to administer, but either policy would make fossil fuel-based power generation more expensive for NSP.

In our analysis, NSP could reduce carbon emissions over the base case by picking the lowest priced combination of:

◆ meeting the gap between demand and existing capacity by adding new natural gas, wind, or biomass generation;

◆ shutting down some existing capacity;

◆ operating relatively low-carbon generators more of the time; and

◆ offering customers incentives to cut back their electricity consumption.

In this analysis, the two investment strategies give rise to four scenarios that represent cost-minimizing investment streams. Each scenario was analyzed with respect to how the mix of power generation types, costs per kWh, and carbon emissions change over time. Also, cumulative carbon emissions, 2050 emissions as a percentage of 1995 emissions, and present value costs were calculated. *(See Figures 5-17 and 5-18.)*

What does the analysis show? First, taking early actions to reduce carbon emissions can be advantageous for NSP. Under an anticipated large carbon tax or equivalent carbon restriction, it is in NSP's interest to begin hedging its carbon emissions bets by adding wind power to its generation portfolio and by reducing customer demand. Specifically, if NSP considers the likelihood of a $100/ton carbon tax being implemented in 2020 is 44 percent or better, the expected net benefits from modifying its investment portfolio in anticipation are greater than from waiting to find out for sure.

The least expensive investment scenario occurs when NSP does not take early action and the tax does not materialize (present value cost of $19.7 billion). The

FIGURE 5-17. Annual Electricity Output by Type (GWh)

A. Act then Learn—No Tax Case

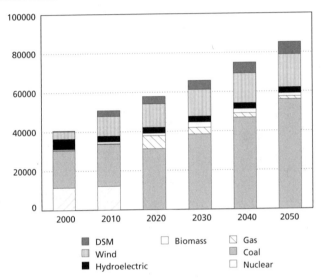

B. Act then Learn—Tax Case

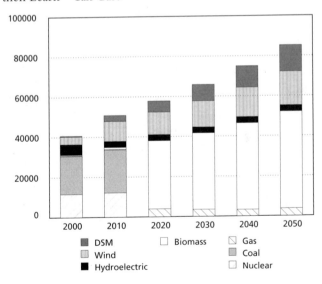

Source: Union of Concerned Scientists.

FIGURE 5-18. **Annual Electricity Output by Type (GWh)**

A. Wait and See—No Tax Case

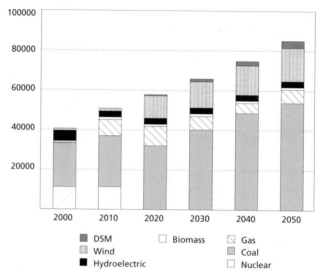

B. Wait and See—Tax Case

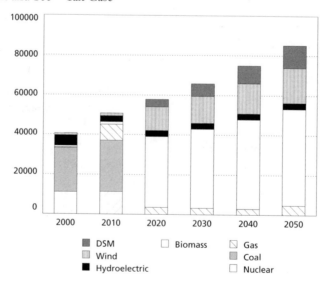

Source: Union of Concerned Scientists.

TABLE **5-1.**   **Scenario Simulation Results**

	Present Value Costs*	Cumulative CO$_2$ Emissions (million tons) 1995–2050	2050 Emissions as % of 1995 Emissions
Wait and See (No Tax)	19.7	2215	290%
Wait and See (Tax)	30.1	446	10.4%
Act Then Learn (Tax)	28.6	356	8.5%
Act Then Learn (No Tax)	20.9	2006	169%

*Including carbon tax payments

Source: World Resources Institute and Union of Concerned Scientists.

most expensive is when NSP follows the same strategy and the tax does get imposed ($30.1 billion). In other words, hedging protects NSP from the worst-case outcome and causes it to forego the best possible outcome. *(See Table 5-1.)*

NSP's "no regrets" strategy might actually lead to big regrets, for both society and the company. By going beyond "no regrets"—for example, by investing in renewables before it is required—the utility would gain valuable resources and experience should carbon taxes later be imposed. From the viewpoint of society more generally, anticipating the carbon constraint—by going beyond "no regrets"—is far more advantageous: in three out of the four scenarios, societal benefits outweigh societal costs. From society's perspective, it would be better in the end for NSP to begin changing its investment portfolio even if there were only a 28 percent chance of a carbon tax being imposed.

The differences between what a utility considers in its financial interest and what is good for society at large gets at the nub of the difficulty private firms face in incorporating sustainability into investment decisions. Here, discount rates, estimated costs, and perceived benefits all come into play, along with two other factors not analyzed here—the ability to spur technological change and to shift risk away from current ratepayers.

Discount Rate

NSP discounts the stream of future costs and benefits at the market cost of capital (6.2 percent). The social discount rate, in contrast, measures society's preference for current over future consumption. The social discount rate chosen here (3

BOX 5-6. Discounting and Long-Term Decision-Making

One rule of thumb in economic analysis is that an investment should be undertaken if the discounted sum of the benefits exceeds the discounted sum of the costs (including social benefits and costs). When an investment has a positive net "present value," society's net gains accruing at one point in time are large enough to fully compensate the net losses occurring at another, regardless of whether those who benefit actually compensate those who lose.

The choice of discount rate is tremendously important in determining the extent to which future costs and benefits affect today's actions. The future is typically discounted to reflect society's preference for the present over the future, as well as inflation and project risk. The discount rate typically used by private investors is higher than when investments are evaluated from society's perspective, and the higher the discount rate, the less the future counts.

The divergence between private and social discount rates grows when an investment has social costs or benefits that occur over many decades. Costs and benefits far in the future become irrelevant because discounting reduces their present value to almost nothing. In the case of greenhouse gas emissions, long time lags mean that the conventionally discounted net present benefits of any actions to reduce emissions are small.

A common justification for discounting is that future generations need not have equal standing with the present generation because advances in technology will make them wealthier than we are today. However, the assumption of increasing wealth in the face of adapting to climate change and energy resource constraints is itself questionable. Even so, public and private decision-makers are unlikely to give up the practice anytime soon.

percent) falls in the middle of the range used in other cost/benefit analyses of climate policy (Toth, 1995). Even at a 3 percent discount rate, $100 of benefits accrued 40 years from now are worth only $31 today. *(See Box 5-6.)*

Differences in the discount rates assumed in decision-making can skew comparisons of alternative investment strategies. Relative to social discount rates, a higher private discount rate will favor investment strategies in which benefits materialize early and costs come due later. In comparing individual technologies, using higher discount rates favors those that can be paid far down the road. But both renewable energy and energy-efficiency equipment typically require more capital than do equivalent fossil fuel capacity and less efficient electricity-using equipment (which cost relatively more to operate). In general, the cost of renewable and fossil generation (levelized over their lifetimes) is particularly sensitive to what discount rate is used in calculations.

Counting Costs and Benefits

Another difference between private and social perceptions of what makes a good investment springs from whether various outcomes are counted as costs or benefits. Most of the environmental and economic benefits of avoiding carbon emissions by choosing carbon-free energy sources would accrue broadly to society and to future generations, not just to NSP or its customers. While NSP would benefit from continued economic growth in its service area should climate stabilize, these benefits are not proportional to NSP's own emission reductions but rather to those made globally. Only if all the world's utilities cut emissions at the same time would private and societal benefits begin to converge.

As for costs, imposing carbon taxes would raise NSP customers' electricity rates. Whether from society's perspective these taxes are seen as real costs or merely transfers depends on how the revenues are used. If used to reduce other taxes to NSP's customers dollar for dollar, the carbon tax would be revenue neutral. (The carbon tax introduced in Minnesota would be offset by dollar-for-dollar decreases in property taxes.)

Technology Commercialization

NSP's incentives to invest in technology commercialization are weaker than society's. Because the costs of many climate-friendly energy technologies are expected to come down as production increases, orders for equipment have benefits for all future users. As an individual firm, NSP cannot capture all or even most of such benefits from commercializing a climate-friendly technology. The returns to society as a whole (or, for that matter, to the electricity industry) from investing in climate-friendly electricity technologies are potentially much higher. From NSP's perspective, waiting until costs of climate-friendly technologies drop because others invest makes sense. Still, if NSP invested in commercializing climate-friendly technologies to contribute to a broad-based coordinated effort, benefits to NSP would probably outweigh the costs.

Shifting Costs Away from Current Ratepayers

Another difference between social and private perspectives is that society has an explicit interest in intergenerational equity. Under the wait-and-see strategy, NSP saves current ratepayers from having to pay extra for something that may never be needed. Rates are about $0.002/kWh lower between now and 2010 than if

NSP makes early investments to cut its carbon emissions. If a tax is imposed in 2010, however, rates for future customers will be about \$0.007/kWh higher than they would have been with early investments.

A related issue is that any costs beyond those incurred from following a "least cost" investment strategy could make it hard for NSP to retain and attract customers as competition among power providers heats up. Since carbon constraints might never be imposed, NSP has a further incentive to choose the investment strategy that shifts costs onto future ratepayers.

Responding to New Information

Investments both in generating capacity and in reducing carbon emissions are typically sunk once equipment is installed. Thus, knowing that information about the need to control carbon emissions will become available at a given date in the future helps determine which "irreversible" investments are worth making today. Conversely, since several climate-friendly technologies need further commercialization, waiting until the "carbon signal" is sounded to take the plunge could mean that they won't be available when they are urgently needed.

From both NSP's and society's perspectives, it is desirable to avoid making irreversible investments to control carbon emissions until shown that they are necessary. If all else is equal, alternative technologies for meeting projected power needs that have short lead times (wind and gas generation or energy-efficiency improvements, for instance) thus have an advantage over those, like coal-fired plants, that take much longer. Of the carbon-control alternatives available to NSP, simply changing the order in which different plants are turned on would reduce carbon emissions without incurring fixed costs.

In other respects too, the standpoints of society and private firms diverge. The incentive to put off irreversible investments in carbon control until better information is available may be outweighed by the public's desire to avoid irreversible climate effects (Schimmelpfennig, 1995). As a private firm, however, NSP has no way to capture the benefits of avoiding climate change, so the prospect of new investment-related information pushes would-be investors to delay irreversible carbon-control investments.

CURRENT POLICIES TO RECONCILE PRIVATE AND SOCIAL INCENTIVES

Public policy can help create incentives for incorporating sustainability into private decisions. Indeed, it has already. But as the following analysis shows many of the policy tools used to incorporate public concerns into private decisions will become less relevant as the industry restructures.

Reforming Price Signals

One way to make private and societal interests converge is to make market price signals reflect marginal social costs. In other words, the prices of making and delivering goods and services should reflect all of the costs of manufacturing and distribution. Presently, they do not. For example, the prices of various energy forms and fuels typically do not include all the environmental costs of the air pollution or climate change that they give rise to. These environmental costs, though real, are not included in market prices; economists have termed them "externalities."

Environmental Externalities

Both state and federal governments have enacted various mechanisms to correct environmental distortions in price signals for electricity or fuel inputs. The corrections are necessary because the price of electricity from various sources does not fully reflect the environmental costs of electricity generation, let alone the risk of climate change. Coal is particularly underpriced in this respect: a price increase of 50 percent appears justified if environmental and human health costs are taken into account (Boyd et al., 1995).

Tradable Emission Permits

Departing from traditional "command-and-control" approaches to pollution control, the 1990 federal Clean Air Act Amendments (CAAA) capped national sulfur-dioxide emissions (SO_2) and established a market for electricity generators to trade emission allowances under that ceiling. The law calls for a 10 million ton reduction in national SO_2 emissions, and each ton avoided is considered one emission allowance. A special bonus pool of 300,000 allowances was created to reward new efficiency-related or renewable-energy initiatives. (For every 500 MWh saved through such measures, a utility earns one allowance.) This market-oriented approach reduces the cost of achieving environmental goals because each polluter can decide on the most efficient way to respond to the cap. So far, the overall concept of emissions trading has been proven feasible and has been proposed for other pollutants besides SO_2.

Social Costing

Even when complying with environmental controls, power plants may still emit damaging amounts of pollutants. Some public utility commissions have tried to internalize the environmental costs of power-supply options, especially in ranking the options for generating power. One such approach uses "environmental adders" to increase the cost of a kilowatt-hour of electricity by an amount estimated to cover the environmental costs. Environmental adders have not greatly influenced new generation choices to date, however, since controversies rage over what monetary values should be used and how to calculate them.

Utilities use environmental adders mainly in their "least cost" planning at the state level. But as sales of power among utilities become more commonplace and electricity suppliers increasingly sell power directly to large customers or power pools, this approach may be retired because identifying sources of power to apply social costs to will be harder.

Other Price Distortions

Government has a long history of intervening in the electricity sector and in the fossil-fuel industry in ways that affect retail electricity prices—subsidies to various energy sources, tax code provisions, and rate setting. The relative competitiveness of alternative ways of meeting energy needs is also influenced by government subsidies.

Fuel Subsidies

Several subsidies to publicly and privately supplied power increase the use of fossil fuels over other sources of electricity. Those that encourage the use of coal have the largest environmental impact—extending low-interest loans to the Rural Utility Service, granting municipal utilities tax-exempt status, allowing coal firms to renege on full support of the Abandoned Mine Reclamation Fund, giving utilities a percentage-depletion allowance, and not requiring coal companies to fully support the Black Lung Trust Fund. While the emissions impacts of such subsidies are modest now, by 2035 extra annual carbon emissions could reach 46 million metric tons (Shelby et al., 1995).

Some studies conclude that our tax code is biased against renewable technologies. According to one, taxes on technologies favor fuel costs relative to capital and labor costs, which gives natural gas an advantage over coal and, especially, over renewable technologies. Others show that solar-thermal electric generation technologies pay significantly higher net taxes than gas-fired plants, depreciation notwithstanding. Still others find no consistent tax code bias in favor of certain technologies and against others.

The 1992 Energy Policy Act called for analysis of energy subsidies, with an eye to removing those that distort energy choices. But subsidies are tar babies:

getting rid of them is politically tougher than creating them, so government often simply creates countervailing subsidies for technologies given short shrift under current tax laws.

Various state and federal tax incentives have been established to, as analysts like to say, "level the playing field" for renewables and energy efficiency. These include property tax exemptions, tax credits, and accelerated depreciation. In stimulating lasting markets for the targeted technologies, their record has been mixed. While tax subsidies for renewables have stimulated investment, early solar subsidies were based on the capital costs of equipment, not on energy output—a signal to manufacturers to make expensive equipment, whether or not it produces much energy. Also, subsidies have been implemented inconsistently, which makes investment planning difficult. Finally, unforeseen factors have diluted subsidies' impact on would-be investors. For example, a wind developer hoping to take advantage of a production tax credit for wind must rely heavily on equity capital, which raises overall financing costs (Kahn, 1996).

Electricity Rate Structures

Public Utility Commissions (PUCs) have also actively regulated utility-rate structures. So-called declining block rates, long used to encourage electricity use by giving heavy users lower rates, have been reformed in most states. PUCs have also required utilities to offer discounts for customers willing to have their service interrupted during periods of peak electricity demand. The Sacramento Municipal Utility District has instituted a feebate program for new electricity hook-ups that reflects the relative cost that new customers impose on the power grid.

In a restructured industry, PUCs will still pass generation and transmission costs through to consumers. But these utilities will establish minimum charges for access to the grid and other distribution system costs.

Leveling the Playing Field

Long before real competition became a prospect for the electricity industry, policy-makers tried to realize the benefits of a more competitive structure, requiring utilities to incorporate diverse energy options into their planning and acquisition decisions.

Least-Cost Planning

Prominent among the regulatory reforms enacted by states in the past decade is "least-cost planning," which requires utilities to examine all resource options (including energy-efficiency measures) and choose the mix that meets energy demand at the lowest long-term social cost. "Integrated resource planning," a more recent term for the same idea, explicitly incorporates social costs, the risks associated with various resource options, and public participation in planning (Hirst, 1992). But even in most states where integrated resource planning is mandatory, utilities do not have to take account of climate risks.

On the positive side, integrated resource planning has slowed down capacity expansion and thus saved ratepayers from paying for unnecessary growth. On the negative side, besides excluding climate risks, this approach to planning tends to be contentious, time-consuming, and complicated.

As wholesale competition for power escalates, integrated resource planning will focus mainly on near-term purchases of alternative types of power. If passels of companies emerge to provide generation, transmission, and distribution services, broad social concerns may lose out to narrower concerns. If, for instance, generation, transmission, and distribution services become organizationally, analytically, and financially unbundled in the new industrial structure, grid-support benefits may not get their due. If that happens, distributed renewable applications will look less attractive and the public benefits of using them—cleaner air, for instance—may be overlooked.

Supply Diversification

The 1978 federal Public Utility Regulatory Policies Act (PURPA) broke utilities' monopoly on power generation. It required utilities to buy power from qualifying renewable power and cogeneration facilities (QFs) and to pay them prices based on the cost saving of not having to build their own plants. PURPA jumpstarted the independent power and renewable energy industry in those states with favorable implementation policies, and ten states now account for 73 percent of the nation's QF capacity (Hamrin and Rader, 1993).

PURPA's mandatory purchase provisions have come under increasing attack, however. Some utilities claim that they are now saddled with contracts representing excessive and expensive capacity and argue that PURPA has fulfilled its purpose of stimulating competition and so should be repealed. If customers can shop for low-price electricity, utilities with expensive QF contractual obligations may be stuck with stranded investments. Once, utilities could roll expensive generation in with less costly generation, but now they are increasingly sensitive to whether each power source is competitive. Renewable technologies with high capital costs may be attractive in the long run, but they hold less appeal to an industry facing near-term immediate pressure.

Another state mechanism to promote diverse power generation bases is a competitive set-aside for renewables. By 1991, concerns about cost-effectiveness and overcapacity had prompted 36 states to implement competitive bidding among all power sources. But the utilities' obsession with lowest near-term cost per kWh meant that renewables accounted for only 2 percent of all capacity acquired under such schemes in 1993. Environmental and other nonfinancial benefits were simply not part of all-source bidding. To make amends, some state regulators began requiring utilities to solicit competitive bids from developers of renewable resources to meet part of their capacity needs.

Under one such set-aside approach, a utility issues a "Green Request for Proposal (RFP)" for new capacity from renewable resource projects only. A Green RFP has the cost-control advantages of competitive bidding but also reduces market risks to renewable resource developers. Pioneers in this concept include New England Power, the Bonneville Power Administration, Portland General Electric, and the Sacramento Municipal Utility District. The PUCs in California and other states have also ordered utilities to acquire specified amounts of renewable capacity (Swezey and Sinclair, 1992). This tool lets developers know that factors other than price per kWh count and spares them the high transaction costs of open-ended bidding. While the amount of capacity acquired under such programs is small relative to the total acquired by utilities over the same period, it is large relative to what the renewable energy industry requires to remain viable.

State-level set-asides are not likely to survive in a competitive environment. FERC voided California's set-aside because it requires utilities to acquire capacity at costs that are above those that the utilities would otherwise have to pay. More generally, in a competitive environment, electricity distributors are likely to combine financial approaches and supply diversification to manage price risks. A hedging strategy could still include developing a diverse power generation portfolio, but individual firms managing energy price risks may also turn increasingly to such financial instruments as long-term gas contracts, a portfolio of purchased power contracts, or options and futures contracts to keep power costs down. If supply isn't diversified as reliance on these financial instruments grows, social benefits may be sacrificed to profitability.

Demand-Side Management

Utilities spend more than $2 billion a year on demand-side management (DSM) programs to lower their operating costs and avoid having to build more capacity. Under today's regulatory regime, the value of investments in such programs is often measured by how much the utility's total costs fall. As long as total costs and customers' electric bills were held down, PUCs have let utilities offset DSM expenditures with small rate increases to make up for foregone kWh sales revenues. Some states have also developed mechanisms to decouple kWh sales from profits. Performance-based regulation (PBR), for instance, ties utility profits to the company's success in meeting certain objectives, including demand management. (PBR has also been used to stimulate renewables. Wisconsin offers utilities a premium rate of return for power generated from renewable energy sources.)

Learning how to make energy production and use more efficient takes time. But utilities have made DSM programs more cost-effective by taking advantage of the turnover of equipment and buildings to introduce changes in power generation portfolios, by increasing the scale of installations, by improving information on how to save energy, by gaining marketing experience, and by evaluating programs.

To keep customers as the industry grows more competitive, utilities will also have to change in their DSM programs. First, measures that control peak demand will be favored over measures that save energy. Programs that do more to reduce overall consumption than to cut peak demand, such as rebates for compact fluorescent lamps, will not fare well while those that optimize the use of the utility's capital assets or that minimize the need to make new capital investments will. Second, DSM programs will focus more on customer service and less on energy and capacity savings. Programs will target the customers that could most easily leave the utility system, such as large industrial customers, and help them become more productive, which may or may not reduce their energy consumption. Similarly, DSM programs will recover costs from participating customers rather than provide "give-aways." Thus, financing some energy-efficiency equipment is a safer bet for utilities than offering rebates for such equipment. Third, DSM programs will become increasingly performance based to increase cost-effectiveness. Selling energy services rather than kilowatt hours is one approach. Already, Wisconsin Electric Power Company offers supermarkets a fixed service fee in exchange for refrigeration and air conditioning services. The utility owns and maintains the equipment, provides the power to run it, and installs efficient equipment, thereby maximizing its profits. Alternatively, utilities may contract with energy service companies to provide some energy savings or to reduce demand at specified prices (Lawrence Berkeley Laboratory, 1995).

Technology Commercialization

The costs of some sustainable energy technologies could come down substantially thanks to technical innovations, experience, or economies of scale. As noted, however, individual firms, the industry as a whole, and society in general all have different incentives to commercialize new technologies. Firms may not be able to fully capture returns related to technical breakthroughs, and renewables' declining costs encourage companies to delay investment until costs fall even farther. To increase companies' incentives to try efficient new technology, government has borne some of the costs of research and development, demonstrations, and market diffusion—key steps in technology commercialization. State and federal governments have boosted overall investment in commercializing sustainable energy technologies through public investment, tax policy, and utility regulation.

Research and Development

Under conventional monopoly regulation, local utility revenues paid for experiments with new technologies as long as Public Utility Commissions agreed that ratepayers would reap environmental or social benefits from such technologies. Electric utilities have conducted their own research or contributed to an electric industry consortium, the Electric Power Research Institute (EPRI). Still, industry spending on R&D has been low relative to that in other industries, partly because equipment vendors have made such investments and partly because they don't feel competitive pressures. Moreover, some consumer advocates claim that ratepayers foot the bill for technology-induced benefits that others reap.

Government programs have complemented private R&D investment by spreading risks among disparate projects and among multiple investors. For emerging sustainable electricity technologies, R&D costs are shared by government and industry, as are the costs of some projects. Some important technological advances, such as aero-derivative turbines, were also developed under government sponsorship.

Unfortunately, technology commercialization requires steady support, and congressional authorizations have been roller coaster rides. Expenditure levels rose in the 1970s, then fell in the 1980s, only to rise again in the early 1990s. In the years ahead, they will probably fall again. These swings have mirrored those in the prevailing philosophy of government's role in technology commercialization and international competitiveness. Meanwhile, in general the United States has not shared R&D investment risks with other countries interested in climate-friendly energy technologies—too bad, since this is one way to buffer downswings.

As the electricity market grows more competitive, utilities try to cut costs for noncritical services, including technology commercialization. As one indicator, EPRI programs have shrunk and have been redirected toward research that addresses clients' immediate needs. Major suppliers of fossil fuel-based power can weather this period of simultaneous cuts in government and in utility investments in technology commercialization, but few producers of renewable generation equipment have the deep pockets needed to do so.

In the long run, competitive pressures facing the electricity industry should induce them to invest in R&D at least as much as under the old regulatory regime. But the market, not societal objectives, will be the driving force. Any surviving vertically integrated utilities will be less interested in sponsoring or using basic research on power generation since they will rely more on power brought from external sources. Only if fossil fuel prices rise to reflect the risk of global warming will industry turn to carbon-free or low-carbon energy sources, and suppliers of these technologies have no guarantee that that day will come or, if it does, when.

Market Diffusion

In recent years, policy-makers have launched various initiatives to transform markets for electricity-using equipment. The goal is to shift the mix of products manufactured, offered, and purchased without having to deal with customers one by one. In market-transformation programs, "trade allies" (contractors, builders, retailers, distributors, and manufacturers) work together to realize larger, faster, longer lasting, and more broadly distributed savings than programs aimed at individual customers could. Such programs may coordinate utility and government efforts to establish energy codes for buildings, equipment-efficiency standards, utility rebates, information dissemination, or guaranteed markets for highly efficient equipment (Hirst, 1994). Refrigerators, fluorescent light ballasts, compact fluorescent lamps, personal computers, and buildings have all become more efficient thanks to these programs (Geller and Nadel, 1994). *(See Box 5-7.)*

Market-transformation techniques have also been used to bring down the costs of photovoltaics, fuel cells, and solar hot-water heaters. Typically, consortia of utilities and equipment suppliers form to aggregate markets for targeted technologies. One such consortium calls for competitive bids for supplying an aggregate quantity of a particular technology with specified performance and other characteristics. This

BOX 5-7. High-Efficiency Refrigerators in the Marketplace

The U.S. Department of Energy is working with a consortium of utilities and other stakeholders—the New York Power Authority (NYPA), the U.S. Department of Housing and Urban Development, the New York City Housing Authority, and appliance manufacturers—to annually bring more than 50,000 super-efficient refrigerators to market. Government agencies will play a key role. The Housing Authority must replace thousands of refrigerators each year for its units and pays for the electricity in those units. Under this program, NYPA will purchase the large number of refrigerators that the Housing Authority would otherwise buy. In turn, HUD will let the Housing Authority pay back the money invested by NYPA with the money saved by lowered electricity bills. The consortium's sponsors hope to change whole product lines and manufacturing processes (Gardner and Foley, 1995; U.S. DOE, 1996).

approach generates enough orders to interest manufacturers in investing in new production capacity. Some trade allies get rewarded for taking risks: if commercialization succeeds, the utilities that placed early orders for equipment receive royalty payments on equipment sales (Kozloff and Dower, 1993).

Market transformation has worked better with some technologies than others. Experience has shown that current U.S. markets for some technologies may not be large enough to increase production or lower unit costs dramatically. For example, demand for photovoltaics among U.S. utilities is slack because current fuel costs are relatively low and electricity demand is growing only slowly. The answer may lie abroad, where fast-growing economies with only nascent power grids will value some of these technologies more. In any case, interest in market transformation is likely to wane among utilities or electricity distributors as the industry changes shape, even though this approach may yield more cost-effective energy savings than the customer-targeted services approach to demand-side management.

While reduced industry spending could delay the commercialization of any technology, the most damage is done to those whose costs stand to fall the farthest. For example, unless aggregated utility demand is high enough, renewable energy equipment suppliers will lack incentives to further reduce costs. More specifically, one manufacturer's proposal to build a photovoltaic facility that can provide 5.5 cent/kWh power in return for minimum guaranteed purchases for ten years will not materialize until local utilities participate in its "forward pricing" scheme.

Promoting Sustainable Energy Investments by Consumers

A wide range of information programs has been implemented by federal, state, and local governments and by utilities to improve the public's understanding of electricity use patterns, energy efficiency, and renewable energy sources. Some information campaigns are targeted to various categories of electricity customers; others focus on training architects, developers, engineers, and builders. Programs to make electrical energy efficiency easier to grasp include energy audits offered by utilities, energy service companies, or government agencies—all of which provide residential and commercial building owners with detailed information on how to reduce energy consumption cost effectively. Another approach is labeling: the federal government requires labels on major appliances that tell customers about the relative efficiency of various models. Similarly, measures of solar energy equipment's performance have been standardized and a certification program set up to reduce buyers' risks. Energy-rating systems for buildings have also been developed, and the U.S. Environmental Protection Agency awards its Energy Star labels to new homes that exceed regular building code standards by 30 percent.

Consumer information goes hand in hand with equipment and building standards. The most inefficient products can be driven off the market where efficiency standards have been established and savvy consumers know how to interpret them. Still, the issue is complicated. Even though some levels of energy efficiency investments are attractive from the vantage of their lifetime costs, incentives for those making capital investment decisions and those paying electricity bills may be mismatched. Indeed, why should landlords who do not pay electricity bills install efficient appliances if initial costs are higher? Responding to this market failure, the federal government has established minimum efficiency standards for a range of electricity-using equipment.

MOVING TOWARD SUSTAINABILITY IN A COMPETITIVE POWER MARKET

The premise embodied in sustainability—that our generation has a moral obligation to the well-being of future generations as well as our own—suggests major shifts in how electricity services are provided. Over the next several decades, such shifts are certainly feasible. Technological capability, affordability, and service reliability are not unrealistic goals, even in the face of climate change. But unfettered market forces alone can't trigger the transformation. The rift between what we as a society might want and how individual decisions are made is likely to grow larger as the electricity industry restructures and the traditional mechanisms for incorporating societal concerns diminish. Ironically, as we enter an era of more decentralized decision-making in the electricity industry, addressing the climate problem requires more coordination than ever.

Yet, competition can be made compatible with sustainability if electricity producers, users, industry regulators, and other policy-makers all rise to the occasion. State and federal regulators should ensure that sustainable energy options can compete fairly in emerging electricity markets by continuing to balance efficiency, affordability, and environmental goals as the industry becomes more competitive. Even with restructuring and the likely demise of integrated resource planning, Public Utility Commissions will retain important regulatory responsibilities and acquire some new ones. And while power generation and even retail services may be provided by a range of firms, transmission and distribution monopolies have a continued responsibility to make their operations more sustainable.

The United States has logged substantial experience with a broad range of policies that increase electricity suppliers' and consumers' incentives to acquire sustainable electricity technologies. To accelerate needed technological transformations, these tools can be adapted to new technologies and industry structures. The targeted investments, regulations, market-based policies, and other interventions needed to bring about technological transformation can be made consistent with the American values of equity, choice, accountability, and common sense.

Of course, intervening in the marketplace will be a political challenge given Americans' current mistrust of government. *(See Box 5-8.)* But with this constraint in mind, policy-makers can still reform price signals, ensure that all technologies get to compete on fair terms in electricity markets, complement private sector efforts to commercialize emerging technologies, and nurture voluntary corporate leadership and individual responsibility.

Reforming Price Signals

State and federal policy-makers should reform pricing policies to reflect more fully the social costs of electricity use. Price signals already influence which resource options are chosen for electricity production and how much electricity is consumed. Better price signals make it more likely that the benefits of using sustainable technologies will be realized. In a more competitive and less regulated industry, price reforms (supported by other policies) to achieve societal objectives become more important than ever.

BOX 5-8. Policy Design Constraints

To help guide the selection and design of policy tools to promote sustainability, WRI conducted several focus groups to solicit the opinions of residential electricity consumers. Some of the results are relevant to the electricity sector (Cosgrove, 1995).

Incrementalism

Participants felt overwhelmed with the magnitude and complexity of sustainability problems. They want meaningful, incremental, and tangible solutions, and they don't want major cost or lifestyle changes. But they are willing to pay a little more to develop and deploy technologies to protect the environment if they can be assured they will get something tangible in return.

Accounting for Uncertainty

Participants expressed skepticism about scientific consensus on climate change and the need to act now to mitigate the risk of global warming. The concept of collective action to avoid risks seems well-accepted in other policy areas, however, such as national security. (In fact, focus group participants viewed the government role in providing national security as one of the most positive government functions.)

Fairness

Participants expressed greater willingness to change from the status quo if resulting economic burdens are shared equitably. There are many notions of fairness, one of which would be satisfied if costs fall proportionately on those who benefit from electricity services. There was a sense that some ratepayers and various classes of customers, as well as shareholders, are not bearing their fair share of electricity service costs.

Choice and Control

Focus group participants expressed a desire to be presented with various choices, including positive incentives for "doing the right thing." Participants were divided on the relative effectiveness of government versus the market in influencing individual behavior, recognizing some shortcomings of relying on either one. Participants professed greater trust in decision-making at the community than at the national level.

Environmental Costs

For environmental threats likely to be regulated in the future, state policy-makers should write rules to explicitly allocate risk among the various industry segments. For example, state policy-makers should prohibit electricity distributors from passing on the costs of complying with future carbon-emission controls (and other environmental regulations) to ratepayers as fuel-price adjustments. In addition, power purchase agreements should specify exactly who will bear the costs of carbon constraints. In general, risks should be assigned to those parties that can manage them most efficiently and are appropriately compensated.

Explicit allocation of risk is likely to increase some suppliers' electricity costs. If regulators allocate electricity price risks properly among generation developers, wholesale power purchasers, and retail electricity customers, some will be winners and some will be losers (Yardley, 1995). Some portion of the risk of carbon-related or new environmental regulations will be shifted from ratepayers to other market participants. By requiring the industry to become explicit about who bears carbon control risks, policy-makers will create ripple effects. Capital markets will adjust bond ratings of carbon-dependent utilities. Other large-scale sources of capital may start to consider climate risks, as the vanguard of the insurance industry does now.

Once long-term goals for atmospheric carbon dioxide concentrations become widely accepted, the United States and other signatories to the Climate Treaty will probably agree to binding national carbon emissions constraints. Each signatory country will, in turn, impose them on the electricity industry and on other industries with significant carbon emissions. A tax on the carbon content of fossil fuels imposed at the mine mouth or wellhead would increase costs to all fossil fuel users. In contrast, the same tax imposed on carbon emissions at the point of wholesale power sales would affect power generation only. In any case, policy-makers will need a good sense of how the industry will respond to the tax, and they should announce the phase-in of taxes well in advance and stick to the levels and dates announced. Alternatively, a carbon emissions cap with tradable permits—like the present regime for sulfur emissions—could be imposed on wholesale power generators. Under this approach, no revenues are collected, though allocating emission rights among power generators will be complicated because ownership of power plants changes. In response to either a cap or a tax, independent or utility power generators can choose among highly efficient fossil-generation technologies, generation with no carboniferous by-products, or carbon offsets.

Electricity Rates

Public Utility Commissions should ensure that electricity rate structures reflect the true costs of providing electricity services to various kinds of customers. Rates should give consumers price signals to use energy efficiently. For example, area rates (common early in the industry's history) reflect differences in transmission and distribution system costs related to the customer's location. Returning to this rate structure would provide more accurate incentives for demand-side management and make it clear where on site generation is economically competitive. Similarly, variable hook-up fees for new customers should reflect the costs that the building's electricity demand will impose on the distribution, transmission, and generation system.

Time-of-use rates, under which customers are charged according to the costs they impose on the system at any point in time, should also be offered. Such rates reward on site generation and energy efficiency measures that coincide with high periods. Distribution companies with purchased power contracts specifying that higher capacity charges will be incurred during peak periods already have some incentive to implement time-of-use rates. While few have experimented with such rates to date because costly special meters are needed to make the system work, these costs are no longer too steep for utilities.

The competitive pressure to reduce cross-subsidies among and within customer classes may give distribution companies an incentive to implement area and time-of-use rates. But other rate options tied explicitly to sustainable power development should be tested, too. For example, while some households and businesses may enter spot markets for power, others will want to lock in stable long-term rates based on renewable power sources with low operating costs. Others may opt for discounted rates for service that can be interrupted when system demand is high or renewable power flows are low.

Energy Subsidies

Congress should pass legislation eliminating energy subsidies that bias electricity investments toward unsustainable technologies. According to one study, getting rid of these federal subsidies would prevent total carbon emissions from growing after 2015 and reduce emissions by 2 percent in the year 2035 (Shelby et al., 1995).

Congress should also eliminate those subsidies in nonenergy sectors that indirectly bias decisions against sustainable energy technologies. For example, subsidies to farmers to grow "program" food crops raise the opportunity cost of growing biomass energy crops. If there were no subsidies for program crops, "fuel crops" would be more attractive.

Leveling the Playing Field in Emerging Electricity Markets

Appropriate price signals may be necessary, but they are not enough to open electricity markets to sustainable service providers. Also key is sweeping aside barriers to new market entrants and to consumer choice.

Retail Electricity Services

Public Utility Commissions (PUCs) should ensure that entities offering distributed generation, demand-side management, and other retail electricity services can compete with regulated electricity distributors for these services. All states need not follow the same regulatory model. Rather, PUCs should make sure that electricity customers have a wide range of services and prices to choose from, including sustainable options.

Regulators may not foresee all the service options that could be offered, but can rid rules of biases against emerging services. For example, individual households or businesses could generate their own power and sell the surplus to the grid. Many large businesses already generate or cogenerate power. For small customers, micro-cogeneration systems, PV, wind, and fuel cells are possibilities. Off-grid self-generation might also be appropriate for highly specialized end-use markets, such as charging batteries for electric vehicles. For small self-generators, PUCs should provide for backward running meters and net billing, as law requires in California.

While wires for distribution systems are generally assumed to be most efficiently provided by a monopoly, firms can compete for other distribution services (i.e., demand management, on site generation). Energy service companies, cooperatives, local governments, nongovernmental organizations, and others should be able to compete freely against regulated electricity distributors, who should not be allowed to use information on customer demand or other privileges of power to keep new entrants out of the market. On the other hand, established electricity distributors should be allowed to set up service centers, perhaps at the substation level, to compete with independent service providers.

Local government should express citizens' preferences for sustainable electricity services—energy efficiency, renewable supply options, self-generation, and district energy—and even provide them, if necessary. Local governments could work with neighborhood groups and power distributors to provide energy-efficiency services as utilities cut them back. In one California deal between a city and its electricity distributor, the same level of electricity services is provided with fewer kWhs: the utility finances efficiency improvements, electricity bills stay the same until capital is paid off, then bills are permanently reduced. This way, the distributor retains customers who might otherwise shop around in a deregulated retail market. In a related model, groups of consumers align themselves in an energy-efficiency cooperative to free up capacity that they then sell to the utility. In densely populated cities, local government could provide district energy services too.

Another option is to form a municipal utility to respond directly to consumer preferences for sustainability. Yet, because of legal and administrative challenges, relatively few communities are likely to municipalize their utilities (The Quad Report, 1995).

A third option is to act as an agent for constituents who want to shop around for power. By aggregating individual customers, municipalities could negotiate better deals with power generators who, in return, would have a more stable electric load and the reduced transaction costs that volume makes possible. Moreover, small municipalities would not have to go it alone. Associations of local governments within the state, for example, could broker electricity generation services for their members.

A fourth option is to establish service requirements for local distribution franchises. Municipalities already provide similar services in issuing cable TV franchises and negotiating service contracts with private trash collectors.

Local government is not the only way to represent and aggregate the preferences of environmentally conscious energy users. In Denmark, local cooperatives have formed to develop wind power for local use and sale to the grid. Such "communities" need not have traditional political boundaries: as long as they are located in states that allow retail access, individual households or businesses could band together to find a retail electricity supplier, as Bank of America branches have done in this country and McDonald's franchises have done in the United Kingdom.

Wholesale Power Markets

Through their rulings on market operations, the Federal Energy Regulatory Commission (FERC) and state PUCs should ensure that all competitors in the electrical power market face comparable environmental regulations. Without such a principle, plants with higher NO_x, particulate, and CO_2 emissions per kWh might be used more as demand for cheap power rises. Options for implementing so-called environmental comparability include setting emissions caps, using benchmark emissions rates, "de-grandfathering" old plants, and issuing tradable pool-wide emissions permits.

Similarly, FERC should ensure that system operation and dispatch rules and formulas for determining transmission costs do not bias the market against renewables. Transmitting utilities should assess transmission costs to all generating resources on the basis of equivalent capacity and actual energy generated. Transmission costs should be divided between capacity and energy on the same basis as generation costs. Capacity-related costs should be assessed to generating resources in proportion to actual capacity that the resource contributes during usage peaks.

In addition, FERC should not allow the site-specific nature of renewable energy resources to create barriers to transmission access. Under some transmission-pricing proposals, a renewable power generator would have to pay a charge (often unrelated to the actual cost of sending power) to each transmission system owner located between the generation unit and the customers. Such proposals hurt renewable sources' competitiveness.

Commercializing Emerging Technologies

Even if sustainable electricity options are fairly weighed by electricity suppliers and users, decisions about electricity supply and demand are made with current technologies in mind, so in some cases the potential for major cost reductions is overlooked. To make amends, the public sector can complement and leverage private investment to make the costs and quality of electricity services provided by sustainable energy technologies comparable to those of other energy options.

Assuring Adequate Commercialization Investments

Congress should mandate the creation of a Sustainable Energy Trust Fund (SETF) to commercialize sustainable energy technologies and capitalize it through a fee on retail electricity sales. In this "off-budget" electricity industry program, industry would collect fees and administer revenues. Many precedents exist for a trust fund that supplements private investment in commercializing technologies to address a long-term environmental problem, including the Highway Trust Fund and the Nuclear Waste Trust Fund.

SETF would complement public and private sector organizations' efforts to commercialize specific technologies. For some technologies, this would mean ensuring a stream of orders for equipment over time by paying the difference between current market rates for wholesale power and a competitively bid price from renewable power developers. The "Non-Fossil Fuel Obligation," which the United Kingdom operates this way, is credited with helping to reduce the cost of wind power in that country, as well as with increasing overall renewable energy capacity.

SETF could basically be used to cover the difference between market prices and renewable power costs for a specified period of time. In the United States, a renewable energy portfolio standard has been proposed at both state and national levels to ensure that a minimum percentage of all new capacity would be renewable. Under the national version of this proposal, all electricity providers would be required to acquire a minimum percentage, but individual states would be free to set a higher standard. While the portfolio standard does not necessarily increase customer charges, electricity suppliers would pass through any extra costs to customers.

SETF could also facilitate cooperative ventures between utilities and equipment suppliers. One such model is a utility consortium that issues requests for competitive bids for a bulk order of a particular technology with specified performance and other characteristics. By pooling applications within and among utilities, the consortium generates threshold annual sales so manufacturers have some incentive to enlarge their production capacity.

SETF should be able to integrate domestic and international efforts. For some technologies, even aggregated material markets may not be enough to realize economies of scale. Allowing utilities or other technology users from abroad to enter the pool could create a large enough market to reduce unit costs appreciably and reduce the investment risks of scaling up production. Technologies whose costs fall steeply as production or experience rises are excellent candidates for a coordinated multilateral program. Some renewable electricity technologies are already competitive in off-grid markets that are far larger in developing countries than in the United States, and overseas market demand would push costs down and help make such equipment competitive in the United States.

The amount of money required to complement commercialization investments is small relative to both other trust fund revenues and as a percentage of energy bills. SETF could be funded by either a levy on fossil fuel sales to electric power production or on wholesale electricity production from all or only fossil-based generators. Nothing should prevent SETF from seeking other revenues either. At present levels of consumption, a small levy of 0.05 cents/kWh would generate about $1 billion/year, about the amount that the Electric Power Research Institute (EPRI) and the federal government have collectively spent on sustainable energy technologies in recent years. This amount is much smaller than the yield of a carbon tax designed to reduce fossil fuel consumption, and required revenues would decline as the cost of technologies dropped. Large energy users would pay proportionately more since they impose greater costs on future generations.

SETF could either be part of an existing public or private institution or be a new independent federally chartered nonprofit corporation. Either way, its focus would be on commercialization activities that individual firms can't easily carry out. Besides electric generation, storage, and efficiency technologies, its programs could also cover distributed generation.

Transforming Markets for Electricity Uses

As the electricity industry becomes less integrated and more competitive, its incentives to stimulate customer investments in energy efficiency will wane. Whether other private-sector entities will take up this role formerly played by electric utilities is unclear. But electricity users will face the same barriers to making efficiency investments as before, and government will have more incentive than ever to make sure that the mix of electricity-using products being sold is as efficient as possible.

Given the economies of scale associated with national markets for manufactured technologies, the federal government should expand its role in market-transformation activities. Depending on the technology, government should provide information services, engage in public procurement, mandate equipment-performance standards and building codes, and facilitate utility/equipment supplier partnerships to identify and aggregate markets. For example, the energy efficiency of many appliances can be upgraded through competitive procurements similar to those that have worked in the refrigerator industry.

Government policy must also ensure that one-time opportunities to reduce the U.S. economy's carbon intensity are not lost. Many decisions made outside of the electricity sector affect long-term electricity-consumption patterns. For instance, to avoid unnecessary electricity consumption in new construction, state and local governments should implement and enforce standards and codes for minimum efficiency levels, building orientation, and other land-use controls. Similarly, new urban infrastructure should be designed so that district energy systems can later be installed.

Helping Consumers to Make Sustainable Energy Decisions

Another step that must be taken if the electricity sector is to become competitive and sustainable is to help change consumers' behavior. Opinion polls conducted over the past two decades indicate high and stable public support for continued development of renewable energy and energy-efficiency technologies. Yet, individual behavior often diverges from expressed values. To close this gap, credible and accurate information about energy choices would help.

Improving Information

PUCs should require retail electricity providers to provide consumers with meaningful information on prices and services in a format that allows comparison with other providers. Consumer surveys, including WRI's own focus groups, indicate that most residential customers have little idea about what mix of sources generates the power they use or about the relative amount of power needed for different uses. Clearly, energy consumers' understanding of the sustainability implications of supply and demand decisions must be improved. Regulators should review the information that utilities give their customers and make sure that the links between electricity use and sustainability are spelled out clearly. As competition increases within the electricity industry, the quantity but not the quality of information will likely increase if the comparable advertising campaigns by long distance telephone companies are any indication.

Local governments can also help fill the information gap. Through bill stuffers about the sustainability-related risks associated with their choice of electricity provider, they can provide citizens with objective credible information about the characteristics of various power suppliers. Local government could also offer information on investments in energy-using equipment, especially that geared to locking in long-term energy savings when equipment is replaced.

Past experience shows that linking regulations with information programs or economic incentives makes them more effective. For example, building regulations would better reduce the energy requirements of new buildings if paired with economic incentives (such as utility feebates for hook-ups) or information programs (such as building-energy ratings). Similarly, localities should combine "minimum solar access" provisions in subdivision ordinances with training programs on passive solar design for developers.

Leading the Way

As part of the restructuring process, the electricity industry should develop and implement model sustainability principles. A few utility leaders have already embraced sustainability to some extent on the road to greater competition. Examples include the Sacramento Municipal Utility District, Waverly (Iowa) Municipal Utility, Pacificorp, Ontario Hydro, New England Electric System, Seattle City Light,

Portland General Electric, Austin (Texas) Municipal Utility, Los Angeles Department of Water and Power, Southern California Edison, Eugene (Oregon) Water and Electric, Green Mountain Power, and Bonneville Power Administration. *(See Boxes 5-9 and 5-10.)* Most such leaders have been public utilities, but the preferences shown by public utility ratepayers are probably similar to those of their investor-owned utility counterparts.

Differing circumstances aside, there is no substitute for high-level impassioned leadership. CEOs should spearhead the development and implemention of sustainability principles within their firms and reward bottom-up initiatives that serve these principles. The principles that are adopted should cover all unregulated subsidiaries, such as power generation developers operating here or abroad. In the case of climate-related risks, the modest voluntary measures implemented to date are not enough. The electricity industry has been a pioneer in applying risk analysis and management techniques to other investment decisions and now needs to adapt these techniques to sustainability risks, perhaps by evaluating the risk benefits of a portfolio of sustainable technologies and plans for their rapid diffusion.

Another shrewd leadership move for the electricity industry is to collaborate with environmental groups to develop a labeling program that designates those electricity suppliers receiving some minimum percentage of their power from renewable sources. The percentage could begin at 10 percent and grow over time. Utilities in Sweden recently became eligible to compete for a similar eco-label (Cutter Information Corp., 1996).

As the country with the largest total and per capita emissions, the United States will need to set an example in reducing emissions if other key countries are expected to make globally coordinated cuts. The electricity industry should also demonstrate international leadership. Already, utility consortia for technology commercialization are being expanded to the international level. For example, in 1995 the Utility Photovoltaic Group, a U.S. consortia created to stimulate utility applications of photovoltaic cells, approved a new membership class for foreign utilities. U.S. utilities have also been teaming up with foreign utilities through other organizations (such as the Electric Power Research Institute, Edison Electric Institute, and the E7 group). However, U.S. utilities' wealth of expertise in renewable energy, demand-side management, and environmental protection should be more effectively conveyed to utilities in other countries, especially those where the industry is being restructured.

BOX 5-9. New England Electric System/Conservation Law Foundation

"Grand Bargain"

Under this proposal, transmission assets are sold to an independent company (TRANSCO) for a market-based price, ideally reflecting their replacement value. To the extent that above-market generation liabilities are not directly assumed by TRANSCO, cash proceeds would be used to write down the stranded investment to market price as far as possible. The sale price is recovered through a transmission-access contract between TRANSCO and the distribution entity; all users connected to the distribution company pay their share of this contract. This fixed fee contract stands independent of any commodity-based transmission usage charges.

This proposal preserves momentum to clean up power plant air emissions. Under the old regime, utilities could re-power plants with cleaner fuels and recover their costs with rate increases. That option would be gone with deregulation. In return for recovering all stranded assets through the sale of transmission assets, utilities would bring all generation facilities up to the standards required of new generation but would have the latitude to trade off among various plants to meet the requirements of an overall ceiling on emissions.

A demand-side management funding mechanism is applied to all distribution customers. Investment in current programs and delivery mechanisms is continued at current levels, though programs evolve to emphasize lost-opportunity markets, market transformation, T&D cost reduction, and the reduction of financial intervention.

As important as the substantive aspects of the "Grand Bargain" is the process for reaching it. The proposal has been hammered out collaboratively by key stakeholders, notably the largest New England utility holding company and a major environmental group. While many details of the proposal and closure on the deal remain to be worked out among all stakeholders, the progress achieved so far suggests that closure will be easier than if a more confrontational process had been used. As of July 1995, some 19 environmental, utility, consumer, and government organizations in the region had filed a single set of principles on utility restructuring (O'Driscoll, 1995).

Timing the Transition to Sustainability

The threats posed to sustainability by continuing business as usual in the electricity sector are serious, but the United States does have some breathing space in which to work out solutions. Unlike the energy crises of the 1970s, the time frame for fully addressing these threats is in decades rather than weeks. Electricity producers and consumers can begin the technological transforma-

BOX 5-10. Ontario Hydro's Sustainable Energy Development (SED) Strategy

Ontario Hydro is one of the world's largest utilities. Until recently, it was characterized by poor environmental performance, high rates, and lack of accountability to stakeholders. But in 1994, this public utility's stated mission was to become "a leader in energy efficiency and sustainable development, and to provide customers with safe and reliable energy services at competitive prices" (Ontario Hydro, 1994). Under the utility's new Sustainable Energy Development Strategy, principles and practices in support of sustainable development include practicing eco-efficiency, taking a precautionary approach to human health risks and environmental damage, integrating environmental and social factors into decision-making, participating in the development of public policies promoting sustainable development, encouraging employees to conduct activities sustainably, and monitoring progress toward sustainable development.

Investment decision criteria developed in 1994 require the following to be considered:

- full life-cycle costs, from design to decommissioning and disposal;
- expected damage to ecosystems, communities, and human health;
- potential environmental impacts of alternatives;
- quantification and monetization of these impacts, if possible; and
- the trade-offs made in selecting the preferred alternative (Ontario Hydro, 1994).

An internal assessment of this new strategy during the first year of implementation concluded that significant progress had been made, particularly in programs relating to ozone-depleting substances, renewable energy, greenhouse gases, and in-house energy efficiency; criteria for decision-making; and tracking personnel performance through indicators. Progress has been deemed slow in internal motivation and education, funding for initiatives, moving from environmental compliance to leadership, customer energy-efficiency programs, reforming R&D programs, and introducing full-cost accounting. Because the Sustainability Energy Development (SED) principles were adopted during a period of substantial budget cuts and corporate restructuring, staff in the various operating divisions have been hard-pressed to embrace their new responsibilities. In addition, a change in Ontario Hydro's CEO prompted staff to question the continued importance of the strategy. To reestablish priority, momentum, and openness in implementing the strategy, the assessment team recommended holding monthly meetings with the president, preparing quarterly progress reports on the strategy, highlighting activities in business plans, and enhancing the financial resources available to leverage SED activities in the company's business units.

tions without prematurely retiring capital stock. Only modest amounts of new electrical production capacity are needed in the next several years, and the market cost of fuels that power existing capacity is expected to remain relatively stable. These factors afford the industry and policy-makers a perfect opportunity to make progressively larger investments of time and money in new supply- and demand-side technologies.

The policy package needed to usher in sustainability can be phased in over time to coincide with expected future opportunities and needs. One goal should be to ensure that a portfolio of cost-competitive sustainable technologies and associated business plans is available when the time comes to replace retired nuclear and fossil capacity. Another should be to prepare for the likely adoption of international carbon constraints, perhaps starting between 2000 and 2010.

Some policies make sense to implement immediately because they don't require resolving uncertainties about the nature of future carbon restrictions and because they are insensitive to the nature and pace of market reforms. Such policies include incorporating environmental risks into wholesale power trans-actions, reforming retail electricity rates, removing subsidies that distort decisions among energy choices, and creating a Sustainable Energy Trust Fund. The cutbacks in public and private R&D lend particular urgency to the creation of the Trust Fund, while opportunities for competition in wholesale power markets created by recent FERC rulings make policies that promote fair competition in such markets urgent too.

Policy-makers should also ensure that we don't lose no-cost or cheap one-time opportunities to make our electricity infrastructure and electricity-consuming equipment less sensitive to climate risks. By making the entire electricity-generation and distribution system more resilient as capital stock is replaced, we can protect ourselves against the high costs of making dramatic reductions in carbon emissions in the future.

Other policies depend on the pace of restructuring at the state and federal levels. Ensuring that a wide range of sustainable electricity service providers can enter and compete in retail electricity markets will be critical as retail competition and vertical de-integration evolve in various states over the next decade or so. Similarly, improving information to customers should go hand in hand with efforts to make the best choices. Finally, the electricity industry should immediately adopt sustainability principles and incorporate them into any proposals that it makes to Congress or individual states related to its preferred structure and operations, as well as to its overseas operations.

TABLE 5-2. Net Present Values (Benefits—Costs) in $1995 Billions

NSP Perspective: Benefits = 0; Discount Rate = 6.2%

	Tax	No Tax	Expected Value (P=.5)
WAS	-10.4	0	-5.2
ATL	-8.9	-1.2	-5.05

Society Perspective: Benefits = $100/ton; Discount Rate = 3.0%

	Tax	No Tax	Expected Value (P=.5)
WAS	33.7	0	16.85
ATL	44.1	-2.8	20.65

Society Perspective: Present Value Benefits = $26.5 billion; Discount Rate = 3.0%

	Tax	No Tax	Expected Value (P=.5)
WAS	0	0	0
ATL	3.9	-2.8	.55

Sources: World Resources Institute and Union of Concerned Scientists

The probable lead time for stabilizing atmospheric carbon concentrations allows the United States time enough to try out and then incrementally refine various policy responses. Federal initiatives notwithstanding, the 50 states and many communities have acted as policy laboratories. Public and private policymakers have learned much over the past two decades about what works and what doesn't.

CONCLUSION

The U.S. electric power sector is, by far, the largest and fastest growing consumer of fuels in the nation, a fact at the heart of its nonsustainability. In 1950, U.S. electric utilities accounted for only 15 percent of total primary fuels consumption. By 1995, the figure had reached 36 percent. Without doubt, this historical trend toward electrification of the U.S. economy will continue. Americans are not alone. Similar trends toward electrification are occurring worldwide.

Faced with the long-term movement toward electrification, the electric power sector finds itself at the focus of profound, and sometimes conflicting, social, regulatory, scientific, and technological forces that will require the industry to reinvent itself. Some of these forces—such as the need to reduce air pollution emissions—lend themselves readily to technical fixes, such as fuel switching or pollution scrubbers. But the most important of these factors is the threat of global climate change, an issue that is at the center of U.S. power production's nonsustainability.

In 1992, the United States ratified the UN climate treaty. Eventually, it will have to drastically reduce its carbon dioxide emissions, the largest source of which is electric power production (currently, a source of 35 percent of total U.S. emissions). While national CO_2 emissions grew by 7 percent between 1980 and 1993, utility emissions rose by 17 percent. These emissions trends—the result of burning more fossil fuels—are clearly in conflict with long-term needs to curb carbon emissions.

Utility greenhouse gas emissions can be reduced. The key is an integrated strategy for improving end-use efficiency and changing generating technologies to more efficient ones that require less carbon to run. To some extent, both of these options are already being pursued. Thanks to dramatic improvements in the efficiency of lights, motors, heat pumps, industrial processes, and so forth, growth in electricity consumption—which historically grew by about 6 percent per year for much of the century—has averaged only 2.2 percent since 1973.

But efficiency improvements, welcome as they may be, will not be sufficient over the long term. Decentralized, more sustainable sources of power will be needed. Unfortunately, these more environmentally friendly technologies tend to have higher capital costs and their clear environmental benefits—reducing CO_2 emissions, for instance—currently have little value in the marketplace. Moreover, changes in the structure and competitiveness of the industry now under way are likely to change the incentives for investing in various supply and demand technologies. Competition is likely to favor technologies whose investment costs are lower and can be recouped relatively quickly; and these facilities can have higher emissions.

Still, there are undeniable signs that a transition is under way as more sustainable technologies begin to penetrate the market. These technologies generate power from fossil fuels more efficiently, reduce the amount of electricity needed to accomplish a given task, or provide power from renewable technologies. The electric utility industry has an incentive to begin investing in these technologies as a way to hedge climate risks. Even though the details of a carbon-restriction policy cannot be spelled out precisely at this time, electricity suppliers may find it in their economic interest to begin shifting their mix of resources. Yet, they still do not have strong enough incentives to invest in renewable energy sources. Under past practices, state public utility commissions have tried through regulation to give utilities economic incentives to do what is best for society. While some of these measures worked, many points of leverage will no longer be available as the electric power sector is restructured and deregulated. In their absence, new policy tools will be needed to promote sustainability as new players (consumers, communities, progressive industry leaders, entrepreneurs marketing green power services, etc.) become involved. As federal and state governments change laws to deregulate, "unbundle" important utility functions, and introduce competition into the industry, the public sector needs to explicitly incorporate financial and regulatory provisions to ensure that the new, emerging power sector will continue moving toward a more sustainable industry.

REFERENCES

Aspen Institute. "The Fading Influence of Government and National Boundaries as the Electric Utility Industry Restructures: How Far Will It Go?" presented to Energy Policy Forum, Aspen, Colorado, July 9-12, 1995.

Awerbuch, Shimon. "The Surprising Role of Risk in Utility Integrated Resource Planning," *The Electricity Journal*, April 1993, Vol. 6, no. 3: 20-33.

Bauer, D., E. Hirst, and B. Tonn. IRP and the Electric Utility Industry of the Future: Workshop Results. August 1994. From July 1994 Workshop by Oak Ridge National Laboratory.

Boyd, Roy, Kerry Krutilla, and W. Kip Viscusi. "Energy Taxation as a Policy Instrument to Reduce CO_2 Emissions: Net Benefit Analysis," *Journal of Environmental Economics and Management*, July 1995, vol. 29, no. 1: 1-24.

Brower, Michael C. et al. *Powering the Midwest: Renewable Electricity for the Economy and the Environment.* Washington, D.C.: Union of Concerned Scientists, 1993.

Bruce, James P., Hoesung Lee, and Erik F. Haites, eds. *Climate Change 1995—Economic and Social Dimensions of Climate Change*, New York, NY: Cambridge University Press, 1996, 362.–EMF (Energy Modeling Forum), 1993: *Reducing global carbon emissions—Costs and policy options*, EMF-12, Stanford University, Stanford, CA. [As cited in IPCC.]

Burtraw, Dall and Pallavi R. Shah. *Fiscal Effects of Electricity Generation: A Full Fuel Cycle Analysis.* Washington, D.C.: Resources for the Future, 1994.

Byrne, John, Young-Doo Wang, and Steven M. Hoffman. *Utility and Commission Attitudes Towards Photovoltaic Technology and Demand-Side Management in the Utility Sector.* Newark, Delaware: University of Delaware, 1992.

Casten, Thomas. "Whither Electric Generation? A Different View," *The Energy Daily's Environment News Extra*, September 7, 1995.

Cavallo, Alfred J. "Replacing Baseload Power Plants with Wind Plants," presented to Windpower '95 Conference, Washington, D.C., March 27-30, 1995.

Center for Clean Air Policy, "Emissions Impacts of Competition: Further Analysis in Consideration of Putnam, Hayes & Bartlett, Inc's Critique," June 1996.

Cavallo, Alfred J. and Matthew D. Keck, "Cost Effective Seasonal Storage of Wind Energy," *Wind Energy*, 1995, vol. 16., no. H00926.

Center for Energy Efficiency and Renewable Technologies (CEERT), "The CEERT Proposal: Clean and Economical Electricity," *Coalition Energy News*, Fall 1994/Winter 1995: 1+.

Chamberlin, John H. and Patricia Herman. "The Energy Efficiency Challenge: Save the Baby, Throw Out the Bathwater," *The Electricity Journal*, December 1995, vol. 8, no. 10: 38-47.

Chestnut, Lauraine. *Dollars and Cents: The Economic and Health Benefits of Potential Particulate Matter Reductions in the United States.* New York, NY: American Lung Association, 1995.

Cosgrove, Tom. *Sustaining the American Dream or a Pipe Dream?* Boston, Massachusetts: MacWilliams Cosgrove Snider Smith Robinson, July 1995.

Cutter Information Corp. "Swedish NGO Awards Green Electricity Label." *Global Environmental Change Report.* February 9, 1996, vol. 8, no. 3: 7.

Daycock, Suzanne. "The Restructuring of Electric Power Industry," *National Coal Leader*, February 1996: 13.

Dockery, D., and C. Pope. "Acute Respiratory Effects of Particulate Air Pollution." *Annual Review of Public Health,* 15, 1994, pp. 107–43.

Dockery, D., and C. Pope. "An Association Between Air Pollution and Mortality in Six U.S. Cities." *New England Journal of Medicine.* Dec. 9, 1993, Vol. 329:24, pp. 1753–1759.

East-West Center. "Energy Outlook to 2010: Asia-Pacific Demand, Supply, and Climate Change Implications," *AsiaPacific Issues*, April 1995, no. 19.

Edison Electric Institute (EEI), Statistical Committee. *Statistical Yearbook of the Electric Utility Industry 1993.* Washington, D.C.: 1994.

Electric Power Research Institute et al. (EPRI). *Distributed Utility Valuation Project Monograph.* San Francisco, CA: July 1993.

Energy Information Administration (EIA). *Annual Energy Outlook 1996 with Projections to 2015.* Washington, D.C.: U.S. Department of Energy, 1996a.

———. "Operable and Future Electric Units Statistics as of 12/31/93: EIA-860," unpublished computer data. June 21, 1995.

———. *Annual Energy Review 1994.* Washington, D.C.: U.S. Department of Energy, July 1995a.

———. *International Energy Annual 1993.* Washington, D.C.: U.S. Department of Energy, May 1995b.

———. *International Energy Outlook 1995.* Washington, D.C.: U.S. Department of Energy, June 1995c.

———. *Inventory of Power Plant Data Bases,* unpublished computer data. Washington, D.C.: U.S. Department of Energy, 1995d.

———. *Monthly Energy Review.* Washington, D.C.: U.S. Department of Energy, 1995e.

———. *Electric Power Annual 1993.* Washington, D.C.: U.S. Department of Energy, December 1994.

———. *Federal Energy Subsidies: Direct and Indirect Interventions in Energy Markets.* Washington, D.C.: U.S. Department of Energy, 1992.

Fenichel, Anita. *Impacts of Demand-Side Management Programs on the Environment: An Analytical Approach and Case Study Application.* Washington, D.C.: Alliance to Save Energy, 1993.

Galen, Paul S. et al. "Electric Utility Industry Restructuring: Strategic Issues Paper for DOE's Office of Energy Efficiency and Renewable Energy." Golden, Colorado: National Renewable Energy Laboratory, May 1995.

Gardner, Margaret and Thomas Foley. "The Role of Federal and State Government in Market Transformation," *Energy Services Journal,* 1995, vol. 1, no. 2: 119–128.

Geller, Howard and Steven Nadel. "Market Transformation Strategies to Promote End-Use Efficiency," *Annual Review of Energy and the Environment,* 1994, vol. 19: 301–346.

Gellings, Clark W. "Potential Energy Savings from Efficient Electric Technologies," *Energy Policy,* April 1991, vol. 19, no. 3.

Gielecki, Mark, and James G. Hewlett. "Commercial Nuclear Electricity Power in the United States: Problems and Prospects," *Monthly Energy Review,* Washington, D.C.: Government Printing Office, Energy Information Administration, 1994: 1–14.

Hadley, Stanton W., Lawrence J. Hill, and Robert D. Perlack. *Report on the Study of the Tax and Rate Treatment of Renewable Energy Projects.* Oak Ridge, Tennessee: Oak Ridge National Laboratory, December 1993.

Hamrin, Jan, and Nancy Rader. *Investing in the Future: A Regulator's Guide to Renewables.* Washington, D.C.: National Association of Regulatory Utility Commissions, 1993.

Hirsh, Richard F. *Technology and Transformation in the American Electric Utility Industry.* Cambridge, United Kingdom: Cambridge University Press, 1989: 131–192.

Hirst, Eric. "A Bright Future: Energy Efficiency Programs at Electric Utilities," *Environment,* November 1994, vol. 36, no. 9: 10–15+.

Hirst, Eric. *A Good Integrated Resource Plan: Guidelines for Electric Utilities and Regulators.* Oak Ridge, Tennessee: Oak Ridge National Laboratory, 1992.

Hoff, Tom and Howard Wenger. "Evaluating Distributed Generation and Targeted DSM: A Simplified Approach," presented to National Renewable Energy Laboratory and Electric Power Research Institute Utility Renewable Energy Planning and Modeling Workshop, Newton, Massachusetts, April 20, 1995.

Holmes, Christopher and Kevin Neal. *Using DSM to Help Meet Clean Air Act Targets: A Case Study of PSI Energy.* Washington, D.C.: American Council for an Energy-Efficient Economy, October 1994.

Intergovernmental Panel on Climate Change (IPCC), *Climate Change 1995: The Science of Climate Change,* Cambridge, United Kingdom: Cambridge University Press, 1996.

Intergovernmental Panel on Climate Change (IPCC). *Radiative Forcing of Climate Change: The 1994 Report of the Scientific Assessment Working Group of IPCC.* Geneva, Switzerland: 1994.

Johansson, Thomas B. et al. *Renewable Energy: Sources for Fuels and Electricity.* Washington, D.C.: Island Press, 1993.

Kahn, Edward. "The Production Tax Credit for Wind Turbine Powerplants Is an Ineffective Incentive." *Energy Policy,* 1996, vol. 24, no. 5: 427–435.

Kahn, Edward. Presentation at Fall 1994 NARUC/DOE Conference.

Kidney, Steve. "Uncertain Science May Lead to New Particulate Regs." *The Energy Daily's Environment News Extra,* May 23, 1996: 1-3.

Kozloff, Keith, and Roger Dower. *A New Power Base: Renewable Energy Policies for the Nineties and Beyond.* Washington, D.C.: World Resources Institute, 1993.

Krause, Florentin et al. *Fossil Generation: The Cost and Potential of Low-Carbon Resource Options in Western Europe.* El Cerrito, California: International Project for Sustainable Energy Paths, vol. 2, part 3C, 1994a.

———. *Nuclear Power: The Cost and Potential of Low-Carbon Resource Options in Western Europe.* El Cerrito, California: International Project for Sustainable Energy Paths, vol. 2, part 3E, 1994b.

Krause, Florentin, Jonathan Koomey, and David Olivier. *Renewable Power: The Cost and Potential of Low-Carbon Resource Options in Western Europe.* El Cerrito, CA: International Project for Sustainable Energy Paths, vol. 2, part 3D, 1995.

Lamarre, Leslie. "Assessing the Risks of Utility Hazardous Air Pollutants," *EPRI Journal,* January/February 1995, vol. 20, no. 1: 6-15.

Larson, Ronald W., Frank Vignola, and Ron West, eds. *Economics of Solar Energy Technologies.* Boulder, Colorado: American Solar Energy Society, 1992.

Lawrence Berkeley Laboratory, Center for Building Science News. "Center Research Supports Electric Utility Restructuring." Berkeley, CA: Fall 1995.

Lee, Henry and Negeen Darani. *Electricity Restructuring and the Environment.*Cambridge, Massachusetts: Harvard Electricity Group, 1995.

Maycock, Paul D. *Photovoltaics as a Customer Peaking and Demand Management Option.* Casanova, Virginia: PV Energy Systems, 1992.

Moore, Taylor. "Repowering as a Competitive Strategy," *EPRI Journal*, September/October 1995, vol. 20, no. 5: 6-13.

National Biofuels Roundtable, "Principles & Guidelines for the Development of Biomass Energy Systems," May 1994.

National Wind Coordinating Committee. "Utility Restructuring and Wind Energy." NorthBridge, 1994.

New Directions: International Symposium on the Electrical Power Industry Proceedings. Ontario, Canada: Ontario Hydro, April 1994.

New England Electric. *Neesplan 4: Creating Options for More Competitive and More Sustainable Electric Service.* Westborough, Massachusetts: November 1993.

North American Electric Reliability Council. "Electricity Supply and Demand (ES&D), Version 1.2," unpublished computer data. Princeton, New Jersey: 1994.

Northern States Power Company. *Application for Resource Plan Approval 1996–2010.* Minneapolis, Minnesota: 1995.

O'Driscoll, Mary. "Grand Bargain' Advances in Massachusetts," *The Energy Daily*, July 18, 1995, vol. 23, no. 135: 1+.

Office of Technology Assessment (OTA). *Renewing Our Energy Future.* Washington, D.C.: U.S. Government Printing Office, September 1995.

Ontario Hydro Annual Report. Ontario, Canada: Ontario Hydro, April 1994.

Palmer, Karen et al. *Contracting Incentives in Electricity Generation Fuel Markets.* Washington, D.C.: Resources for the Future, 1992.

Powell, Stephen G. and Shmuel S. Oren. "The Transition to Nondepletable Energy: Social Planning and Market Models of Capacity Expansion," *Operations Research*, May-June 1989, vol. 37, no. 3: 373-383.

The Quad Report. "Customer Choice Driving New Electric Industry." Washington, D.C.: Consumer Energy Council of America Research Foundation, July 1995, vol. 3, no. 6: 1+.

RCG/Hagler Bailly, Inc. *Profile IX: U.S. Independent Power Market: 1994 Status and Trends.* Arlington, VA: May 2, 1996.

"Renewables Could Be Helped by Plan to Restructure Wisconsin Utility Field," *The Solar Letter*, April 28, 1995a, vol. 5, no. 11: 136.

Repetto, Robert. "Environmental Productivity and Why It Is So Important," *Challenge*, September-October 1990: 33-38.

Rohrbach, John. "U.S. Nuclear Decommission Trust Planning: Romancing a Millstone?" *The Electricity Journal*, July 1995, vol. 8, no. 5: 56-61.

Rudin, Andrew. "Deficient Efficiency," *Public Power*, May-June 1995, vol. 53, no. 3: 19-23.

Sant and Naill. "Let's Make Electricity Generation Competitive." Aspen, Colorado: The Applied Energy Services Corporation, Aspen Institute, June 23, 1994.

Schimmelpfennig, David. "The Option Value of Renewable Energy: The Case of Climate Change," *Energy Economics*, 1995, vol. 17, no. 4: 311-317.

Seeley, Robert S. "District Energy Growth," *Independent Energy*, November 1995: 24+.

Shelby, Michael et al. *The Climate Change Implications of Eliminating U.S. Energy (and Related) Subsidies.* Washington, D.C.: U.S. Environmental Protection Agency, 1995.

Siemens Solar Industries. "Study Results from the First "Grid Support" Solar Plant Validates Benefits of Solar Power for Utilities," press release. Kerman, California: Siemens Solar Utilities and Pacific Gas & Electric (PG&E), April 11, 1995.

Sindelar, Roger A. "Feds Dictate Competition in Electricity Industry." *National Coal Leader*, May 1996: 1+.

Southeastern Regional Biomass Energy Program (SERBEP). "Alfalfa to Electricity," *SERBEP Update*, January 1996: 2-3.

Spurr, Mark. "Community Energy Systems in U.S. Climate Change Strategy," unpublished paper. September 26, 1994.

"STAPPA/ALAPCO, Utilities Divided Over Risk Findings in EPA Study," *Environment Week*, July 31, 1995, vol. 33, no. 31: 249-50.

"States Struggling to Restructure Utility Industry with Varying Results." *The Solar Letter*, June 9, 1995b, vol. 5, no. 14: 193.

Stein, Jay. "A Study to Determine the Cost-Effectiveness of Active Solar Water Heating as a Demand-Side Management Measure," presented to American Solar Energy Society Annual Conference, Boulder, Colorado, June 13-18, 1992.

Swezey, B. and D. Sinclair. *Status Report on Renewable Energy in the States.* Golden, Colorado: National Renewable Energy Laboratory, 1992.

Tellus Institute. *Energy Report: The Newsletter of the Energy Group at the Tellus Institute.* Boston, MA: February 1994.

Teotia, A.P.S. et al. *District Heating and Cooling Market Assessment.* Argonne, Illinois: Argonne National Laboratory, April 1994.

Toth, Ferenc L. "Discounting in Integrated Assessments of Climate Change," *Energy Policy*, April/May 1995, vol. 23, no. 4/5: 403-409.

Trexler, Mark, private communication, 1995.

U.S. Bureau of Census. "Table 2: Selected Average Annual Expenditures of all Consumer Units, Consumer Expenditure Survey." Washington, D.C.: U.S. Department of Commerce, 1993.

——. "Table 2: Selected Average Annual Expenditures of all Consumer Units Classified by Income Before Taxes, Consumer Expenditure Survey, 1984." Washington, D.C.: U.S. Department of Commerce, 1984.

——. "Table 2: Selected Average Annual Expenditures of all Consumer Units, Interview Survey, 1980." Washington, D.C.: U.S. Department of Commerce, 1980.

——. "Table 2: Selected Average Annual Expenditures of all Consumer Units Classified by Income Before Taxes, Interview Survey, 1972-1973." Washington, D.C.: U.S. Department of Commerce, 1973.

U.S. Bureau of Economic Analysis. *Survey of Current Business.* Washington, D.C.: U.S. Department of Commerce, September 1993.

U.S. Department of Energy (U.S. DOE). "Public Power and Housing Authorities Make a Difference by Commercializing More Efficient Technologies Through Mass Purchasing," *Energy Partnerships Update,* April 1996: 5+.

——. *Energy Conservation Trends: Understanding the Factors Affecting Energy Conservation Gains and Their Implications for Policy Development.* Washington, D.C.: April 1995.

——. *Clean Coal Technologies.* Washington, D.C.: November 1993.

U.S. Environmental Protection Agency (U.S. EPA). *Acid Deposition Standard Feasibility Study Report to Congress, Draft for Public Comment.* Washington, D.C.: February 1995a.

——. *Acid Deposition Standard Feasibility Study: Report To Congress.* Washington, D.C.: October 1995b.

——. *National Air Pollutant Emission Trends, 1900-1994.* Washington, D.C.: October 1995c.

——. *Draft Mercury Study Report to Congress, Vol. 2.* Washington, D.C.: 1994a, ES-7.

——. *Energy Efficiency and Renewable Energy: Opportunities from Title IV of the Clean Air Act.* Washington, D.C.: February 1994b.

——. *National Air Pollutant Emission Trends.* Washington, D.C.: October 1993.

Viscusi, W. Kip et al. "Environmentally Responsible Energy Pricing," *The Energy Journal,* 1994, vol. 15, no. 2: 23-42.

Wamsted, Dennis. "Cleveland Electric Unveils Bonus Program for KWH Consumption." *The Energy Daily,* June 21, 1996, vol. 24, no. 119: 3.

——. "Stranded Costs: The IOS's $135 Billion Headache," *The Energy Daily,* August 9, 1995: 3.

Washington International Energy Group. *1994 Electric Utility Outlook.* Washington, D.C.: January 1994.

Yardley, Robert. "Quotable," *The Energy Daily,* March 7, 1995: 4.

INDEX

364

toxics in the environment, 72-76
visions of the future, 48
Peters, R. L., 231
Peterson, J., 245
Petroconsultants, 131-32
Pharmaceutical companies and developing countries, 106
Phillips, A. F., 200, 237
Phosphorous, 64
Photovoltaic cells, 303, 304-5, 343
Pierce, R. R., 69, 70
Pimentel, D., 74, 93
Pinchot, Gifford, 198, 227
Pittsburgh's (PA) freeway for buses, 158-59
Plater-Zyberk, Elizabeth, 161
Plywood, 205
Point source emissions, 61-62
Policymaking, *see* Environmental policy
Pollution:
    abatement, 1-2, 5-6, 94, 102, 244, 247
    industrial, 207-8
    polluter-pays principle, 100, 109
    prevention, 2, 195, 242-48
Population figures:
    farming and mining, areas dependent on, 94
    forestry, 201
    pressures of population growth, 8-9
    traffic congestion, 148-50
    transportation, 166, 181-82
    vehicle miles traveled, 152-53, 154
Portland (OR), 158, 211
Postel, S., 231
Potato famine, Irish, 79
Poterba, James, 169
Poverty, ix, 7-8, 27-29
Powell, D. S., 193, 197-99, 221, 224, 226
Power producers, private independent, 295
Precautionary principle, 245-46
Precipitation, 50, 56, 81, 83, 84, 210-11
Predators, 233
President's Council on Sustainable Development (PCSD), ix-x, 3, 9, 28, 121-22, 181
Pretty, J. N., 73, 76
Price(s):
    agricultural commodities, 88-89
    congestion pricing, 178-79
    electric rates, 288, 289, 299, 321-22, 336
    fuels, 169-72
    guarantees, 97
    reforming electrical price signals, 333, 335-36
    signals, 33, 34-35
    social costs reflected in, 323-25
    softwood saw timber, 226
Prime farmland, 96
Private forest lands, tax reforms and stewardship of, 263-65
Private independent power producers, 295
Private-sector actions, 111-12
Productivity, forest, 194, 220-30, 265
Project XL (for Excellence and Leadership), 248
Property tax informs, 264-65
PRT, *see* Personal Rapid Transit
Public attitudes and barriers to change:

common sense test, 26
economic system, changing the, 25
fairness test, 26-27
gas tax, 24
justice test, 27-29
Public education, 168, 332, 341-42
Public scrutiny, forest companies open to, 268-69, 270
Public transit, 159, 163-65, 180-81
Public Utility Commissions (PUCs):
    consumer information, 342
    declining block rates, 325
    demand-side management, 328
    electric vehicles, 174-75
    renewable energy, 327
    retail sales, 295, 337-38
Public Utility Regulatory Policies Act of 1978 (PURPA), 326
Pulp and paper mills:
    demand for wood products, 205-8
    pollution, 11, 12, 37, 193, 243-47
    technology aiding sustainable development, 252

Qualifying renewable power and cogeneration facilities (QFs), 326
Quality of life, 97, 249

Rader, N., 326
Raeburn, P., 79, 80
Ranganathan, J., 38, 244
Raphael, M. G., 211
Rayburn, A. L., 75
Raytheon, 165, 180
Recreation:
    driving, 147-48
    forest-based, 212-13
    water supply, 69
Recycling, 11, 19, 193, 206-7, 209
Reformulated gasoline, 169
Refrigerators, high-efficiency, 331
Regional forest planning, 253-54
Regional sustainable forest sector plans, 261-62
Regulations, *see* Standards/regulations
Reid, W. V., 40
Reilly, J. M., 81, 82
Renewable energy:
    agriculture, 48
    discount rates, 320-21
    fossil fuels, displacing, 39
    intermittent renewables, 305-6
    Public Utility Commissions, 327
    supply diversification, 326
    Sustainable Energy Trust Fund, 339-40
    tax codes, 324-25
    transmission capacity, 306-7
    visions of the future, 19
Renner, R., 248
Repetto, R., 35, 37, 38, 76, 245, 289
Reproductive health services, 9
Research and development (R&D):
    agriculture, 31-32, 109-11
    electricity industry/power, 329-30

ABOUT THE AUTHORS

DARYL DITZ is an Associate in WRI's Technology and Environment Program. Previously, Dr. Ditz worked on solid and hazardous waste issues at the Center for the Environment at Cornell University.

ROGER C. DOWER directed the Climate, Energy, and Pollution Program at WRI from 1989 to 1996. Before that, he was Chief of the Energy and Environment Unit at the Congressional Budget Office of the U.S. Congress. Currently, he is Director of Federal and Regional Project Development for Sycom Enterprises in Washington, D.C.

PAUL FAETH is a Senior Associate in WRI's Program in Economics and Technology, where he directs the Economics of Sustainable Agriculture Project. Previously, he worked at the International Institute for Environment and Development and the USDA's Economic Research Service.

NELS C. JOHNSON is a Senior Associate in the Biological Resources and Institutions Program at WRI, where he works on biodiversity conservation planning and forest management policy. Previously, he worked for the International Institute for Environment and Development in Washington and Oregon State University's cooperative forest research program.

KEITH LEE KOZLOFF was a Senior Associate in WRI's Climate, Energy, and Pollution Program from 1991 through 1996. Currently, Dr. Kozloff is a Senior Associate with Hagler Bailly Consulting, Inc., in Arlington, Virginia.

JAMES J. MACKENZIE is Senior Associate in WRI's Program in Climate, Energy, and Pollution. Formerly, Dr. MacKenzie was Senior Staff Scientist at the Union of Concerned Scientists and a senior staff member for energy at the President's Council on Environmental Quality.

WALTER V. REID is an ecologist and Vice President for Program at WRI. Before that, Dr. Reid was a Senior Associate in WRI's Program in Forests and Biodiversity.

Frontiers of Sustainability
Advisory Panel

Mr. David Burwell
President
Rails-to-Trails Conservancy

Mr. Robert Grady
Principal
Robertson, Stephens and Co.

Mr. Hank Habicht
Senior Vice President
Safety Kleen Corporation

Dr. Deborah Jensen
Vice President for Science
The Nature Conservancy

Dr. Kai Lee
Professor and Director
Center for Environmental Studies
Williams College

Ms. Catherine Mater
Vice President
Mater Engineering

Ms. Kathleen Merrigan
Senior Analyst
Henry A. Wallace Institute for Alternative
  Agriculture

Dr. Wallace Oates
Professor of Economics
University of Maryland

Dr. Susan Offutt
Administrator
Economic Research Service
U.S. Department of Agriculture

Mr. David Richards
Consultative Group on Environment,
Development and Social Progress

Mr. Edward Strohbehn, Jr.
Attorney
McCutchen, Doyle, Brown and Enerson

Mr. Mason Willrich
EnergyWorks

Island Press Board of Directors

SUSAN E. SECHLER, *Chair,* Executive Director, Pew Global Stewardship Initiative

HENRY REATH, *Vice-Chair,* President, Collector's Reprints, Inc.

DRUMMOND PIKE, *Secretary,* President, The Tides Foundation

ROBERT E. BAENSCH, *Professor of Publishing,* New York University

PETER R. BORRELLI, Executive Director, Center for Coastal Studies

CATHERINE M. CONOVER

LINDY HESS, Director, Radcliffe Publishing Program

GENE E. LIKENS, Director, The Institute of Ecosystem Studies

JEAN RICHARDSON, Director, Environmental Programs in Communities (EPIC), University of Vermont

CHARLES C. SAVITT, President, Center for Resource Economics/ Island Press

VICTOR M. SHER, President, Sierra Club Legal Defense Fund

PETER R. STEIN, Managing Partner, Lyme Timber Company

RICHARD TRUDELL, Executive Director, American Indian Resources Institute